S0-CAS-423

Chemiluminescence and Photochemical Reaction Detection in Chromatography

Chemiluminescence and Photochemical Reaction Detection in Chromatography

John W. Birks, Editor

Department of Chemistry and Biochemistry and Cooperative Institute for Research in Environmental Sciences (CIRES) University of Colorado, Boulder, CO 80309

John W. Birks
Department of Chemistry and Biochemistry
and Cooperative Institute in Environmental
Sciences (CIRES)
Campus Box 449
University of Colorado at Boulder
Boulder, CO 80309-0449

Library of Congress Cataloging-in-Publication Data

Chemiluminescence and photochemical reaction detection in
chromatography/John Birks, editor

p. cm.
Includes index.
Bibliography: p.
1. Chromatographic analysis. 2. Chemiluminescence.
3. Photochemistry. I. Birks, John W.
QD79.C4C44 1989
543′.089—dc20 89-5734
CIP

British Library Cataloguing in Publication Data

Birks, John W.
Chemiluminescence and photochemical reaction
detection in chromatography.
1. Chromatography
I. Title
543′.089

ISBN 3-527-26782-4

Printed in the United States of America.
ISBN 0-89573-281-5 VCH Publishers
ISBN 3-527-26782-4 VCH Verlagsgesellschaft

Printing History:
10 9 8 7 6 5 4 3 2 1

Published jointly by:

VCH Publishers, Inc.
220 East 23rd Street
Suite 909
New York, New York 10010

VCH Verlagsgesellschaft mbH
P.O. Box 10 11 61
D-6940 Weinheim
Federal Republic of Germany

VCH Publishers (UK) Ltd.
8 Wellington Court
Cambridge CB1 1HW
United Kingdom

This book is dedicated to the memory of our friend and colleague Roland Frei

Contents

Preface

I have been fascinated by chemiluminescence since my childhood. I can remember sitting in front of a campfire at an age of nine or ten and thinking that it must be impossible to understand what fire is and how it can produce light. I was similarly fascinated with the glow of fireflies (or lightning bugs as we called them). In retrospect, it is perhaps even more fascinating to realize how much is known or at least knowable concerning these chemiluminescent and bioluminescent reactions. This fascination certainly contributed to my selection of a thesis topic as a graduate student and to many of the areas of research I choose to pursue today. Fortunately for myself and many others, this "playing with fire" can be perfectly justified as a means of developing ultrasensitive and ultraselective methods of chemical analysis.

Photochemical reactions and chemiluminescence are opposite processes. In the former, light is absorbed by molecules to cause chemical reactions to take place. In the latter, chemical reactions produce light. Although both phenomena have been known for centuries, it is only during the past few years that they have been applied to any appreciable extent to chemical analysis, and it is only very recently that cold (non-flame) chemiluminescence and photochemical reactions have been applied to detection in chromatography. Commercialization of the resultant technologies has been slow, however, probably reflecting a lack of confidence by instrument manufacturers in detectors based on chemical rather than physical interactions. Fortunately, some new startup companies employing chemists for instrument design, in addition to electrical engineers, are beginning to find unique applications for these detectors and to break into the chromatography detector market.

This book is addressed to the adventurous analytical chemists who are willing to try something a little out of the ordinary to solve their analysis problems and to those researchers who would like to contribute to this very fertile field. There is still much room for fundamental studies of photochemical and chemiluminescent reactions, as well as the invention of new detection schemes. This book details the theory of chemiluminescence and photochemical reaction detection, describes and characterizes various detection schemes, and provides reviews of the applications published to date.

I thank the various chapter authors for their diligence and willingness to continue to update their chapters, because of the rapid advances in the field, right up to the final galleys. My graduate students certainly deserve more credit than myself for the three chapters they co-authored. Finally, I would like to thank another graduate student, Thomas (Red) Chasteen, for proofreading and for preparing the word and chemical indexes.

We are dedicating this book to the memory of a wonderful friend and colleague, Roland Frei, who passed away suddenly and unexpectedly this year. Roland pioneered the field of post-column reaction detection in HPLC and made major contributions to both chemiluminescence and photochemical reaction detection. He had a great enthusiasm for both life and science. From his post as Professor of Analytical Chemistry at the Free University of Amsterdam, he promoted and fostered many international collaborations. I personally spent many months as a guest in his townhouse "Hotel Frei," a short bus ride from the lab. "Well gentlemen, shall we open another bottle of wine?"

John W. Birks
Boulder, Colorado
May 11, 1989

Chemiluminescence and Photochemical Reaction Detection in Chromatography

1

Photophysical and Photochemical Principles

John W. Birks

Department of Chemistry and Biochemistry and Cooperative Institute for Research in Environmental Sciences (CIRES), University of Colorado, Boulder, Colorado

The absorption of light to initiate photochemical reactions and the emission of light in chemiluminescent reactions both involve transitions between the quantized electronic–vibrational–rotational states of molecules. The energies of these transitions are typically in the range 20 to 200 kcal/mol (1.4 μm to 140 nm), corresponding to near infrared, visible and ultraviolet wavelengths. Photochemical and chemiluminescent reactions nearly always involve transitions between different electronic states of molecules; exceptions may include such processes as photochemical reactions induced by multiphoton absorption and chemiluminescence arising from vibrational transitions (e.g., HF vibrational overtone emission in the reaction of F_2 with organosulfur compounds discussed in Chapter 2). We will limit our discussion in this chapter to the photophysical and photochemical properties of nonadiabatic processes, i.e., processes involving more than one potential energy surface. There are excellent books devoted to the subject of photochemistry, and the reader is referred to these for a more detailed discussion.[1-3] The purpose of this chapter is to review the most general principles of photochemistry and chemiluminescence in order to provide a background for the chapters to follow.

Electronic States and Chromophores

Molecular orbital theory provides a basis for predicting the occurrence of electronic states of molecules, their relative energies and a useful means of designating various types of electronic transitions. In this theory, a linear combination of two atomic orbitals, centered on adjacent atoms, results in two molecular orbitals. One is of lower energy than either of the two atomic orbitals and results in bonding, and one is of higher energy than either of the two atomic orbitals and is antibonding. If the two atomic orbitals overlap along the axis connecting the two atomic centers, the resulting bonding and antibonding molecular orbitals are designated σ and σ^*, respectively. If the atomic orbitals overlap above and below a plane containing the two centers, then the bonding and antibonding orbitals are designated π and π^*. Single bonds are σ bonds, while double and triple bonds arise from additional stabilization resulting from sharing of electrons in π molecular orbitals. In addition, atoms such as oxygen, sulfur, nitrogen and the halogens may have valence electrons that do not participate in bonding. These "lone pair" electrons and their electronic states are designated by the letter n for "nonbonding." In general, a molecule will have a number of σ bonding orbitals at very low energy, π orbitals at higher energy (if there are double or triple bonds in the molecule) and still higher energy n orbitals (if lone pairs are present in the molecule). Electronic transitions may result from promotion of electrons from one molecular orbital to another, giving rise to electronic configuration changes such as $\sigma \rightarrow \sigma^*$, $\pi \rightarrow \pi^*$, $n \rightarrow \sigma^*$ and $n \rightarrow \pi^*$. Excited electronic states are designated by the transition from the ground state that gives rise to them. For example, an n, π^* state refers to the electronic state that arises when a ground-state molecule is excited by promotion of a lone pair electron to the π^* orbital.

It is the usual case that stable molecules have an even number of electrons and that in the ground electronic state all electrons are paired. This is particularly true of organic compounds. There are some stable free radicals, especially notable are NO and NO_2, and the ground state of O_2 has two unpaired electrons, making it a diradical. Prediction of the latter has been heralded as one of the great triumphs of molecular orbital theory. Figure 1-1 shows the molecular orbital diagram and electron configurations for the ground electronic state of O_2 and its first two excited electronic states. In filling the 16 electrons of oxygen into the molecular orbital diagram we must obey the

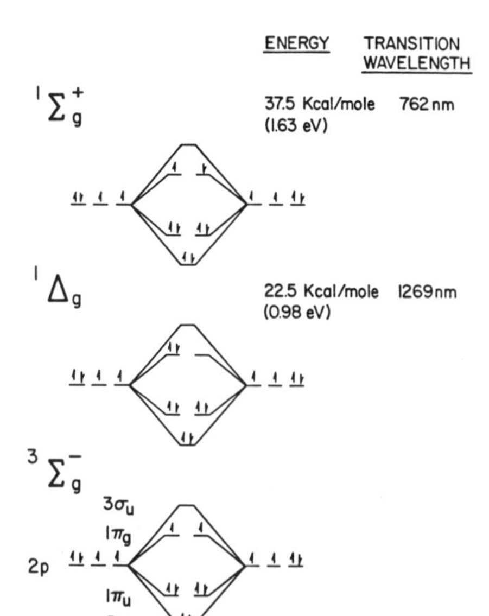

Figure 1-1 ■ Molecular orbital diagrams and electron configurations for the ground and two lowest electronic states of the oxygen molecule. From B. Shoemaker, "Singlet-Delta Molecular Oxygen as a Tool for Selective Chemiluminescent Detection in Chromatography," Ph.D. Thesis, University of Illinois, 1981.

Pauli principle, i.e., as we fill lower to higher energy orbitals we may place up to two electrons in each orbital, and these two electrons must have opposite spin. When we add the final two electrons to the energy level diagram, we have three choices. There are two degenerate π^* orbitals into which we may place the electrons. The ground-state configuration $^3\Sigma_g^-$ results when we place the two electrons in different orbitals with their spins unpaired, as required by Hund's rule. Pairing the electrons in the same orbital results in the first excited electronic state $^1\Delta_g$, which lies 22.5 kcal/mol above the ground state. A still higher excited electronic state of O_2, $^1\Sigma_g^+$, lying 37.5 kcal/mol above the ground state, results when the two electrons are placed in different orbitals, but with their spins paired.

The oxygen example also illustrates the spectroscopic designation of electronic states for diatomic molecules.[4] As in atoms, the angular momentum of electron spin is quantized. The total orbital angular momentum, however, is no longer a constant of the motion. As a result of the reduction in symmetry of the electric field from spherical (for atoms) to axial (for diatomic molecules), only the component of orbital angular momentum projected along the internuclear axis provides a good quantum number, M_L. In an electric field (unlike a magnetic field) reversing the direction of motion of all electrons does not change the energy of the system, although it does change M_L into $-M_L$. Therefore, in diatomic molecules states differing only in the sign of M_L are degenerate. On the other hand, states having different values of $|M_L|$ tend to have widely different energies resulting from the internal electric field of the molecule. Thus, it is appropriate to define electronic states of diatomic molecules in terms of the quantum number Λ given by

$$\Lambda = |M_L| = \left| \sum_{\text{all electrons}} m_l \right| \tag{1-1}$$

The term symbol used to designate electronic states of diatomic molecules

$$^{2S+1}\Lambda_{g/u}^{+/-} \tag{1-2}$$

specifies the value of Λ (Σ for $\Lambda = 0$, Π for $\Lambda = 1$, Δ for $\Lambda = 2$ and Φ for $\Lambda = 3$; analogous to S, P, D and F electronic states of atoms). The spin multiplicity, given as $2S + 1$, may be used to deduce the number of unpaired electrons in the molecule. The value of the spin quantum number S is given by the sum

$$S = \left| \sum_{\text{all electrons}} m_s \right| \tag{1-3}$$

The singlet states ($S = 0$, $2S + 1 = 1$) arise in O_2 when the spins of all electrons are paired. For the ground state of O_2, $S = \frac{1}{2} + \frac{1}{2} = 1$ and $2S + 1 = 3$; i.e., the ground state is a triplet. Note that for radical species such as NO, which have one unpaired electron, the ground state is a doublet ($2 \times \frac{1}{2} + 1 = 2$).

The g and u designations specify whether or not the wave function for that electronic state is symmetrical with respect to inversion through a center of symmetry and therefore only applies to homonuclear diatomic molecules (and also molecules having equal nuclear charge, e.g., HD and HT). If the electron coordinates x_i, y_i, z_i are replaced by $-x_i$, $-y_i$, $-z_i$ and the wave function does not change sign, then the state is said to be *gerade* (from German denoting even) and the symbol g is attached to the term symbol. If the wave function does change sign under this symmetry operation, the symbol u is used to indicate an *ungerade* (odd) state. The $+/-$ designation further describes the symmetry of Σ states. Here the symmetry operation is reflection through a plane containing the two nuclei. The $+$ sign indicates that the wave function does not change sign for this operation, while the $-$ sign is used when the wave function does change sign.

Selection Rules

The term symbols just defined are very useful in evaluating what electronic transitions are likely to take place in molecules. It can be shown that the intensity of an electronic transition resulting from the interaction of a photon with the electric dipole of a molecule is proportional to the square of the integral

$$\int \Psi_{\text{initial}} \mu \Psi_{\text{final}} \, d\tau \tag{1-4}$$

where Ψ_{initial} and Ψ_{final} are the wave functions describing the initial and final electronic states, respectively, and μ is the dipole moment operator. Quantum mechanical group theory may be used to determine under what conditions this integral is nonzero.[5] The greater the symmetry of the molecule, the greater the number of selection rules that may be derived for electronic transitions. Some of the more important selection rules for electronic transitions in diatomic molecules are

$$\Delta S = 0 \tag{1-5}$$

$$\Delta \Lambda = 0, \pm 1 \tag{1-6}$$

$$u \leftrightarrow g, \qquad u \not\leftrightarrow u, \qquad g \not\leftrightarrow g \tag{1-7}$$

$$\Sigma^+ \leftrightarrow \Sigma^+, \qquad \Sigma^- \leftrightarrow \Sigma^-, \qquad \Sigma^+ \not\leftrightarrow \Sigma^- \tag{1-8}$$

As an example, it is interesting to note that transitions (either absorption or emission of electromagnetic radiation) between the ground $^3\Sigma_g^-$ and the $^1\Delta_g$ lowest excited electronic state of O_2 are forbidden by the first three of the preceding selection rules (see Fig. 1-1 for the electronic configurations resulting in these states). Transitions between the $^3\Sigma_g^-$ and $^1\Sigma_u^+$ state violates the first, second and fourth of these rules. In agreement with these selection rules, the radiative lifetimes of the $^1\Delta_g$ and $^1\Sigma_g^+$ states have been determined by experiment to be ~ 45 min and ~ 75 s, respectively. These lifetimes are many orders of magnitude longer than the theoretical limit of $\sim 10^{-9}$ s for an allowed transition.

It should be emphasized that the selection rules are derived assuming an electric dipole interaction between the electromagnetic wave and the molecule. Magnetic dipole, electric quadrupole and higher moments of interaction also are possible and have their own selection rules. Because the strengths of these interactions are typically several orders of magnitude less than for the electric dipole interaction, absorption and emission based on such interactions can usually be ignored both in solution and in the gas phase. Many electric dipole forbidden transitions, including those of oxygen as discussed, may be observed as emissions by molecules in the near vacuum of the upper atmosphere of the earth and in interstellar space where collisional deactivation is relatively unimportant.

The selection rules (1-5) to (1-8) are very useful in the interpretation of gas-phase chemiluminescence, as discussed in Chapter 2, where diatomic molecules frequently are the emitting species. For molecules of lower symmetry, the selection rules are more limited, and for most organic molecules, the only pertinent selection rule is that for spin multiplicity ($\Delta S = 0$). Excited states of molecules whose radiative relaxation to the ground state is forbidden are said to be "metastable." Metastable states play enormously important roles in the mechanisms of photochemical reactions. Some examples of metastable states include the lowest triplet (T_1) state of most organic molecules, the $^1\Delta_g$ and $^1\Sigma_g^+$ states of O_2 already discussed and collectively called "singlet oxygen," the $^3\Sigma_u^+$ state of N_2 (a component of "active nitrogen") and the $^3P_1$ state of Hg. These metastable states are important in that they are sufficiently long lived to enter into chemical reactions, and their electronic potential energy may be used to overcome the activation barriers of reactions.

Spin-Orbit Coupling

Phosphorescence, the emission of light accompanying the triplet–singlet transition is, according to the $\Delta S = 0$ selection rule, strictly forbidden. Although generally much weaker than fluorescence, for which the spin selection rule is obeyed, phosphorescence is observed to occur to an appreciable extent in many molecules, especially those containing heavy atoms. The explanation for this discrepancy is based on the coupling of spin and orbital angular momenta. The selection rules for changes in spin and orbital angular momenta as previously stated were derived assuming strong coupling between the individual spin angular momenta of the electrons to form a resultant vector S (with component Σ along the internuclear axis of a diatomic molecule) and strong coupling of the individual orbital angular momenta of the electrons to form a resultant L (with component Λ along the internuclear axis of a diatomic molecule). This idealized case (Hund's case a) applies extremely well to molecules containing only light atoms. As atoms become larger, however, the individual spin and individual orbital angular momenta of the electrons begin to couple with one another, and the derived selection rules no longer apply.

From the wave-mechanical point of view, the singlet and triplet states are no longer completely separable. The spin-orbit interaction introduces into the Hamiltonian operator a term which operates on both spin and space variables. As a result, the wave functions that are solutions to the wave equation are no longer pure singlets and triplets, but are mixed to a small extent. The mixed triplet wave function is given by

$$\Psi_T' = \Psi_T^0 + \lambda_{TS}\Psi_S^0 \tag{1-9}$$

where the zero superscript refers to the pure zero-order state and λ_{TS} is the mixing coefficient. As the result of mixing of singlet and triplet states, the transition moment integral [Eq. (1-4)] will be nonzero, and the transition is therefore partially allowed. The mixing coefficient λ_{TS} is found to increase as the energy difference $E_S - E_T$ decreases. In addition to providing an explanation for singlet–triplet absorption and phosphorescence, mixing of states plays an important role in nonradiative processes such as intersystem crossing and internal conversion, as will be discussed.

Table 1-1 ▪ Representative Chromophores for Absorption of UV Radiation

Chromophore	Notation	λ_{max}
σ electrons		
C—C and C—H	$\sigma \rightarrow \sigma^*$	~150
Lone pair electrons		
—O—	$n \rightarrow \sigma^*$	~185
>N—	$n \rightarrow \sigma^*$	~195
—S—	$n \rightarrow \sigma^*$	~205
—I	$n \rightarrow \sigma^*$	~260
>C=O	$n \rightarrow \pi^*$	~300
>C=O	$n \rightarrow \sigma^*$	~300
π electrons		
>C=C<	$\pi \rightarrow \pi^*$	~190

Source: Adapted from B. P. Straughan and S. Walker, *Spectroscopy*, Chapman and Hall, London, 1976, Vol. 3, p. 131.

Chromophores

Empirical relationships between molecular structure and the colors of organic compounds were formulated by the early dye chemists. It was found that characteristic absorption bands could be ascribed to specific groupings of atoms called *chromophores* (color carriers) and that the locations of these bands were virtually independent of the structure of the remainder of the molecule. It also was found that certain substituents, called *auxochromes*, do not confer color themselves, but when present in the molecule substantially intensify and shift the position of the absorption maxima of the chromophores. A shift to the red (longer wavelengths) is said to be *bathochromic* and a shift to the blue (shorter wavelengths) is termed *hypsochromic*. With the advent of quantum mechanics and modern molecular photochemistry, it has become possible to understand the existence of chromophores and the effects of substituents in terms of transitions between the electronic states previously discussed. Examples of representative chromophores are given in Table 1-1. Most of these isolated chromophores result from high energy electronic transitions that fall in the vacuum ultraviolet (< 200 nm), although some, such as the n, σ^* transitions of alkyl sulfides and iodides and a weak n, π^* transition of carbonyls, do occur in the near UV. The presence of auxochromes in the molecule will cause a shift in the value of λ_{max} given in Table 1-1. For example, the halogens result in a bathochromic shift. For CH_3Cl, CH_3Br and CH_3I, the λ_{max} are 173, 204 and 258 nm, respectively.

An important means by which the absorption spectra of molecules are shifted into the near UV and visible wavelength regions is by conjugation of chromophores. A molecular orbital diagram showing the effect of conjugation of two double bonds on the energy levels and electronic transition energy is given in Fig. 1-2. The linear combination of two filled π orbitals on the two ethylene molecules results in two molecular orbitals in butadiene, one of lower energy and one of higher energy. Similarly, a linear combination of the two π^* orbitals results in two molecular orbitals

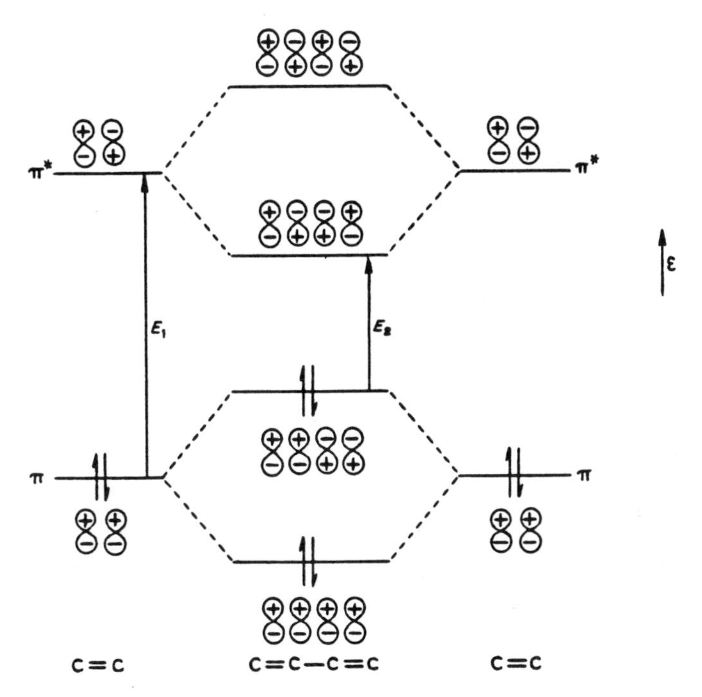

Figure 1-2 ■ Effect of conjugation of two double bonds on the energy levels and electronic transitions. Because E_2 is smaller than E_1, the effect of conjugation is to shift the absorption spectrum to longer wavelengths. From B. P. Straughan and S. Walker, eds., *Spectroscopy*, Chapman and Hall, London, 1976, Vol. 3, p. 132.

of lower and higher energy. Because the energy of the highest occupied molecular orbital is now raised and the energy of the lowest unoccupied molecular orbital is lowered, the energy difference between the two levels is reduced. Thus, the wavelength of the photon required to excite the transition is shifted to the red. (For illustrative purposes, the figure greatly exaggerates the degree of this effect.) The shift to longer wavelengths is illustrated by the series of molecules: ethylene ($\lambda_{max} = 193$ nm), 1,3-butadiene (217 nm), hexatriene (258 nm) and octatetraene (300 nm). Another important result for photochemistry is that the molar absorptivity also increases with an increasing number of conjugated double bonds.

Empirical rules have been established for estimating the longest wavelength λ_{max} of α,β-unsaturated aldehydes and ketones containing various auxochromes (Table 1-2). Applying these rules to the molecule

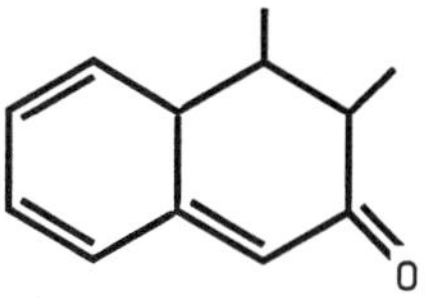

O

Table 1-2 ▪ Empirical Rules for Calculating λ_{max} for α,β-unsaturated Ketones and Aldehydes

$$\overset{\delta}{C}=\overset{\gamma}{C}-\overset{\beta}{C}=\overset{\alpha}{C}-C=O$$

Value assigned to parent α,β-unsaturated six-ring or acyclic ketone		215 nm
Value assigned to parent α,β-unsaturated five-ring ketone		202
Value assigned to parent α,β-unsaturated aldehyde		207
Increments for		
double bond extending the conjugation		30
each alkyl group of ring residue	α	10
	β	12
	γ and higher	18
Auxochromes		
—OH	α	35
	β	30
	δ	50
—OAc	α,β	6
—OMe	α	35
	β	30
	γ	17
	δ	31
—SAlk	β	85
—Cl	α	15
	β	12
—Br	α	25
	β	30
—NR_2	β	95
Exocyclic nature of double bond		5
Homodiene component		39

Source: Adapted from B. P. Straughan and S. Walker, *Spectroscopy*, Chapman and Hall, London, 1976, Vol. 3, p. 134.

we may calculate the λ_{max} as follows[6]:

parent value	215 nm
β substituent	12
ω substituent	18
2 × extended conjugation	60
exocyclic double bond	5
homoannular diene	39
	349 nm

The calculated value of 349 nm is in excellent agreement with the observed value of 348 nm in ethanol. The empirical constants given in Table 1-2 vary from solvent to

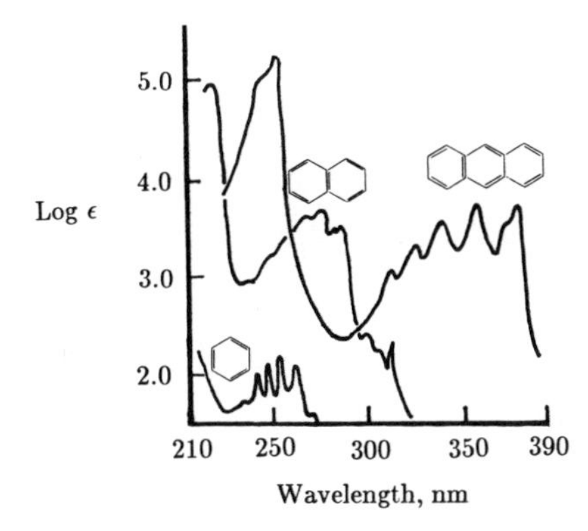

Figure 1-3 ■ Absorption spectra of benzene, naphthalene and anthracene in ethanol. Adapted from W. V. Mayneord and E. M. F. Roe, *Proc. Roy. Soc.* (*London*), *Ser. A* **152**, 299 (1935).

solvent. The solvation of the excited state of a molecule is likely to be different from that of the ground state, because the electronic charge distribution is altered. Polar solvents generally result in a red shift for $\pi \rightarrow \pi^*$ transitions. This amounts to about 10 nm in going from hexane to ethanol as the solvents. For $\pi \rightarrow \pi^*$ transitions the excited state is usually more polar than the ground state, and the dipolar interaction with the solvent molecules is stabilizing, i.e., lowers the energy of the excited state. In contrast, the $n \rightarrow \pi^*$ transition of the carbonyl group undergoes a blue shift as the polarity of the solvent increases. This is explained by a greater degree of hydrogen bonding of the ground state with the solvent as compared to the excited state. Thus, the ground-state energy is lowered relative to that of the excited state. Many solvent shifts are not readily explained, but are almost certainly due to differential interaction of the ground and excited states with the solvent molecules.

As seen in Fig. 1-3, the absorption spectra of polycyclic aromatic compounds, given by the series benzene, naphthalene and anthracene, because of extended conjugation and delocalization of π electrons, also shifts toward longer wavelengths with increasing numbers of conjugated rings. Although the positions of λ_{max} in these complex spectra do not behave in a predictable way, the longest wavelength at which absorption occurs and the integrated extinction coefficient do obey the expected trend for "linear" polycyclic aromatics. The effect of conjugation is illustrated by the comparison of hexacene with 6,15-dihydrohexacene:

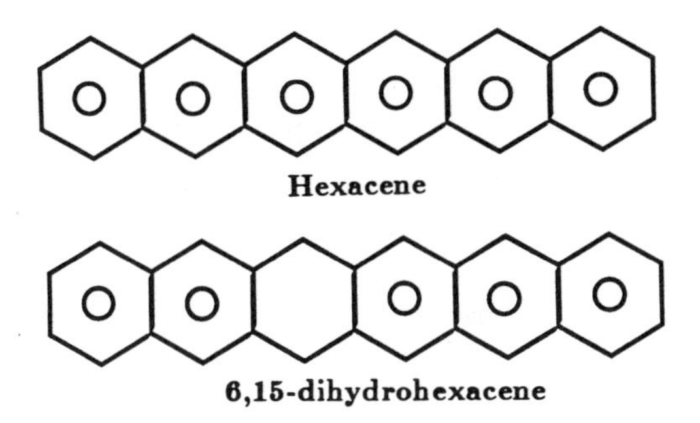

Table 1-3 ▪ Spectra of Monosubstituted Benzenes[a]

	Primary Band		Secondary Band		
Substituent	λ_{max} (nm)	ϵ_{max}	λ_{max} (nm)	ϵ_{max}	$\lambda_{sec}/\lambda_{pri}$
H	203.5	7,400	254	204	1.25
NH_3^+	203	7,500	254	160	1.25
CH_3	206.5	7,000	261	225	1.25
I	207	7,000	257	700	1.24
Cl	209.5	7,400	263.5	190	1.25
Br	210	7,900	261	192	1.24
OH	210.5	6,200	270	1,450	1.28
OCH_3	217	6,400	269	1,480	1.24
SO_2NH_2	217.5	9,700	264.5	740	1.22
CN	224	13,000	271	1,000	1.21
CO_2^-	224	8,700	268	560	1.20
CO_2H	230	11,600	273	970	1.19
NH_2	230	8,600	280	1,430	1.22
O^-	235	9,400	287	2,600	1.22
$NHCOCH_3$	238	10,500	...	...	...
$COCH_3$	245.5	9,800	...	...	...
CHO	249.5	11,400	...	...	...
NO_2	268.5	7,800	...	...	...

[a] Water as solvent, trace of methanol added for solubility where necessary.
Source: H. H. Jaffe and M. Orchin, *Theory and Applications of Ultraviolet Spectroscopy*, Wiley, New York, 1962, p. 257.

Hexacene is a green compound, while 6,15-dihydrohexacene is a colorless compound whose absorption spectrum is approximately the sum of that of naphthalene and anthracene.

The effects of substituents on absorption spectra of aromatic compounds are exemplified by the results for benzene, given in Table 1-3. All substituents, whether electron releasing or electron withdrawing, cause a shift in both the primary and secondary bands of benzene to longer wavelengths. As seen in the table, the ratio of λ_{max} for the two bands, $\lambda_{sec}/\lambda_{pri}$, remains nearly constant at about 1.25.

Potential Energy Surfaces

Molecules placed in excited electronic states can undergo a variety of nonradiative processes, including chemical reactions that are either slow or totally inaccessible from their ground states. These changes in electronic charge distribution and relative positions of the nuclei may be visualized in terms of potential energy surfaces for relevant electronic states of the molecule. A potential energy surface is simply a plot of the total potential energy of a molecule as a function of the nuclear coordinates. If one considers that a linear molecule has $3N$-5 and a nonlinear molecule $3N$-6 vibrational modes (the number of normal coordinates required to relate the relative positions of all the N nuclei), it is clear that for as few as three atoms this "surface" is actually four dimensional (potential energy V and three nuclear coordinates) or higher and not easily visualized. Pictorial representations of these surfaces are usually limited to two- or three-dimensional plots that show how the potential energy of the molecule changes along one or two critical coordinates. The state of a molecule at any particular time is given by a point on this surface. Classically, this point may be thought of as moving on the surface under the influence of forces which are given by the derivatives of V with respect to each of the coordinates,

$$F_i = -\frac{\partial V}{\partial q_i} \tag{1-10}$$

Implicit in this description is the Born–Oppenheimer (or adiabatic) approximation which assumes that the velocities of electrons are so much greater than those of nuclei that the electronic charge distribution in a molecule will rapidly adjust itself to any change in the nuclear coordinates. In the language of wave mechanics this means that the wave functions for nuclear and electronic motion are separable. The Born–Oppenheimer approximation is extremely useful for qualitative considerations in that each electronic state of a molecule ($S_0, S_1, S_2, \ldots; T_1, T_2, \ldots$) is described by a separate potential energy surface. As will be discussed at greater length in the following text, intersections of these surfaces (also called curve crossings) are of fundamental importance to the nonradiative processes of internal conversion and intersystem crossing. At these intersections, the nuclei have the same coordinates, and the potential energy of the molecule is identical for the two electronic states; thus, nonradiative transitions from one surface to another are favored by the Franck–Condon principle.

It is interesting to note that although the Born–Oppenheimer approximation allows us to visualize nonradiative processes in terms of potential energy surfaces, it is the breakdown of the Born–Oppenheimer approximation at certain intersections of these surfaces that allows the molecule to change electronic states. An example of the effect of nuclear motion on curve crossings for n, π^* and π, π^* states of a carbonyl is given in Fig. 1-4. The nonbonding orbitals of oxygen, n, and the π molecular orbital of C=O are orthogonal for all in-plane vibrations of the molecule. This results in adiabatic (zero-order) curve crossings for any change in coordinates for which the molecule remains planar. Thus, in Fig. 1-4 it is seen that for planar vibrations the potential energy curves do not interact in the region of their crossing. A molecule in the n, π^* state, represented by a point moving on the n, π^* curve, will remain in the n, π^* state. On the other hand, nonplanar vibrations of the molecule will cause the p orbitals of the oxygen atom to rehybridize to sp^n where n depends on the degree of out-of-plane bending. Thus, the n, π^* and π, π^* states of the molecule are coupled. In a narrow region near the intersection of the two potential curves the wave function describing the electronic state of the molecule is a mixture of n, π^* and π, π^* wave functions, so that the electronic state of a molecule at this intersection has a finite (often large) probability of switching states. In Fig. 1-4 it can be seen that a point moving from left to right along the n, π^* potenial energy curve can either continue on this curve via the dashed line when it nears the intersection with the π, π^* curve or continue along the solid line and enter the "pure" π, π^* state. Using the familiar paradigm of electrons and molecular orbitals, we say that an electron has jumped from a π orbital to an n orbital.

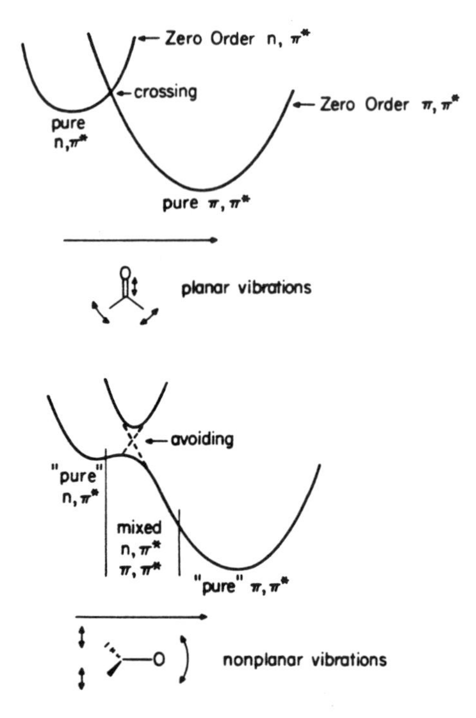

Figure 1-4 ■ Curve crossings for n, π^* and π, π^* states for planar and nonplanar vibrations. Nonplanar vibrations result in mixing of states, thereby allowing a molecule to change electronic states near the intersection of the two zero-order curves without emission of a photon. From N. J. Turro, *Modern Molecular Photochemistry*, Benjamin/Cummings Menlo Park, CA, 1978, p. 163.

Molecular vibrations also are important in inducing transitions between singlet and triplet states at the intersections of their potential energy surfaces. To be effective the vibration must induce spin-orbit coupling in the molecule. As we shall see, potential energy surfaces are a valuable aid in understanding photochemical reactions and their inverse—chemiluminescent reactions.

Relaxation Mechanisms

The energy deposited in a molecule by the absorption of a photon is rapidly redistributed by a variety of photophysical porocesses illustrated in the Jablonski diagram of Fig. 1-5. The potential energy curves for the S_0, S_1 and T_1 states of a typical organic molecule are shown in Fig. 1-6. This figure also shows the processes of absorption, fluorescence and phosphorescence to and from various quantized vibrational levels of the electronic states and the resulting spectra. As can be seen in Fig. 1-6, the fluorescence emission spectrum is shifted to longer wavelengths than the absorption spectrum. This is because absorption arises from the lowest vibrational levels of S_0 and

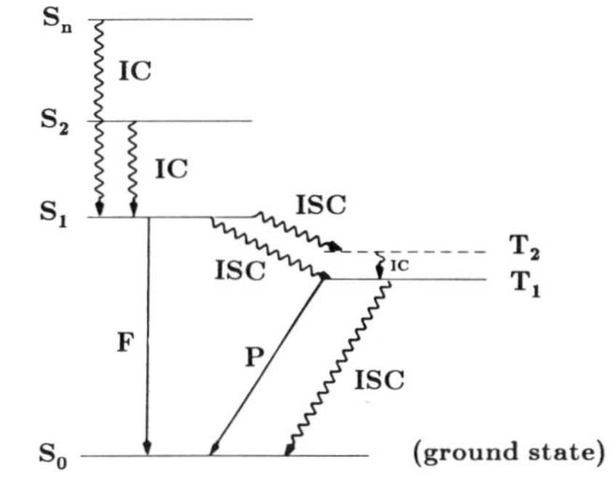

Figure 1-5 ■ Jablonski diagram showing the processes defined as absorption (A), fluorescence (F), internal conversion (IC), intersystem crossing (ISC) and phosphorescence (P). Processes involving absorption or emission of a photon are shown as straight lines and nonradiative processes are shown as wavy lines.

terminates in many excited vibrational levels of S_1, while fluorescence originates in the lowest vibrational levels of S_1 and terminates in many excited vibrational levels of S_0. Since the T_1 electronic state always lies lower than S_1, the phosphorescence spectrum is shifted even further to the red. The various photophysical processes described by Figs. 1-5 and 1-6 are defined as follows:

Absorption—capture of a photon by a molecule to produce an excited state, e.g., $S_0 + h\nu \rightarrow S_1$.

Vibrational relaxation—radiationless deactivation of vibrational energy levels within a given electronic state.

Internal conversion—radiationless transitions between states of the same spin, e.g., $S_1 \rightarrow S_0$ + heat, rate constant k_{IC}.

Intersystem crossing—radiationless transitions between states of different spin, e.g., $S_1 \rightarrow T_1$ + heat, rate constant k_{ISC}.

Fluorescence—radiative emission connecting states of the same spin, e.g., $S_1 \rightarrow S_0 + h\nu$, rate constant k_F.

Phosphorescence—radiative emission connecting states of different spin, e.g., $T_1 \rightarrow S_0 + h\nu$, rate constant k_P.

Energy transfer—transfer of electronic energy from one molecule to another, e.g., $D(T_1) + A(S_0) \rightarrow D(S_0) + A(T_1)$, rate constant k_{ET}.

For organic molecules, which generally have a singlet ground state, S_0, the allowed absorptions are to excited-state singlets, $S_1, S_2, \ldots, S_n$. (Using powerful light sources such as lasers, it is possible to excite triplet states as well.) Once a molecule has been activated by absorption of light, it may undergo a variety of deactivation processes. To quantify the various deactivation pathways, it is useful to define the quantum yield

$$\phi = \frac{\text{number of events}}{\text{number of photons absorbed}} \tag{1-11}$$

As will be discussed, once a molecule has absorbed light to form an excited singlet state, it may undergo internal conversion to the ground state, intersystem crossing to a triplet state, fluorescence, energy transfer to another molecule or chemical reaction, and the sum of the quantum yields for all of these processes must be unity:

$$\phi_{IC} + \phi_{ISC} + \phi_F + \phi_{ET} + \phi_R = 1 \tag{1-12}$$

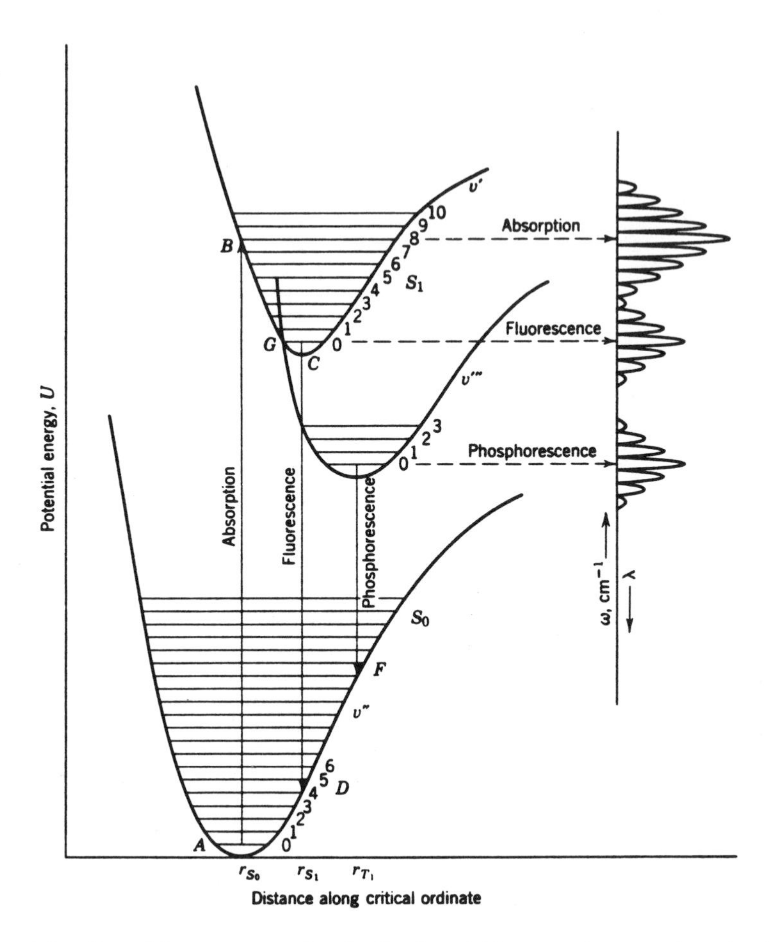

Figure 1-6 ■ Potential energy curves for the S_0, S_1 and T_1 states of an organic molecule. The critical coordinate is a vibrational mode of the molecule for which quantized vibrational levels are indicated. The point G is the intersection between S_1 and T_1 where intersystem crossing may occur. The equilibrium distances r were chosen such that $r_{S_0} < r_{S_1} < r_{T_1}$ so that the spectra are spread out and better visualized. In complex molecules that are fairly symmetrical, $r_{S_0} \simeq r_{S_1} < r_{T_1}$ and the (0, 0) bands of absorption and fluorescence nearly coincide, but phosphorescence bands are significantly displaced to longer wavelengths. Also, spectra are usually more complex, as several vibrational modes may be active. From J. G. Calvert and J. N. Pitts, Jr., *Photochemistry*, Wiley, New York, 1966, p. 274.

Here, ϕ_{IC}, ϕ_{ISC}, ϕ_F, ϕ_{ET} and ϕ_R are the quantum yields for internal conversion, intersystem crossing, fluorescence, energy transfer and chemical reaction, respectively.

Vibrational Relaxation

In the condensed phase, vibrational relaxation of a molecule, once formed in a particular electronic state, occurs on the same time scale as nuclear motion, i.e., $\sim 10^{-13}$ to 10^{-12} s. In nearly all cases, it can be assumed that once a molecule enters a new electronic state, interaction of the molecule with the solvent will result in complete deactivation of the vibrational modes of the molecule to establish a Boltzmann distribution before other photophysical processes take place, especially radiative processes. A possible exception is internal conversion, which can take place on the same time scale as vibrational relaxation. An important result of this principle is that fluorescence and phosphorescence spectra both originate in the $v = 0$ vibrational levels of the molecule, so that both spectra are red-shifted from the absorption spectrum.

Internal Conversion

With very few exceptions, in the condensed phase the fate of any excited singlet above S_1 is rapid internal conversion to S_1:

$$S_n \rightarrow S_{n-1} \rightarrow \cdots \rightarrow S_1 \tag{1-13}$$

This is evidenced by many experimental observations that the quantum yield for fluorescence and phosphorescence is independent of the wavelength of excitation and that the fluorescence spectra obtained are characteristic of S_1 and independent of which singlet state is originally excited. It appears that for most organic compounds $k_{IC}(S_n \rightarrow S_1)$ falls in the range 10^{11} to 10^{13} s^{-1}, i.e., the lifetime for deactivation to S_1 is of the order of 0.1 to 10 ps. This is the same time scale as that of nuclear motion (vibration). In contrast, internal conversion from S_1 to S_0 is often so slow as to be negligible in comparison to the other possible fates of S_1. The rapidity of $S_n \rightarrow S_1$ internal conversion as compared to $S_1 \rightarrow S_0$ is best explained in terms of the Franck–Condon principle and potential energy curve crossings.

The probability of a nonradiative process such as internal conversion and intersystem crossing (to be discussed) has been derived in quantum mechanical terms using first-order perturbation theory and may be expressed in the form known as the Fermi golden rule:

$$k_{nr} = \frac{2\pi}{\hbar} \beta^2 \rho \langle v' | v'' \rangle^2 \tag{1-14}$$

Here, $\langle v' | v'' \rangle^2$ is the Franck–Condon factor, i.e., the square of the overlap integral between the vibrational wave functions of the initial and final states, ρ is the density of vibrational levels in the final electronic state that is coupled to the initial state and β^2 is the electronic integral containing the appropriate operator for the electronic transition. β^2 is large for singlet–singlet and triplet–triplet (spin allowed) nonradiative transitions, but small for singlet–triplet (intersystem crossing) transitons. For internal conversion the dominant factor in Eq. (1-14) is the Franck–Condon factor. In general, in the absence of any potential energy surface crossings, internal conversion is nearly forbidden by the Franck–Condon principle in that the nuclei must undergo rather drastic changes in positions and momentum (i.e., the overlap between vibrational wave functions of the initial and final states is small). This can explain the very slow rate constant usually observed for $S_1 \rightarrow S_0$ when S_1 lies more than ~ 50 kcal mol^{-1} above S_0. For most molecules, the spacings between successive states in the singlet and triplet manifolds become smaller and the probability of potential energy surface crossings increases. Near the intersection of two potential energy surfaces a molecule can change its electronic charge distribution (electronic state) without making large changes in the positions of its nuclei, so that the nonradiative transition is favored. This *energy gap*

Table 1-4 ■ Quantum Yields for Fluorescence and Intersystem Crossing of Organic Molecules

Molecule (configuration of S_1)	ϕ_F	ϕ_{ISC}	$1-(\phi_F+\phi_{ISC})^a$	E_{S_1} (kcal/mol)
Benzene (π, π^*)	0.05	0.25	0.70	110
1,4-dimethylbenzene (π, π^*)	0.35	0.65	< 0.05	100
Naphthalene (π, π^*)	0.20	0.80	< 0.05	92
Anthracene (π, π^*)	0.70	0.30	< 0.05	76
Tetracene (π, π^*)	0.15	0.65	0.20	60
Pentacene (π, π^*)	0.10	0.15	0.75	50
Azulene (π, π^*)	0.000	...	Low	50
Acetone (n, π^*)	0.001	~ 1.0	< 0.05	85
Biacetyl (n, π^*)	0.002	~ 1.0	< 0.05	65
Benzophenone (n, π^*)	0.000	~ 1.0	< 0.05	75
Cyclobutanone (n, π^*)	0.000	0.00	1.0	80
1,3-pentadiene (n, π^*)	0.000	0.00	1.0	100

[a]Represents an upper limit to ϕ_{IC}.
Source: Adapted from N. J. Turro, *Modern Molecular Photochemistry*, Benjamin/Cummings, Menlo Park, CA, 1978, p. 181.

law thus provides an explanation for both the slow rate of internal conversion for $S_1 \rightarrow S_0$ and the fast rate of internal conversion of $S_n \rightarrow S_1$.

Exceptions to the generalization that $S_n \rightarrow S_1$ internal conversion completely dominates fluorescence and intersystem crossing are extremely rare. Azulene and its derivatives are the most cited examples.[7-10] Because of an unusually small value of k_{IC} of $\sim 7 \times 10^8$ s^{-1}, azulene has a small but significant quantum yield for fluorescence from S_2 of ~ 0.03. This is explained by the exceptionally large energy gap of ~ 40 kcal mol^{-1} between S_2 and S_1. Azulene also is characterized by an exceptionally rapid rate of $S_1 \rightarrow S_0$ internal conversion of ~ 10^{12} s^{-1}, suggestive of a small gap between S_0 and S_1 and/or a surface crossing between S_0 and S_1.

Although internal conversion between S_1 and S_0 is often insignificant as compared to fluorescence or intersystem crossing, for molecules with small energy gaps between S_1 and S_0, ϕ_{IC} can be quite large. Tetracene and pentacene, for example, have low S_1 energies of ~ 60 and ~ 50 kcal mol^{-1} and large values for ϕ_{IC} of 0.20 and 0.75, respectively (Table 1-4). Molecules undergoing photochemical reaction in their excited singlet states also may exhibit large quantum yields for internal conversion. For example, in Table 1-4 it is seen that cyclobutanone and 1,3-pentadiene both have $\phi_F + \phi_{ISC} = 0$. In the case of cyclobutanone a cleavage reaction occurs with a quantum efficiency of ~ 0.3, while 1,3-pentadiene undergoes an isomerization with an efficiency of ~ 0.1. By inference, the quantum efficiencies of internal conversion are ~ 0.7 and ~ 0.9, respectively. This internal conversion is probably due to a reversal in the photochemical reaction that leaves the molecule in S_0. Note in Table 1-4 that benzene has a quantum yield for internal conversion of ~ 0.70. This may be due to a reversible photochemical reaction or a curve crossing between S_0 and S_1.

Intersystem Crossing

Intersystem crossing from S_1 to T_1 plays a critical role in determining the emission spectroscopic and photochemical properties of molecules. Molecules with small values

Table 1-5 ▪ Rate Constants for Intersystem Crossing

Molecule	k_{ISC} (s^{-1})	ΔE_{ST} (kcal/mole)	Transition
Naphthalene	10^6	30	$S_1(\pi,\pi^*) \rightarrow T_1(\pi,\pi^*)$
Anthracene	10^8	2–3	$S_1(\pi,\pi^*) \rightarrow T_2(\pi,\pi^*)$
Pyrene	10^6	30	$S_1(\pi,\pi^*) \rightarrow T_1(\pi,\pi^*)$
Triphenylene	5×10^7	20	$S_1(\pi,\pi^*) \rightarrow T_1(\pi,\pi^*)$
1-bromonaphthalene	10^9	30	$S_1(\pi,\pi^*) \rightarrow T_1(\pi,\pi^*)$
9-acetoanthracene	$\sim 10^{10}$	~ 5	$S_1(\pi,\pi^*) \rightarrow T_2(n,\pi^*)$
Acetone	5×10^8	5	$S_1(n,\pi^*) \rightarrow T_1(n,\pi^*)$
Benzophenone	10^{11}	5	$S_1(n,\pi^*) \rightarrow T_2(\pi,\pi^*)$
Benzil	5×10^8	5	$S_1(n,\pi^*) \rightarrow T_1(n,\pi^*)$
Biacetyl	7×10^8	5	$S_1(n,\pi^*) \rightarrow T_1(n,\pi^*)$
9,10-dibromoanthracene	$\sim 10^8$	30	$S_1(\pi,\pi^*) \rightarrow T_1(\pi,\pi^*)$
			$S_1(\pi,\pi^*) \rightarrow T_2(\pi,\pi^*)$
[2.2.2]-diaza-bicyclooctane	10^6	25	$S_1(n,\pi^*) \rightarrow T_1(n,\pi^*)$

Source: Adapted from N. J. Turro, *Modern Molecular Photochemistry*, Benjamin/Cummings, Menlo Park, CA, 1978, p. 186.

of k_{ISC} are usually highly fluorescent; exceptions are those compounds that undergo fast S_1 to S_0 internal conversion, as previously discussed. Molecules having high values of k_{ISC} are usually phosphorescent at low temperatures and in organized media where the rate of $T_1 \rightarrow S_0$ intersystem crossing is slow. The T_1 state of organic molecules is responsible for several classes of photochemical reactions, and many of the post-column photochemical reaction detection methods developed for chromatography are based on specific reactions of the triplet state. The high reactivity of the triplet state is partly due to the relatively long lifetime in solution of this metastable species, partly due to its high internal energy content and partly due to its diradical character.

The rate constants for intersystem crossing, singlet–triplet energy gaps (ΔE_{ST}) and electronic configurations of initial and final states are given for a variety of compounds in Table 1-5. The first-order rate constant for $S_1 \rightarrow T_1$ intersystem crossing varies over several orders of magnitude from compound to compound. However, nearly all measurements of k_{ISC} fall in the range 10^6 to 10^8 s^{-1} for unsubstituted aromatic hydrocarbons. This range is comparable to that for fluorescence, $k_F \sim 10^6$ to 10^9 s^{-1}, so that T_1 is populated by intersystem crossing to a significant extent in all but the most highly fluorescent compounds. The energy gap law (based on the Franck–Condon principle) previously discussed for internal conversion also applies to intersystem crossing, but as seen in Table 1-5 intersystem crossing from S_1 can terminate in more than one triplet state. For example, when the $S_1 \rightarrow T_2$ gap is small, the overall $S_1 \rightarrow T_1$ transition can take place via $S_1 \rightarrow T_2$ intersystem crossing followed by rapid $T_2 \rightarrow T_1$ internal conversion. Thus, anthracene has a value of k_{ISC} that is about 2 orders of magnitude greater than that for either naphthalene or pyrene, in part because its $S_1(\pi,\pi^*)$ state can couple with a proximate $T_2(\pi,\pi^*)$ state.

As discussed earlier, singlet–triplet transitions are forbidden in the absence of spin-orbit coupling, and heavy atoms are found to be strong promoters of spin-orbit coupling. Thus in Table 1-5 it is seen that k_{ISC} is about 3 orders of magnitude larger for 1-bromonaphthalene as compared to naphthalene. Besides this intramolecular "heavy-atom" effect, it is found that spin-orbit coupling may be induced by heavy atoms in the solvent. For example, it is common to enhance phosphorescence spectra by the use of solvents containing heavy atoms and/or by the addition of large ions such as Tl(II) into the solvent. The intermolecular heavy-atom effect enhances the quantum yield for intersystem crossing and shortens the phosphorescence (also spin forbidden) lifetime. Heavy atoms also enhance $T_1 \rightarrow S_0$ intersystem crossing, which competes with phosphorescence. Thus, the overall effect of heavy atoms on phosphorescence is variable and depends on its relative ability to stimulate phosphorescence vs. nonradiative quenching to the ground state.

Aside from the heavy-atom effect, it has been found that spin-orbit coupling is much greater between electronic states having different electronic configurations than for those of the same configuration. In Table 1-5 it is seen that the rate constants for intersystem crossing for $S_1(n,\pi^*) \rightarrow T_n(\pi,\pi^*)$ is much greater than that of either $S_1(\pi,\pi^*) \rightarrow T_n(\pi,\pi^*)$ or $S_1(n,\pi^*) \rightarrow T_n(n,\pi^*)$. Benzophenone is the premiere example of this; it has $k_{\text{ISC}} = 10^{11}\ \text{s}^{-1}$ for its $S_1(n,\pi^*) \rightarrow T_2(\pi,\pi^*)$ nonradiative transition with the result that $\phi_{\text{F}} \simeq 0$ and $\phi_{\text{ISC}} \simeq 1$. Chapter 7 describes post-column photochemical reaction techniques for liquid chromatography based on quinone photochemistry. Like benzophenone, quinones experience rapid intersystem crossing from $S_1(n,\pi^*)$ to $T_2(\pi,\pi^*)$. The $T_2(\pi,\pi^*)$ state subsequently undergoes rapid internal conversion to the reactive $T_1(n,\pi^*)$ state.

Fluorescence

Within a few picoseconds following the absorption of a photon, a condensed phase molecule typically has been deactivated to the lowest vibrational levels of the S_1 state. Its fate will now be either internal conversion to the ground state, intersystem crossing to the lowest triplet state, chemical reaction, energy transfer to another molecule or fluorescence. Fluorescence can hardly be discussed in the absence of these other processes, since they are all competitive. In terms of first-order rate constants, the quantum yield for fluorescence is given by

$$\phi_{\text{F}} = \frac{k_{\text{F}}}{k_{\text{F}} + k_{\text{IC}} + k_{\text{ISC}} + k_{\text{ET}} + k_{\text{R}}} \tag{1-15}$$

Obviously, the quantum yield for fluorescence increases and approaches unity as k_{F} becomes large compared to the first-order rate constants for other processes. Since the radiative lifetime is the reciprocal of k_{F}, high fluorescence efficiency is favored by a short radiative lifetime. Fluorescence is the reverse process of absorption, and it is therefore possible to relate the rate constant for fluorescence to the oscillator strength of the transition, which is given by the integral under the absorption curve. This relationship, derived by Strickler and Berg,[11] is given by

$$k_{\text{F}} = 2.88 \times 10^{-9} n^2 \langle \nu^{-3} \rangle_{\text{av}}^{-1} \int \frac{\epsilon}{\nu}\, d\nu \tag{1-16}$$

where n is the refractive index, $\langle \nu^{-3} \rangle_{\text{av}}$ is the averaged frequency of the fluorescence and $\int (\epsilon/\nu)\, d\nu$ is the area under the absorption curve. Molecules that absorb strongly to S_1 have large values of k_{F} and short radiative lifetimes. Thus, in order to be highly fluorescent a molecule also must have a high integrated extinction coefficient. A useful approximation to the radiative lifetime for molecules that absorb in the near UV is given by[12]

$$\tau_F(s) \simeq \frac{10^{-4}}{\epsilon_{max}} \tag{1-17}$$

Thus, for anthracene whose lowest excited singlet is a π, π^* state, $\epsilon_{max} \simeq 10^4$ M^{-1} cm^{-1}, and τ_F is calculated to be $\sim 10^{-8}$ s. A molecule may have a large value for λ_{max} and integrated extinction coefficient but a long radiative lifetime if the strongly absorbing band is due to S_2 or a higher singlet state. This is common for aldehydes, quinones and nitroaromatics, all of which have an n, π^* state for S_1. The more strongly absorbing π, π^* singlet states are rapidly deactivated to S_1 by internal conversion. Thus, although the molecule may absorb very strongly to S_2 or higher singlet states, its fluorescence lifetime is determined by its weak, long wavelength absorption to S_1. The λ_{max} of anthraquinone, for example, is $\sim$ 54,000 M^{-1} cm^{-1} at 251 nm, but the extinction coefficient at 424 nm for absorption to its lowest $^1(n, \pi^*)$ state is only $\sim$ 60 M^{-1} cm^{-1}, and the fluorescence lifetime is approximated by Eq. (1-17) to be 1.7×10^{-6} s. The much longer radiative lifetime of anthraquinone as compared to anthracene allows other deactivation routes to compete more favorably. In particular, anthraquinone has a ϕ_F of $\lesssim 10^{-5}$ but a ϕ_{ISC} of ~ 1.0. Other factors, such as the small energy gap between the S_1 and one or more triplet states also have the effect of enhancing intersystem crossing in anthraquinone and other carbonyl and nitro compounds.

Structural and Substituent Effects. Every molecule that absorbs light would fluoresce if it were not for the competing processes of internal conversion, intersystem crossing, electronic quenching and chemical reaction. Thus, the effects of molecular structure on fluorescence intensity must be explained in terms of the relative rates of the various deactivation processes. Table 1-6 summarizes the effects of substituents on the fluorescence of aromatic compounds. In general, electron releasing groups such alkyl, hydroxyl, methoxyl and amino groups tend to increase fluorescence intensity by increasing the absorption coefficient and decreasing the radiative lifetime so that

Table 1-6 ▪ Effects of Substituents on Fluorescence of Aromatic Compounds

Substituent	Effect on Emission Wavelength	Effect on Intensity
—R (alkyl)	None	Slight increase
$-OH$, $-OCH_3$, $-NH_2$, $-NHR$, $-NR_1R_2$	Decrease	Increase
$-COOH$, $-SH$	Decrease	Decrease
$-NO$, $-NO_2$	...	Total quenching
$-SO_3H$	None	None
$-F$, $-Cl$, $-Br$, $-I$	Decrease	Decrease

Source: Adapted from B. L. Van Duuren and T.-L. Chan, "Fluorescence Spectrometry," in *Spectrochemical Metholds of Analysis*, J. D. Winefordner, ed., Wiley-Interscience, New York, 1971, Vol. 9, p. 419.

fluorescence competes better with other deactivation processes. Conversely, electron withdrawing groups such as nitroso, nitro, carboxyl, sulfhydryl and halogens tend to decrease fluorescence. Nitroaromatic compounds are well known for their total lack of fluorescence. This is due to a very rapid rate of intersystem crossing to the triplet state in these compounds whose S_1 has an n,π^* configuration. In general, compounds having $S_1(n,\pi^*)$ states, such as aldehydes and ketones, usually are at best weakly fluorescent as a result of a long radiative lifetime for $S_1 \rightarrow S_0$ and rapid intersystem crossing to a n,π^* or π,π^* triplet. Halogen substitution usually results in decreased fluorescence efficiency in the order F < Cl < Br < I. Here, not only is there an electron withdrawing effect, but more importantly, the heavy atom may induce intersystem crossing by spin-orbit coupling as discussed previously. Sulfonic acid functional groups are neither electron releasing nor electron withdrawing and have very little effect on the photophysical properties of aromatic compounds. Incorporation of a sulfonate group into a molecule provides a means of increasing its solubility in polar solvents without substantially changing its emission or photochemical properties. Sulfonate groups also provide a means of immobilizing fluorophores or photochemically active species either on anion exchange resins or by covalent bonding through the sulfonate group.

Competing nonradiative transitions are reduced for compounds that have planar and rigid structures. Nonrigid and nonplanar molecules can access regions of their potential energy surfaces that have good Franck–Condon factors for transitions to other electronic states. An example of this principle is given by the comparison of phenolphthalein and fluorescein:

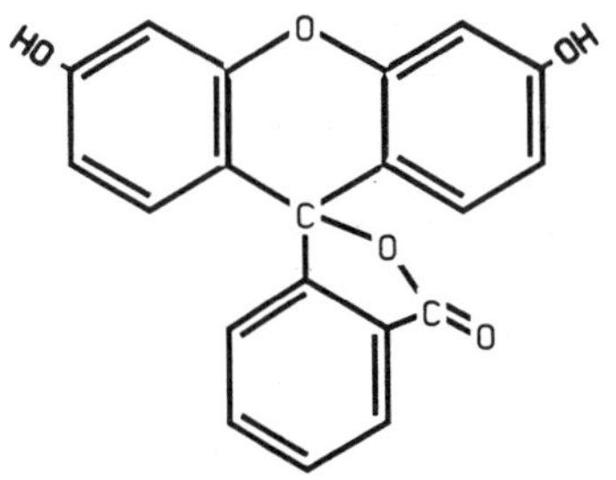

HO O OH C O C=O

Fluorescein

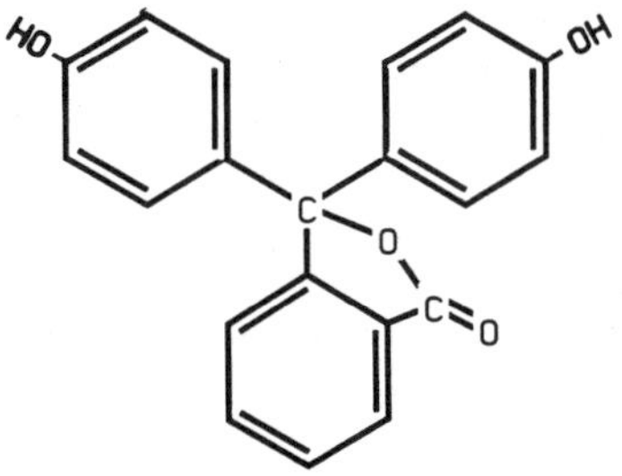

HO OH C O C=O

Phenolphthalein

In phenolphthalein there is free rotation about the C—C bond for each of the phenol groups. In fluorescein, the two phenol groups are held in a rigid, coplanar configuration by a bridging oxygen. The result is that fluorescein is highly fluorescent, while phenolphthalein is nonfluorescent.

These generalizations concerning the properties of fluorescent molecules may be used in the design of post-column reaction systems to enhance detection in chromatography. For example, a common use of photochemical reaction detection in liquid chromatography is to photochemically convert a nonfluorescent compound to a fluorescent compound. Stilbenes undergo ring closure and, in the presence of an oxidizing agent, elimination of two hydrogen atoms, to form phenanthrene derivatives:

cis-Stilbene $\xrightarrow{h\nu}$ (H H) $\xrightarrow{\text{oxidizing agent}}$ Phenanthrene (1-18)

This rigid, planar molecule exhibits much greater fluorescence than either the *trans-* or *cis-*stilbene precursor. Thus, the detection sensitivity of stilbenes, such as the important synthetic hormone diethylstilbestrol, can be greatly enhanced by use of this photochemical reaction following separation via HPLC. The selectivity also is enhanced in that very few potential interferences are likely to have their fluorescence enhanced by photochemical reaction. Another example, discussed in detail in Chapter 6, is the photoreduction of quinones in the absence of O_2 and the presence of a hydrogen-atom donor such as an alcohol. The reaction for anthraquinone is given by:

$$\xrightarrow[\text{RH}]{h\nu} \quad (1\text{-}19)$$

As already discussed, anthraquinone and other quinones undergo rapid intersystem crossing with a quantum yield of ~ 1.0 and thus are nonfluorescent. The dihydroxyanthracene product, as a result of the electron-donating hydroxyl groups, is even more fluorescent than anthracene. This photochemical reaction may be used not only to detect quinones, but also to detect hydrogen donors such as aliphatic aldehydes, ketones, ethers, sugars and steroids, and has the advantage that the analyte need not absorb light itself.[13-15] Thus, it is a photochemical reaction applicable to many compounds not directly detectable in liquid chromatography by UV absorption or fluorescence.

Phosphorescence

Those molecules that undergo intersystem crossing from S_1 also will experience rapid internal conversion within the triplet manifold and rapid vibrational deactivation to the lowest vibrational levels of the metastable T_1 state. Because absorption from S_0 to T_1 is spin forbidden in the absence of spin-orbit coupling and thus very weak, the radiative decay of T_1 to S_0, known as phosphorescence, also is a weak transition characterized by a long radiative lifetime. The first-order rate constant for phosphorescence (whose reciprocal is the radiative lifetime) can, in principle, be calculated by integration under the $S_0 \rightarrow T_1$ absorption curve using the Strickler–Berg equation [Eq. (1-15)]. In practice, the absorbance is seldom large enough to be easily measured. Because of the characteristically long phosphorescent lifetimes of $\sim 10^{-3}$ to ~ 10 s or longer, deactivation by intersystem crossing from T_1 to S_0 dominates in solution at room temperature for nearly all organic compounds. To obtain phosphorescence spectra, the sample is often dispersed in a rigid matrix at low temperature such as a glass-like solid composed of a mixture of ether, pentane and alcohol (EPA) at liquid nitrogen temperature.

A few compounds do exhibit room temperature phosphorescence, biacetyl being the most notable example, and many compounds will phosphoresce at room temperature when incorporated into micelles or other organized media.[16] Micelles have been added to the mobile phase of liquid chromatography to allow detection of phosphorescent compounds.[17–19] Unfortunately, micelles degrade the performance of liquid chromatography columns by reducing the rates of mass transfer between the mobile and stationary phases, and the detection limits using this technique are not competitive with ordinary UV absorption detection. The room temperature phosphorescence of biacetyl also has been used in HPLC to detect nonfluorescent compounds via an energy transfer mechanism to be discussed in the following section.[20–23]

Electronic Energy Transfer

Triplet–triplet intermolecular energy transfer,

$$\mathrm{D}^*(T_1) + \mathrm{A}(S_0) \rightarrow \mathrm{D}(S_0) + \mathrm{A}^*(T_1) \tag{1-20}$$

occurs at a nearly diffusional rate between donor (D) and acceptor (A) molecules provided that the triplet energy of the donor is greater than that of the acceptor. This process is of great importance to photochemistry, and has been used to advantage in mechanistic studies, photochemical synthesis and the development of several detection methods for analytical chemistry. An example of the latter is that of sensitized room temperature phosphorescence. In this technique biacetyl, one of the rare compounds that exhibits room temperature phosphorescence, is added to the mobile phase of a liquid chromatograph. As the analytes exit the chromatographic column, they are excited by the light source of a conventional fluorescence detector. Those compounds that undergo intersystem crossing to a triplet state having an energy greater than the 56 kcal/mol T_1 state of biacetyl can transfer energy to biacetyl, which in turn phosphoresces. Thus, compounds that do not themselves undergo room temperature phosphorescence may be detected. The phosphorescence and fluorescence spectra of biacetyl are shown in Fig. 1-7. In this scheme the analyte serves as a sensitizer.[20] Its

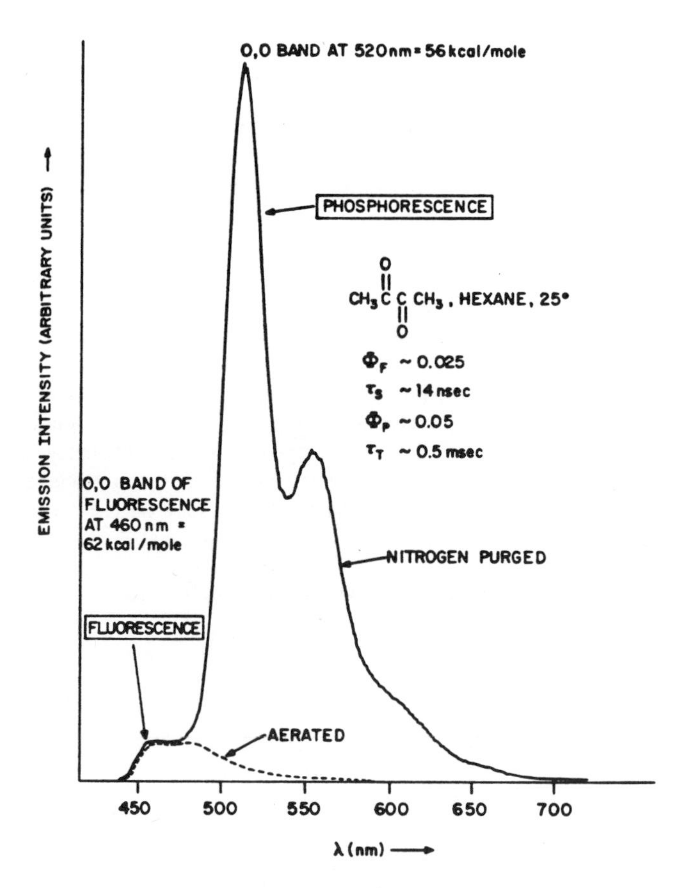

Figure 1-7 ■ Phosphorescence and fluorescence spectra of biacetyl in hexane at room temperature for aerated and deaerated solutions. Oxygen completely quenches the phosphorescence, but has little effect on fluorescence. From N. J. Turro, *Modern Molecular Photochemistry*, Benjamin/Cummings, Menlo Park, CA, 1978, p. 117.

principal limitation is that only compounds having T_1 greater than that of biacetyl can be detected. In order to detect compounds having triplets below 56 kcal/mol, the scheme may be reversed. The biacetyl is excited to establish a standing phosphorescence, and the quenching of that phosphorescence by energy transfer to the analyte is detected as a negative peak.[21–23]

A major disadvantage of both sensitized phosphorescence and micelle-stabilized phosphorescence previously discussed is that the T_1 state of virtually all organic molecules is rapidly deactivated by energy transfer to dissolved oxygen. The effect of oxygen on the biacetyl phosphorescence spectrum may be seen in Fig. 1-7. In both systems oxygen must be rigorously removed from the HPLC mobile phase. Another approach to detecting compounds that undergo rapid intersystem crossing takes advantage of energy transfer to oxygen. Quenching of triplets by oxygen also may be thought of as singlet oxygen sensitization in that the triplet ground state of oxygen is promoted to one of its low-lying singlet states,

$$D^*(T_1) + O_2(^3\Sigma_g^-) \rightarrow D(S_0) + O_2(^1\Sigma_g^+) \text{ or } O_2(^1\Delta_g) \tag{1-21}$$

The $O_2(^1\Sigma_g^+)$ and $O_2(^1\Delta_g)$ states (see Fig. 1-1) lie 37.5 and 22.5 kcal/mol above the ground state, respectively. Those compounds having $T_1 > 37.5$ kcal/mol appear to sensitize formation of the $^1\Sigma_g^+$ state, while those having lower triplets (but greater than 22.5 kcal/mol) form $^1\Delta_g$. In either case, internal conversion is so rapid in solution that the net effect is the formation of $^1\Delta_g$ in either case. In common HPLC solvents such as mixtures of acetonitrile and water or mixtures of methanol and water, the singlet oxygen lifetime with respect to intersystem crossing to the ground state is only a few microseconds. The singlet oxygen may be "trapped" with high efficiency, however, by its reaction with furans spiked into the mobile phase. Changes in spectral properties (either absorption or fluorescence) of the furan "trap" may be used to quantify the amount of singlet oxygen produced. The sequence of reactions

$$\text{analyte } (S_0) + h\nu \rightarrow \text{analyte* } (S_n) \rightarrow \text{analyte* } (T_1) \tag{1-22}$$

$$\text{analyte* } (T_1) + O_2(^3\Sigma_g^-) \rightarrow \text{analyte } (S_0) + O_2(^1\Delta_g) \tag{1-23}$$

$$O_2(^1\Delta_g) + \text{furan} \rightarrow \text{oxidized product} \tag{1-24}$$

$$\text{Net:} \quad \text{furan} + O_2(^3\Sigma_g^-) \rightarrow \text{oxidized product} \tag{1-25}$$

is photocatalytic in that the analyte may be excited repeatedly by the light source with no net loss of analyte (the catalyst). Completion of each cycle results in the oxidation of one furan molecule. Depending on the light intensity, for each analyte molecule hundreds of furan molecules may be oxidized to form products within a short time. Thus the signal, which may be based on either the loss of the furan or appearance of the product, is amplified by the photosensitized reaction.[24] A variety of post-column reaction schemes based on singlet oxygen sensitization are discussed in Chapter 7.

Other types of electronic energy transfer also may occur, but are much less frequently encountered. Transfer of energy from an excited singlet is rare, because the lifetime of the state is very short. Triplet–triplet annihilation is an interesting form of energy transfer that results in *delayed fluorescence*. Here two triplet state molecules may pool their energy to produce an excited singlet and a ground state singlet:

$$B(T_1) + B(T_1) \rightarrow B(S_1) + B(S_0) \tag{1-26}$$

$$B(S_1) \rightarrow B(S_0) + h\nu \tag{1-27}$$

The spectrum of light emitted is identical to that of fluorescence, but the observed lifetime of the luminescence is much longer than for fluorescence and is determined by the lifetime of the triplet state. If the triplet is produced by light excitation, the intensity of luminescence will vary as the square of the light intensity, since the rate of reaction (1-26) is second order in triplet concentration. This experimental observation distinguishes this form of delayed fluorescence from thermally activated delayed

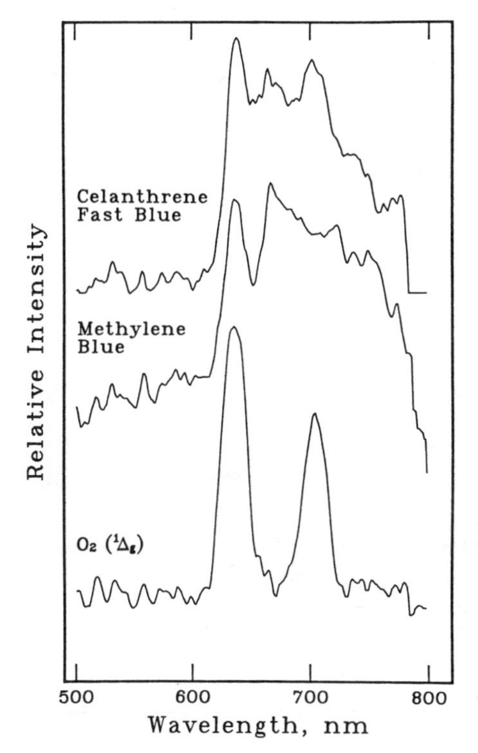

Figure 1-8 ■ Uncorrected diode-array spectra from $O_2(^1\Delta_g)$ chemiluminescence produced in an aerosol spray of H_2O_2 and NaOCl. Lower trace: Dimol emission in the absence of an energy acceptor. The 634 nm emission band corresponds to a transition terminating in the lowest vibrational levels ($v = 0$) of the ground electronic states of both oxygen molecules. The 703 nm emission band results from a transition in which the final state of one of the oxygen molecules contains one quantum of vibrational energy ($v = 1$). Upper traces: Celanthrene fast blue and methylene blue present as energy acceptors. From B. Shoemaker and J. W. Birks, *J. Chromatogr.* **209**, 251–263 (1981).

fluorescence, such as that displayed by pyrene, in which S_1 and T_1 are sufficiently close in energy that thermal activation will occassionally promote a T_1 molecule to S_1 where it can fluoresce.

Singlet oxygen undergoes an interesting electronic energy pooling reaction referred to as *dimol emission*.[25] Two singlet oxygen molecules form excited-state dimers that emit a single photon having twice the energy of the 1.27 μm spin-forbidden transition of individual $O_2(^1\Delta_g)$ molecules as follows:

$$O_2(^1\Delta_g) + O_2(^1\Delta_g) \rightarrow O_2(^3\Sigma_g^-) + O_2(^3\Sigma_g^-) + h\nu\ (\sim 635\ \text{nm}) \tag{1-28}$$

The spectrum of the emission is given in Fig. 1-8. Two emission bands are seen; one band corresponds to (0,0) vibrational transitions for both O_2 molecules, and the second band results when one of the O_2 molecules terminates in the $v = 1$ level of the ground state. The dimols also may transfer energy to fluorophores having low-lying singlet states to induce fluorescence,

$$O_2(^1\Delta_g) + O_2(^1\Delta_g) + F \rightarrow O_2(^3\Sigma_g^-) + O_2(^3\Sigma_g^-) + F^*(S_1) \tag{1-29}$$

$$F^*(S_1) \rightarrow F(S_0) + h\nu \tag{1-30}$$

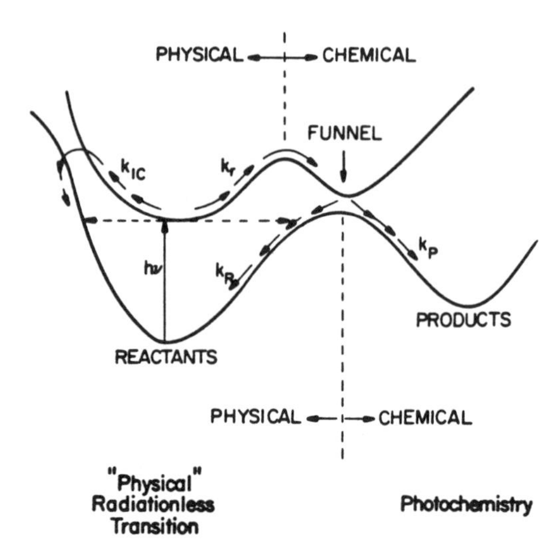

Figure 1-9 ■ Representation of photophysical and photochemical processes occurring on the same potential energy surfaces. From N. J. Turro, *Modern Molecular Photochemistry*, Benjamin/Cummings, Menlo Park, CA, 1978, p. 194.

Also shown in Fig. 1-8 is the emission spectrum that results when methylene blue is present as an energy acceptor. Singlet oxygen can be produced in high concentrations by the chemical reaction[26]

$$H_2O_2 + ClO^- \rightarrow H_2O + Cl^- + O_2\left(^1\Delta_g\right) \tag{1-31}$$

This reaction has been used in an aerosol spray detector for liquid chromatography to produce singlet oxygen;[27] the singlet oxygen in turn excites fluorescent molecules via reaction (1-25). However, this form of energy transfer chemiluminescence is limited to the detection of colored compounds, since S_1 must be less than 45 kcal/mol. The more efficient peroxyoxalate chemiluminescence reaction has subsequently been applied to energy transfer chemiluminescence detection in HPLC,[28] as discussed in Chapters 4 and 5. The peroxyoxalate reaction can transfer up to ~ 105 kcal/mol of energy to certain classes of fluorescent compounds.[29]

Photochemical Reactions

By chemical reaction we mean that the average relative positions of nuclei in the ground state of a molecule or molecules are changed. This change in nuclear geometry may be relatively small, as in the isomerization of $A{=}B{-}C$ to $A{-}B{=}C$, or it may be so large that chemical bonds are broken. The relationship between photophysical and photochemical processes is illustrated in Fig. 1-9. The process is considered to be photophysical when, after being excited to a higher potential energy surface, the molecule finds its way back to the same potential well it was excited from. This may occur by a variety of paths described earlier such as fluorescence, internal conversion (shown in Fig. 1-9), intersystem crossing followed by phosphorescence or two consecutive intersystem crossing steps. In some instances, however, the excited-state molecule may wander onto a portion of the excited surface that undergoes radiationless processes to a region of the ground-state surface that we identify as products. Figure 1-9 is drawn with a "funnel" in the excited state surface that may spill molecules into either ground-state reactants or products in order to show how the same internal conversion process can potentially result in either a photophysical or photochemical event. Of course, in many photochemical reactions this funnel lies entirely on the side of products.

Molecules may undergo a rich variety of primary photochemical reactions, and there are many schemes by which one may classify them. The approach of Calvert and Pitts[1] follows:

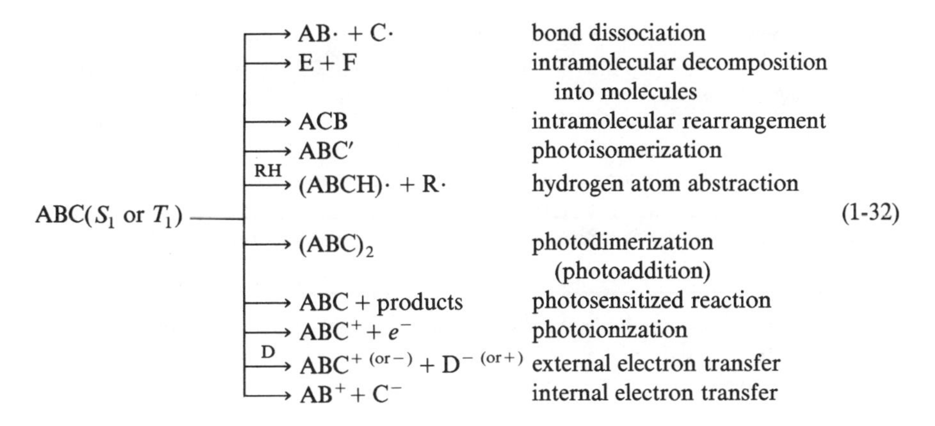

A primary photochemical reaction may be followed by any number of thermal reactions to produce an even greater variety of products. Some aspects of each of the classes of primary photochemical reactions given in (1-32) will be discussed briefly.

Bond Dissociation

In the gas phase this is often the most important type of photochemical reaction encountered. Atmospheric chemistry may be considered to be driven by photodissociation processes. Ozone, for example, is formed in the stratosphere (~ 12 to 50 km altitude) by the sequence of reactions[30]

$$O_2 + h\nu \rightarrow O + O, \qquad \lambda \leq 242 \text{ nm} \tag{1-33}$$

$$O + O_2 + M \rightarrow O_3 + M \tag{1-34}$$

where M is any gas molecule (e.g., N_2 and O_2). In the troposphere nitrogen dioxide photolysis is the principal source of oxygen atoms and thus ozone:

$$NO_2 + h\nu \rightarrow NO + O \tag{1-35}$$

Ozone itself is photolyzed at UV and visible wavelengths. At wavelengths $\lesssim 310$ nm, the products are singlet molecular oxygen and singlet oxygen atoms:

$$O_3 + h\nu \rightarrow O_2(^1\Delta_g) + O(^1D_2), \qquad \lambda \leq 310 \text{ nm} \tag{1-36}$$

The subsequent reaction of $O(^1D_2)$ with water vapor,

$$O(^1D_2) + H_2O \rightarrow 2 \cdot OH \tag{1-37}$$

is the source of OH radicals throughout most of the atmosphere.[31] The hydroxyl radical subsequently enters into a plethora of oxidative reactions. In addition to these photolysis reactions, there are dozens of other photochemical reactions in which chemical bonds are cleaved, all of which have important roles in controlling the chemical composition of the atmosphere. Such photodissociation reactions are typical in the gas phase.

Photodissociation reactions are much less important in the liquid phase, as the rate of vibrational deactivation is much greater and any radicals formed must escape the solvent cage; otherwise, they will recombine. For this reason, quantum yields for photodissociation in the liquid phase are often quite low. Nevertheless, they can be quite important. For example, in Chapter 6, a photochemical scheme for the detection of analytes based on a photooxygenation reaction that produces H_2O_2 is described.[32-34] The H_2O_2 serves as a surrogate for the analyte. The choice of excitation wavelengths for this detection scheme is limited, however, by photodissociation of H_2O_2:

$$H_2O_2 + h\nu \rightarrow 2 \cdot OH \qquad \lambda \leq 320 \text{ nm} \tag{1-38}$$

A low-pressure, 254 nm lamp cannot be used in conjunction with photochemical reaction schemes based on the production of H_2O_2, as the H_2O_2 is totally destroyed during the residence time of the photochemical reactor.

Intramolecular Decomposition into Molecules

The photodissociation of a molecule into two nonradical molecules is best illustrated by the photolysis of aldehydes to produce CO in one of its reaction channels:

$$RCHO + h\nu \rightarrow RH + CO \tag{1-39a}$$

$$RCHO + h\nu \rightarrow R\cdot + HCO\cdot \tag{1-39b}$$

The quantum yield for CO production is found to be highly wavelength-dependent and favored by the absorption of higher energy photons. A variety of other radical and nonradical products also may occur. For example, if the aldehyde contains a γ hydrogen, the photolysis may form an enol-aldehyde and olefin via the Norrish type II split,

$$R_2CHCR_2CR_2CHO + h\nu \rightarrow R_2C = CR_2 + CR_2C(OH)H \tag{1-40}$$

where the Rs are some combination of hydrogen atoms and alkyl or aryl organic functional groups. The small molecules N_2 and CO_2 also are sometimes eliminated in primary photochemical events. For example,

$$C_6H_5{-}\overset{|}{\underset{|}{C}}{-}N{=}N{-}\overset{|}{\underset{|}{C}}{-}C_6H_5 \xrightarrow{h\nu} 2\,C_6H_5{-}\overset{|}{\underset{|}{C}}\cdot + N{\equiv}N \qquad (1\text{-}41)$$

$$C_6H_4{<}(O)(C{=}O){>} \xrightarrow{h\nu} C_6H_4 + CO_2 \qquad (1\text{-}42)$$

Benzyne

The formation of strong bonds in CO (256 kcal/mol), CO_2 (126 kcal/mol) and N_2 (225 kcal/mol) provide the driving force for reactions in which these small molecules are eliminated.

As discussed in Chapter 3, CO is a good analyte for the redox chemiluminescence detector in which the analyte reduces NO_2 to NO at a gold surface. The NO is subsequently detected by its chemiluminescence reaction with O_3. The photoelimination of CO from aldehydes might provide a selective means for detection of these compounds with high sensitivity.

Photoisomerization

Cis–trans isomerization is one of the most general photoreactions of olefins and other compounds having double bonds (e.g., C=N, N=N). The absorption of a photon to

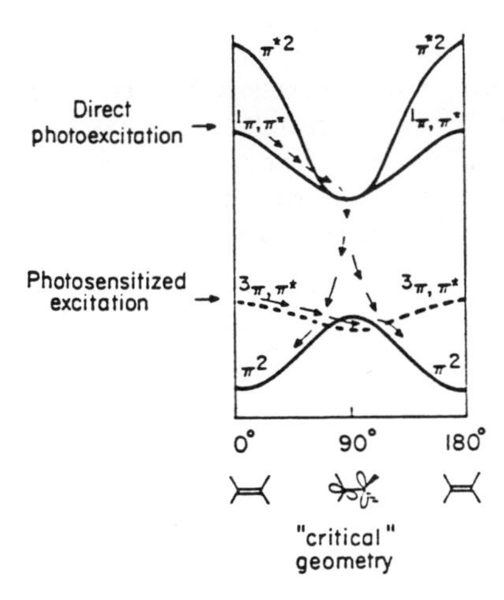

Figure 1-10 ■ Potential energy surface diagram for cis–trans isomerization of olefins. From N. J. Turro, *Modern Molecular Photochemistry*, Benjamin/Cummings, Menlo Park, CA, 1978, p. 475.

remove an electron from a π bonding orbital and place it in a π^* antibonding orbital results in an electron configuration in which the minimum of the potential energy surface occurs at a nonplanar geometry, intermediate between cis and trans, as shown in Fig. 1-10. When the molecule returns to the ground state by internal conversion, it may form either the cis or trans isomer. The reaction also may proceed on the $^3(\pi, \pi^*)$ surface, also shown in Fig. 1-10, and thus is promoted by sensitization reactions as well. In the case where only one isomer, cis or trans, absorbs a particular wavelength of light it is possible to completely convert one isomer into the other. This also is possible using sensitizers if only one of the isomers is a good energy acceptor. For example, *trans*-ionol is excited by lower energy triplet sensitizers ($E_T \leq 65$ kcal/mol) to produce *cis*-ionol,

$$\xrightarrow[^3\text{Sens}]{h\nu} \qquad (1\text{-}43)$$

OH OH

and after sufficient irradiation time the reaction goes to completion even though the cis isomer is less stable. The mechanism of vision involves linear polyenes related to ionol, and it is believed that vision is triggered by a cis–trans isomerization.[35] The detection of analytes based on a similar photosensitized cis–trans isomerization is a possibility that has not yet been explored.

Intramolecular Rearrangement

Photochemical reactions may result in rather extensive changes in molecular structure that are extremely useful in organic synthesis. However, the various classes of rearrangements and their mechanisms are too numerous to describe here. Unfortunately, not many of these have proven useful for enhancing chemical analysis. The rearrangement of *trans*-stilbenes to form the more highly fluorescent phenanthrenes, as described in Eq. (1-18), is one of the few examples of the application of photochemically initiated intramolecular rearrangement to enhance detection. In this case a cis–trans photoisomerization is followed by electrocyclic ring closure and finally oxidation.

Hydrogen Atom Abstraction

The abstraction of hydrogen atoms from hydrogen-atom-donating (HAD) substrates provides one of the more generally useful photochemical reactions for detection in liquid chromatography. Aromatic ketones, especially quinones, are particularly adept at this reaction, which is initiated by an n, π^* triplet state. The reaction may proceed as either an electron transfer followed by a proton transfer or the direct abstraction of a hydrogen atom to produce the semiquinone radical:

$$^{3}Q^{*} + RH \rightarrow QH\cdot + R\cdot \tag{1-44}$$

In the absence of oxygen the semiquinone radical disproportionates to form a quinone molecule and a fully reduced dihydroquinone, QH_2:

$$QH\cdot + QH\cdot \rightarrow Q + QH_2 \tag{1-45}$$

In contrast to quinones, which are nonfluorescent, the dihydroquinone is highly fluorescent, as discussed earlier. In the presence of dissolved oxygen the semiquinone radical donates its hydrogen atom to oxygen to form the hydroperoxyl radical

$$QH\cdot + O_2 \rightarrow Q + HO_2\cdot \tag{1-46}$$

The principal fate of HO_2 is disproportionation to form hydrogen peroxide:

$$HO_2\cdot + HO_2\cdot \rightarrow H_2O_2 + O_2 \tag{1-47}$$

A variety of post-column photochemical reaction-detection schemes have been developed based on the fluorescence of QH_2 or the chemiluminescence detection of H_2O_2 in the anaerobic and aerobic systems, respectively. Either quinones or HAD substrates (alcohols, aldehydes, ketones, amines, saccharides, steroids, etc.) may serve as the analyte. For the detection of quinones in the aerobic system, the reaction is photocatalytic in that up to 100 H_2O_2 molecules may be produced for each analyte molecule during its residence time in the photochemical reactor. These reaction schemes based on hydrogen atom abstraction are discussed further in Chapter 6.

Photodimerization and Photoaddition

This important class of photochemical reactions is exemplified by the photocyclodimerization of α,β-unsaturated carbonyl compounds and the cycloaddition of carbonyls to olefins. Coumarin, for example, photodimerizes in ethanol according to the reaction

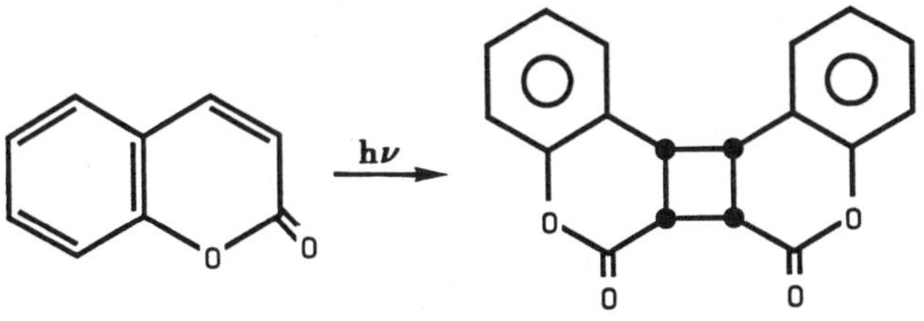

hν (1-48)

to form the cyclobutane derivative. This type of photodimerization causes mutation and death of plant and animal cells; adjacent thymines and to a lesser extent cytosines in DNA dimerize to form stable cyclobutane derivatives upon absorption of UV light:

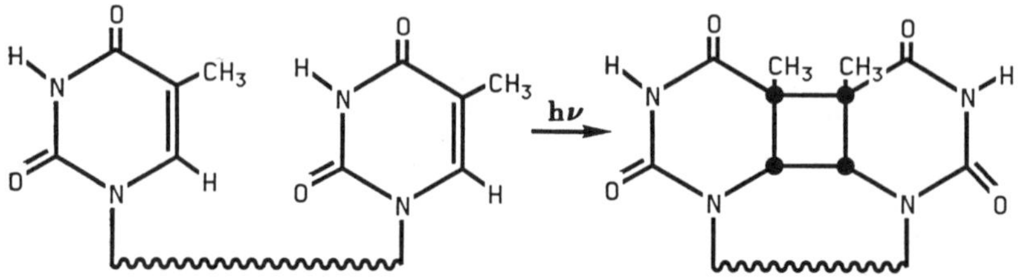

(1-49)

The rippled lines at the bottoms of the structures represent the deoxyribose-phosphate backbone of DNA. These dimers block normal replication and transcription of DNA. UV light in the wavelength region 280 to 320 nm that passes through the Earth's imperfect ozone shield results in sunburn and skin carcinoma in humans. Many types of cells contain a photoreactivating enzyme which selectively binds to these dimers and has a chromophore which absorbs light. The energy derived from the absorbed photon is used to separate the dimer into monomers and thus repair the DNA.

Olefins may add to α,β-unsaturated carbonyls

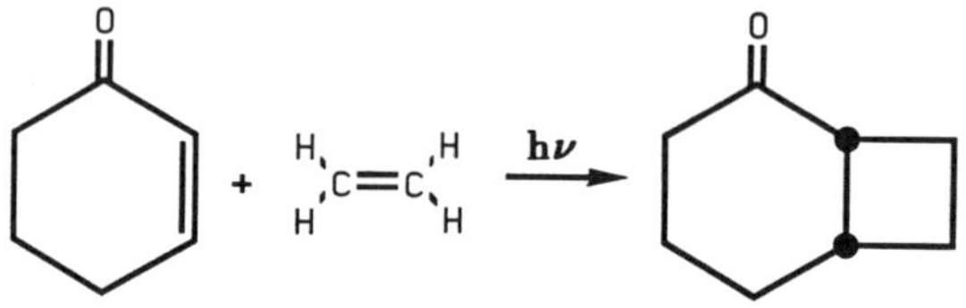

(1-50)

and to aromatic rings

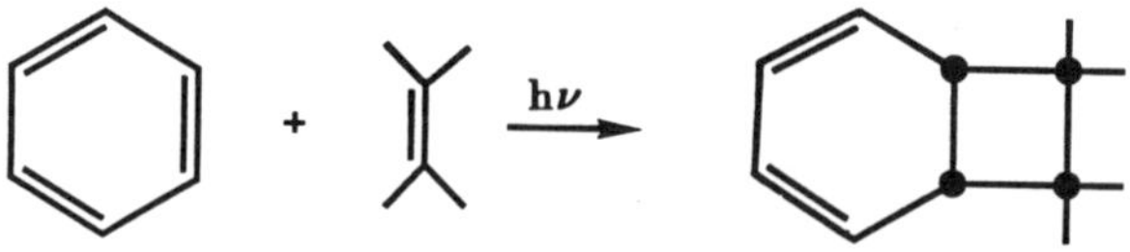

(1-51)

The cycloaddition of a carbonyl to an olefin produces an oxetane. For example, *n*-butyraldehyde adds to trimethyl ethylene as follows

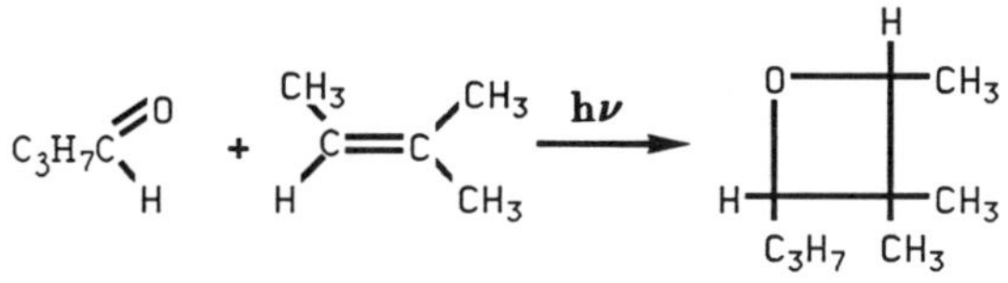

$$C_3H_7CHO + (CH_3)(H)C{=}C(CH_3)_2 \xrightarrow{h\nu} \text{oxetane: O–C(H)(CH}_3\text{)–C(CH}_3\text{)(CH}_3\text{)–C(H)(C}_3\text{H}_7\text{)} \quad (1\text{-}52)$$

In all of these reactions two double bonds are lost, and thus the detectability of the products by either UV absorption or fluorescence is decreased. In fact, photodimerizations are most often a nuisance in post-column photochemical reaction detection. An interesting possibility, however, is the use of photoaddition reactions to photochemically derivatize either carbonyl, olefin or enone compounds in a pre-column step. The reagent could be sufficiently conjugated to absorb UV light efficiently and contain functionalities that are selectively detected, e.g., halogens (for the electron capture detector) or sulfur (for the sulfur chemiluminescence detector). For example, it has been reported that irradiation of hexafluorocyclobutanone in the presence of hexafluoroproylene gives a bicyclic oxetane in 33% yield[36]:

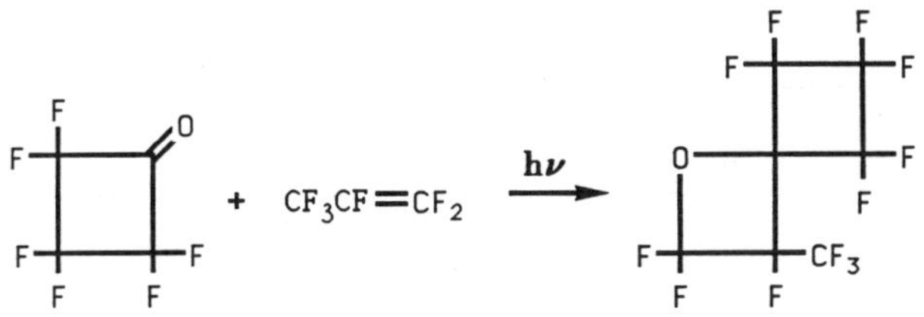

$$\text{hexafluorocyclobutanone} + CF_3CF{=}CF_2 \xrightarrow{h\nu} \text{bicyclic oxetane} \quad (1\text{-}53)$$

Furthermore, thioketones undergo analogous cycloaddition reactions to ketones.[37]

Photoionization

The absorption of light by a molecule to produce a free electron and positive ion,

$$A + h\nu \rightarrow A^+ + e^- \qquad (1\text{-}54)$$

is the basis of a commercially available detector for gas chromatography having high sensitivity. Selectivity may be adjusted via the choice of excitation wavelength, as various classes of molecules have different ionization potentials. Tables 1-7 and 1-8 contain the ionization potenials for selected atoms and compounds. Commercial detectors are provided with lamps having photon energies in the range 8.4 to 11.7 eV. For example, the nominal 8.3 eV (actually 8.44 eV Xe discharge lamp[38]) of HNU Systems, Inc. is highly selective; the only compounds ionized by this source are some amines and sulfur compounds, polynuclear aromatics and a few other large molecules.

Table 1-7 ▪ Ionization Potentials for Some Atoms and Simple Molecules

Molecule	I.P. (eV)	Molecule	I.P. (eV)
H	13.60	I_2	9.28
C	11.26	HF	15.77
N	14.54	HCl	12.74
O	13.61	HBr	11.62
Si	8.15	HI	10.38
S	10.36	SO_2	12.34
F	17.42	CO_2	13.79
Cl	13.01	COS	11.18
Br	11.84	CS_2	10.08
I	10.48	N_2O	12.90
H_2	15.43	NO_2	9.78
N_2	15.58	O_3	12.80
O_2	12.08	H_2O	12.59
CO	14.01	H_2S	10.46
CN	15.13	H_2Se	9.88
NO	9.25	H_2Te	9.14
CH	11.1	HCN	13.91
OH	13.18	C_2N_2	13.8
F_2	15.7	NH_3	10.15
Cl_2	11.48	CH_3	9.84
Br_2	10.55	CH_4	12.98

Source: HNU Systems, Inc., Newton, MA.

An 11.7 eV lamp (11.62 and 11.82 eV Ar discharge) provides an almost universal response. It allows for detection of aliphatics down to C_2, chlorinated hydrocarbons including CCl_4, formaldehyde, alkenes, alkynes, aromatics and many inorganics. The 10.2 eV lamp is most often used, as it is the most intense and provides the greatest sensitivity with moderate selectivity. The Lymann α line produced in a H_2 discharge is a convenient source of 10.2 eV photons; however, the nominal 10.2 eV lamp of HNU is actually a krypton discharge lamp (10.03 and 10.64 eV).[38]

The ionization potential of an excited-state molecule is reduced by the amount of its internal electronic energy. Thus, the lowest singlet state of molecular oxygen, $O_2(^1\Delta_g)$, has an ionization potential of only 11.08 eV, as compared to 12.06 eV for the ground state of O_2. This allows the selective detection of $O_2(^1\Delta_g)$ in the presence of a large excess of ground-state oxygen by use of an 11.7 eV argon discharge lamp.[39] The potential for photochemical amplification in gas chromatography based on singlet oxygen sensitization combined with photoionization of the singlet oxygen produced is discussed in Chapter 7.

Internal and External Electron Transfer

The photochemical production of ions, either directly as indicated in Eqs. (1-32) or as a result of photooxidations to produce acids such as HNO_3 and H_2SO_4 which dissociate into ions, is the basis of the photoconductivity detector for liquid chromatography.[40]

Table 1-8 ▪ Ionization Potentials for Selected Organic Molecules

Molecule	I.P. (eV)	Molecule	I.P. (eV)
Methane	12.98	*n*-butane	10.63
Cyclopropane	10.06	Cyclohexane	9.88
Methyl chloride	11.28	1-chlorobutane	10.67
Methyl bromide	10.53	1-bromobutane	10.13
Methyl iodide	9.54	1-iodobutane	9.21
Methyl alcohol	10.58	*n*-butyl alcohol	10.04
Methanethiol	9.44	1-butanethiol	9.14
Dimethyl sulfide	8.69	Di-*n*-propyl sulfide	8.30
Dimethyl disulfide	8.46	Diethyl disulfide	8.27
Formaldehyde	10.87	*n*-butyraldehyde	9.86
Acetone	9.69	2-heptanone	9.33
Formic acid	11.05	*n*-butyric acid	10.16
Methyl formate	10.82	*n*-butyl formate	10.50
Methyl amine	8.97	*n*-butyl amine	8.71
Dimethyl amine	8.24	Di-*n*-butyl amine	7.69
Trimethyl amine	7.82	Tri-*n*-propyl amine	7.23
Formamide	10.25	Acetamide	9.77
Nitromethane	11.08	2-nitropropane	10.71
Acetonitrile	12.22	*n*-butyronitrile	11.67
Methyl thiocyanate	10.07	Ethyl thiocyanate	9.89
Ethylene	10.52	1-butene	9.58
Furan	8.89	2-methyl furan	8.39
Benzene	9.25	Toluene	8.82
Fluorobenzene	9.20	Chlorobenzene	9.07
Bromobenzene	8.98	Iodobenzene	8.73
Ethylene oxide	10.57	Propylene oxide	10.22
Naphthalene	8.12	1-methylnaphthalene	7.69
Phenol	8.50	Aniline	7.70
Pyrrole	8.20	Pyridine	9.32

Source: HNU Systems, Inc. Newton, MA.

Chemiluminescence

For highly exoergic chemical reactions, one or more of the products is often formed in an excited electronic state. Relaxation of the excited-state molecule by radiative emission (either fluorescence or phosphorescence) completes the reaction sequence referred to as chemiluminescence. One of the most common examples is the gas-phase reaction between nitric oxide and ozone:

$$NO + O_3 \rightarrow NO_2^* + O_2 \tag{1-55}$$

$$NO_2^* \rightarrow NO_2 + h\nu \tag{1-56}$$

This reaction is the source of the signal in the Redox Chemiluminescence Detector™ for gas chromatography described in Chapter 3.

A necessary condition for chemiluminescence is that the reaction be sufficiently exoergic to populate the excited state of the emitting species. For example, for the $NO + O_3$ reaction the enthalpy change for reactants going to products is -200 kJ/mol at 298 K. This reaction is sufficiently exoergic to form NO_2 in both the $^2B_1$ and $^2B_2$ excited electronic states.

Many gas-phase chemiluminescent reactions involve complex mechanisms in which atomic or radical species are formed as intermediates. For example, the $I_2 + F_2$ reaction, which forms IF in the $^3\Pi_{0^+}$ and $^3\Pi_1$ excited electronic states,[41,42] has been demonstrated by molecular beam experiments to involve IIF and F as reaction intermediates[43,44]:

$$I_2 + F_2 \rightarrow IIF + F \quad (1\text{-}57)$$

$$F + IIF \rightarrow IF + IF^* \quad (1\text{-}58)$$

$$IF^* \rightarrow IF + h\nu \quad (1\text{-}59)$$

Several highly efficient chemiluminescent reactions occur in solution and have been applied to chemical analysis. The reactions that have so far proven most useful include those of luminol, lucigenin and oxalate esters. These chemiluminescent reactions and their applications to liquid chromatography are thoroughly discussed in Chapters 4 and 5.

References

1. Calvert, J. G., and Pitts, J. N., Jr. *Photochemistry*. Wiley, New York, 1966.
2. Birks, J. B. *Photophysics of Aromatic Molecules*. Wiley, New York, 1970.
3. Turro, N. J. *Modern Molecular Photochemistry*. Benjamin/Cummings, Menlo Park, CA, 1978.
4. For a detailed discussion of term symbols for diatomic molecules and selection rules for radiative transitions, see: Herzberg, G. *Molecular Spectra and Molecular Structure. I. Spectra of Diatomic Molecules*, 2nd ed. Van Nostrand, Princeton, NJ, 1950.
5. Cotton, F. A. *Chemical Applications of Group Theory*. Interscience, New York, 1963.
6. Burdett, J. K. In *Spectroscopy*, B. P. Straughan and S. Walker, eds. Wiley, New York, 1976, Vol. 3, Chap. 4, p. 133.
7. Beer, M., and Longuet-Higgins, H. C. *J. Chem. Phys.* **23**, 1390 (1955).
8. Viswath, G., and Kasha, M. *J. Chem. Phys.* **24**, 757 (1956).
9. Birks, J. B. *Chem. Phys. Lett.* **17**, 370 (1972).
10. Murata, S., Iwanga, C., Toda, T., and Kohubun, H. *Ber. Bunsen. Gesell.* **76**, 1176 (1972).
11. Strickler, S. J., and Berg, R. A. *J. Chem. Phys.* **37**, 874 (1972).
12. Calvert, J. G., and Pitts, J. N., Jr. *Photochemistry*. Wiley, New York, 1966, p. 280.
13. Gandelman, M. S., and Birks, J. W. *Anal. Chem.* **54**, 2131 (1982).
14. Gandelman, M. S., Birks, J. W., Brinkman, U. A. T., and Frei, R. W. *J. Chromatogr.* **282**, 193 (1983).
15. Gandelman, M. S., and Birks, J. W. *Anal. Chim. Acta* **155**, 159 (1983).
16. Turro, N. J., Chiang, L. K., Ming-Fea, C., and Lee, P. *Photochem. Photobiol.* **27**, 523 (1978).
17. Cline Love, L. J., and Skrilec, M. *Anal. Chem.* **52**, 1559 (1980).
18. Cline Love, L. J., and Skrilec, M. *Anal. Chem.* **53**, 1872 (1981).
19. Weinberger, R., Yarmchuk, P., and Cline Love, L. J. *Anal. Chem.* **54**, 1552 (1982).
20. Donkerbroek, J. J., van Eikema Hommes, N. J. R., Gooijer, C., Velthorst, N. H., and Frei, R. W. *Chromatographia* **15**, 218 (1982).

21. Donkerbroek, J. J., Veltkamp, A. C., Praat, A. J. J., Gooijer, C., Frei, R. W., and Velthorst, N. H. *Applied Spectr.* **37**, 188 (1983).
22. Donkerbroek, J. J., Veltkamp, A. C., Gooijer, C., Velthorst, N. H., and Frei, R. W. *Anal. Chem.* **55**, 1886 (1983).
23. Gooijer, C., Markies, P. R., Donkerbroek, J. J., Velthorst, N. H., and Frei, R. W. *J. Chromatogr.* **289**, 347 (1984).
24. Shellum, C. L., and Birks, J. W. *Anal. Chem.* **59**, 1834 (1987).
25. Bader, L. W., and Ogryzlo, E. A. *Discuss. Faraday Soc.* **37**, 46 (1964).
26. Khan, A. U., and Kasha, M. *J. Am. Chem. Soc.* **92**, 3293 (1970).
27. Shoemaker, B., and Birks, J. W. *J. Chromatogr.* **209**, 251 (1981).
28. Kobayashi, S., and Imai, K. *Anal. Chem.* **52**, 424 (1980).
29. Lechtken, P., and Turro, N. *J. Mol. Photochem.* **6**, 95 (1974).
30. Chapman, S. *Phil. Mag.* **10**, 369 (1930).
31. Brasseur, G., and Solomon, S. *Aeronomy of the Middle Atmosphere*, 2nd ed. Reidel, Boston, 1986.
32. Gandelman, M. S., and Birks, J. W. *J. Chromatogr.* **242**, 21 (1982).
33. Poulsen, J. R., Birks, J. W., Gübitz, G., van Zoonen, P., Gooijer, C., Velthorst, N. H., and Frei, R. W. *J. Chromatogr.* **360**, 371 (1986).
34. Poulsen, J. R., Birks, J. W., van Zoonen, P., Gooijer, C., Velthorst, N. H., and Frei, R. W. *Chromatographia* **21**, 587 (1986).
35. Zubay, G. *Biochemistry*. Addison-Wesley, Reading, MA, 1983, Chap. 33.
36. Harris, J. F., Jr., and Coffman, D. D. *J. Am. Chem. Soc.* **84**, 1553 (1962).
37. Turro, N. J. *Modern Molecular Photochemistry*. Benjamin/Cummings, Menlo Park, CA, 1978, pp. 465–467.
38. Davenport, J. N., and Adland, E. R. *J. Chromatogr.* **290**, 13 (1984).
39. Clark, I. D., and Wayne, R. P. *Proc. Roy. Soc. (London) Ser. A* **314**, 111 (1969).
40. Popvich, D. J., Dixon, J. B., and Ehrlich, B. J. *J. Chromatogr.* **17**, 643 (1979).
41. Durie, R. A. *Can. J. Phys.* **44**, 337 (1966).
42. Birks, J. W., Gabelnick, S. D., and Johnston, H. S. *J. Molec. Spectr.* **57**, 23 (1975).
43. Kahler, C. C., and Lee, Y. T. *J. Chem. Phys.* **73**, 5122 (1980).
44. Trickl, T., and Wanner, J. *J. Chem. Phys.* **74**, 6508 (1981).

2

Chemiluminescence Detection in Gas Chromatography

Andrew A. Turnipseed and John W. Birks

Department of Chemistry and Biochemistry and Cooperative Institute for Research in Environmental Sciences (CIRES), University of Colorado, Boulder, Colorado 80309

The principal advantages of capillary gas chromatography (GC) over its competitor, high performance liquid chromatography (HPLC), are its greater resolving power and the availability of a wider variety of sensitive and selective detectors. The flame ionization detector (FID) is sufficiently sensitive (limit of detection $\simeq 1$ ng) for most applications and is nearly universal. The thermionic emission detector or nitrogen–phosphorus detector (NPD), a variant of the FID in which an alkali metal bead is placed in the flame, may be made selective for either nitrogen- or phosphorus-containing compounds. The electron capture detector (ECD) combines high sensitivity (typically low picogram detection limits) with a high degree of selectivity for compounds having high electron affinities such as halogenated compounds. The photoionization detector (PID) is another highly sensitive detector capable of detecting low picogram quantities of analytes; here the selectivity is directed toward those compounds having low ionization potentials. Selectivity toward sulfur or phosphorus heteroatoms is achieved in the flame photometric detector (FPD). In addition to several other ionization and spectroscopic detectors still considered somewhat exotic, gas chromatography is conveniently interfaced to both mass spectrometry and Fourier transform infrared spectrometers. Considering the smorgasbord of detectors already available to GC, one might ask whether there is any point in exploring still new means of detection. The answer is clear to anyone who is developing a GC method for determination of a new analyte or class of analytes in a complex matrix. Too often the available detectors are still lacking in sufficient sensitivity and/or selectivity. It is here that chemiluminescence detectors are finding their niches.

In a chemiluminescence detector, a reagent gas is mixed with the effluent of the gas chromatograph in a reaction cell monitored by a photomultiplier tube (PMT). If the reaction is sufficiently exothermic, one or more of the products may be formed initially in an excited electronic state, or in some cases products are formed in high vibrationally excited levels of the ground state. Loss of this excess internal energy by emission of light produces a signal at the photomultiplier tube. Chemiluminescence detection can be highly sensitive, because the light emitted is detected against a dark background and individual photons are counted with a high quantum efficiency. By using different reagent gases, a plethora of chemiluminescent reactions may be induced, resulting in a wide variety of selectivities. Additional selectivity may be achieved by use of an optical filter between the chemiluminescence cell and PMT to allow only light from a particular emitting species to reach the detector.

Chemiluminescence reagents are not limited to stable molecules. The highly exothermic reactions of atoms, radicals, ions and electronically excited species are easily produced in gas discharges and transported into the detector cell. Here we review examples of the use of O_3, "active nitrogen" [$N_2(A^3\Sigma_u^+)$ and N atoms], F atoms, F_2, Na atoms and OClO as reagents for chemiluminescence detection in chromatography.

Technically, light emitted from flames is chemiluminescence, and the FPD is a sulfur- or phosphorus-selective chemiluminescence detector whose signal is based on emissions from electronically excited S_2 and POH, respectively. In flames the chemical reactions and excitations are assisted by the high-temperature environment produced by the H_2/O_2 reaction. Here we will only concern ourselves with "cool" chemiluminescent reactions where the temperature in the reaction cell is not substantially altered by the heat of reaction. Such chemiluminescence detectors for gas chromatography have recently been commercialized by Sievers Research, Inc. The Redox Chemiluminescence Detector™ (RCD), which combines a post-column reaction catalyzed on a gold surface with the very well known and well studied reaction of NO with O_3 is a major topic of Chapter 3. The Sulfur Chemiluminescence Detector™ (SCD) is based on light emitted in the reaction of sulfur compounds with molecular fluorine or ozone. To date, these are the only nonflame chemiluminescence detectors having sufficient advantages over existing GC detectors to have become viable commercial products. Many of the chemiluminescent reactions discussed here do have considerable merit, and it is likely that these and other as yet unapplied and even undiscovered reactions will be applied successfully to GC detection.

Active Nitrogen Detector

Although most chemiluminescence detectors exhibit high selectivity, the active nitrogen detector, developed by Sutton *et al.*, provides a nearly universal response to hydrocarbons.[1] Although not developed commercially, this detector has the potential to compete with the FID. Active nitrogen is produced by passing molecular nitrogen through either a microwave or electrical discharge, generally at pressures in the range of 1 to 20 torr. These discharges produce a mixture of nitrogen atoms and highly excited nitrogen molecules, and this "cool plasma" can then be mixed with the effluent from a GC, as shown in Fig. 2-1.

$N_2(A^3\Sigma_u^+)$ is a metastable state with a lifetime of several seconds,[2] as emission from this state to the $X^1\Sigma_g^+$ ground state is spin forbidden and very weak. However, population of the $A^3\Sigma_u^+$ state, either by nitrogen atom recombination or radiative decay from the $B^3\Pi_g$ state, is quite rapid, so that a relatively large concentration of $N_2(A^3\Sigma_u^+)$ metastables can accumulate. The excited metastable state of nitrogen can undergo collisional transfer of energy to analytes when present, providing a pathway to electronic or vibrational excitation of the analytes. In the energy level diagram of

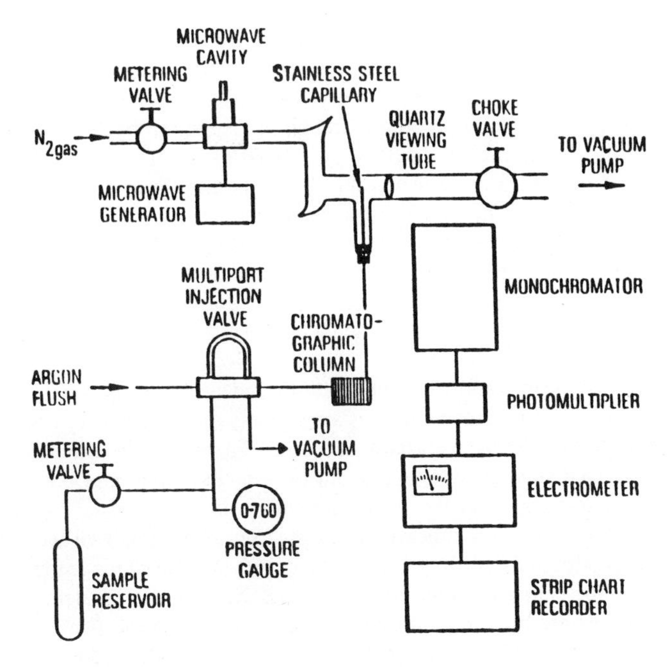

Figure 2-1 ■ Schematic of active nitrogen detector fitted to a capillary gas chromatograph. From D. G. Sutton, K. R. Westberg and J. R. Melzer, *Anal. Chem.* **51**, 1399 (1979).

Fig. 2-2, it can be seen that the metastable $N_2(A^3\Sigma_u^+)$ lies approximately 6.17 eV above the ground state. The production of active nitrogen populates many vibrational levels in the excited metastable state, so that the amount of energy that can be transferred in a collision is actually much higher. Since energy must be transferred to analyte molecules or atoms in exact quanta, the availability of many vibrational levels in both ground- and excited-state nitrogen allows for a wide range of energies to be transferred.

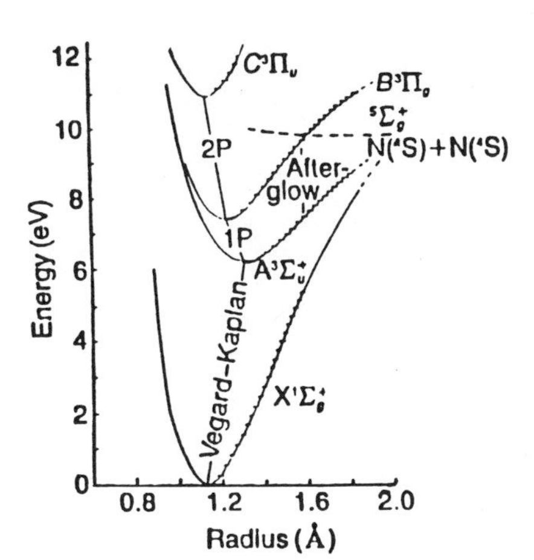

Figure 2-2 ■ Energy level diagram of nitrogen showing the energy levels of importance in the active nitrogen detector. From J. T. Clay and T. M. Niemczyk, *Spectroscopy* **2**, 36 (1987).

Generally, high energy discharges, such as microwave discharges, produce large amounts of nitrogen atoms, which then rapidly recombine to form the metastable $N_2(A^3\Sigma_u^+)$. Electrical discharges produce very few nitrogen atoms. Instead, population of the metastable $(A^3\Sigma_u^+)$ state proceeds through an excitation by the discharge to the $N_2(C^3\Pi_u)$ state. The transition $N_2(C^3\Pi_u) \rightarrow N_2(B^3\Pi_g)$ followed by $N_2(B^3\Pi_g) \rightarrow N_2(A^3\Sigma_u^+)$ can then populate the metastable $(A^3\Sigma_u^+)$ state. In systems using electrical discharges, collisional energy transfer is considered the main mechanism for the excitation of analyte molecules or atoms by active nitrogen. However, energy transfer from metastable nitrogen molecules is not considered to be the complete mechanism when microwave discharges are employed as the source of active nitrogen. The presence of nitrogen atoms in these discharges results in a more reactive plasma. Besides recombining to form metastable nitrogen molecules, these nitrogen atoms can undergo highly exothermic and rapid abstraction and addition reactions with a wide variety of compounds. These reaction rates are generally comparable to or greater than the rate of energy transfer by metastable nitrogen. Therefore, the mechanism for the production of excited-state analytes is thought to be a combination of both reaction by nitrogen atoms and energy transfer by nitrogen metastables.

The universal response to hydrocarbons is based on the fact that virtually all carbon compounds react with active nitrogen to produce excited state cyanogen radicals (CN). Strong chemiluminescent emissions from the $CN(B^2\Sigma^+ \rightarrow X^2\Sigma^+)$ system are observed when hydrocarbons are present in the GC detection cell. The most intense feature of this system is the violet-blue emission at 383 to 388 nm, which corresponds to the $\Delta v = 0$ sequence. The resulting CN spectrum is shown in Fig. 2-3.

The exact source of CN* is unknown at present. Possible chemical reactions involving atomic nitrogen which are sufficiently exothermic to produce excited-state cyanogen are

$$CH_2 + N \rightarrow CN^* + 2H, \qquad \Delta H^0 = -98 \text{ kcal/mol} \tag{2-1}$$

$$CH + N \rightarrow CN^* + H, \qquad \Delta H^0 = -107 \text{ kcal/mol} \tag{2-2}$$

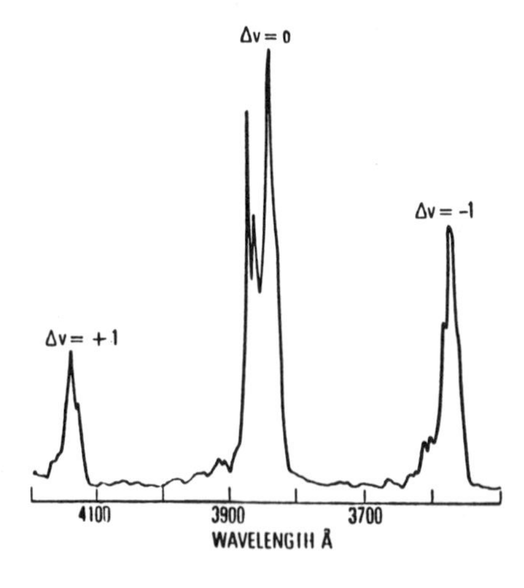

Figure 2-3 ■ Spectrum of $CN(B^2\Sigma^+ \rightarrow X^2\Sigma^+)$ chemiluminescence for hydrocarbons added to the active nitrogen detector. From D. G. Sutton, K. R. Westberg and J. R. Melzer, *Anal. Chem.* **51**, 1399 (1979).

Both pathways require several preceding reactions to produce either CH_2 or CH radicals from the hydrocarbon analyte. The formation of CH radicals by nitrogen atoms has been shown to be endothermic and thus is highly unlikely.[3] Another possibility is that ground-state CN radicals are produced in the reaction, and collisional energy transfer from nitrogen metastables in the active nitrogen plasma provides the necessary excitation energy.[4]

Reactions of unsaturated hydrocarbons with active nitrogen to produce CN* have been proposed to be initiated by reactions of nitrogen atoms. Unsaturated hydrocarbons tend to produce greater CN* emission intensity than saturated alkanes. Reactions with alkenes is thought to occur by the addition of N atoms to the double bond:[2]

$$H_2C{=}CH_2 + N \longrightarrow H{-}\underset{H}{\overset{}{C}}{-}\underset{H}{C}{-}H \text{ (with N bridging the two C atoms)} \quad (2\text{-}3)$$

This is a moderately fast, exothermic reaction. The complex formed in reaction (2-3) can then unimolecularly decay to form HCN and CH_3. The complex is also thought to undergo secondary reactions with excess atoms, but these processes are unconfirmed. In general, once the initiation reaction produces hydrocarbon radicals, these can react extremely fast with nitrogen atoms until virtually all of the initial hydrocarbon is converted to HCN.

In saturated hydrocarbons, addition reactions are not possible, and the abstraction of hydrogen by N atoms is an endothermic process. Initiation of the reactions of saturated hydrocarbons with active nitrogen has been thought to occur by collisional energy transfer from metastable nitrogen[5] or from the reaction of H atoms present in the active nitrogen as impurities.[3] However, these processes are much slower than N-atom reactions with olefins. Therefore, chemiluminescence intensities from saturated hydrocarbons is much less than the corresponding alkene reactions. To overcome the initial barrier to reaction, small amounts of HCl can be added to the detection cell. This catalytically enhances the reactions of saturated hydrocarbons. The catalysis reactions are thought to be

$$N + N(\text{or } N_2^*) + HCl \rightarrow N_2 + H + Cl \quad (2\text{-}4)$$

$$Cl + C_2H_6 \rightarrow HCl + C_2H_5 \quad (2\text{-}5)$$

thereby initiating the reaction by the fast Cl-atom abstraction.[3] Again, once hydrocarbon radicals are produced, they are quickly converted to HCN by nitrogen-atom reactions as described previously for olefins.

The effect of HCl addition aids in the identification of peaks in complex samples. An example is shown in Fig. 2-4. In the initial chromatogram, only the olefinic species can be detected, even though methane is present in the sample. The second chromatogram shows the result of a small addition of HCl. A significant signal is now observed for methane, with virtually no effect on the intensities of the olefins in the sample. With HCl catalysis, a detection limit of 4 ng was obtained for methane, while 200 ng was the least detectable mount without HCl. Similar detection limits were obtained for other saturated hydrocarbons. Unsaturated hydrocarbon detection limits were somewhat better, with vinyl fluoride being detected at 100 pg.[1]

More recently, an atmospheric pressure active nitrogen (APAN) detector has been reported.[6] At atmospheric pressure the high collision rate tends to cause complete

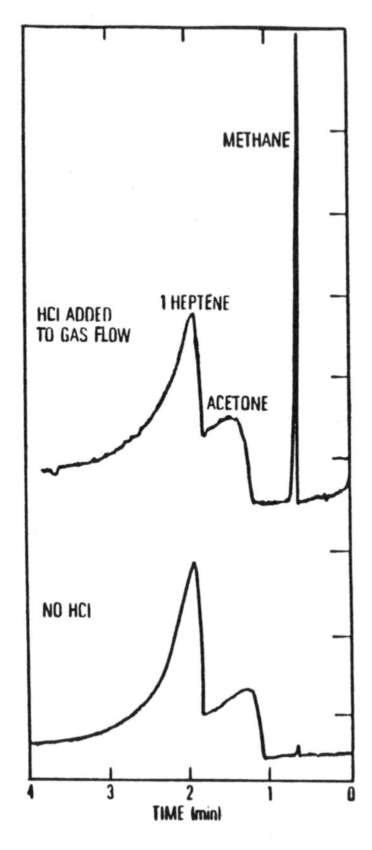

Figure 2-4 ■ Chromatograms of methane, 1-heptene and acetone with and without HCl added to the gas flow. From J. E. Melzer and D. G. Sutton, *Appl. Spectr.* **34**, 434 (1980).

recombination of any nitrogen atoms in the plasma. Therefore, the main reactive species created is metastable $N_2(A^3\Sigma_u^+)$. Although less reactive than atomic nitrogen, the higher concentration allows the overall reaction rates to be comparable to those in the low-pressure, active-nitrogen systems. Therefore, the CL intensities from CN* tend to be similar for the APAN detector. It should also be noted that HCl catalysis is unnecessary since energy transfer rates from metastable nitrogen to both saturated and unsaturated hydrocarbons are nearly identical. Detection limits of hydrocarbons and aromatics were 0.8 and 0.2 ng, respectively.[6]

This detector actually shows a modest increase in sensitivity compared to detection at reduced pressure. This is most likely due to the reduction of background in the APAN. At low pressures, it has been seen that the HCN product tends to polymerize. A polymeric film containing CN would form on the quartz windows, increasing the residual background by further reaction with active nitrogen to produce small amounts of CN*. This buildup of polymer was not observed in the atmospheric pressure version of the detector. The lack of polymer accumulation facilitates operation, since in the low-pressure active nitrogen detector the windows must be periodically removed and cleaned.

Although most of the work with the active nitrogen detector has explored the universal hydrocarbon response, selectivity for other classes of compounds can be accomplished by monitoring different emitters. Organometallics have been found to react with active nitrogen to produce not only CN*, but also excited state metal atoms, which can then chemiluminesce.[1] By monitoring the characteristic atomic emission lines for the metal of interest, selective detection of these compounds can be achieved in the prescence of large excesses of organics. The spectrum of trimethylaluminum is shown in Fig. 2-5 as an example. Detection limits for Pb and Sn are in the nanogram range, while Hg gave a limit of detection of 2 pg in the APAN detector.[6]

Selectivity can be achieved for a variety of other elements as well. Generally, emissions from diatomic molecules consisting of nitrogen and the analyte element are

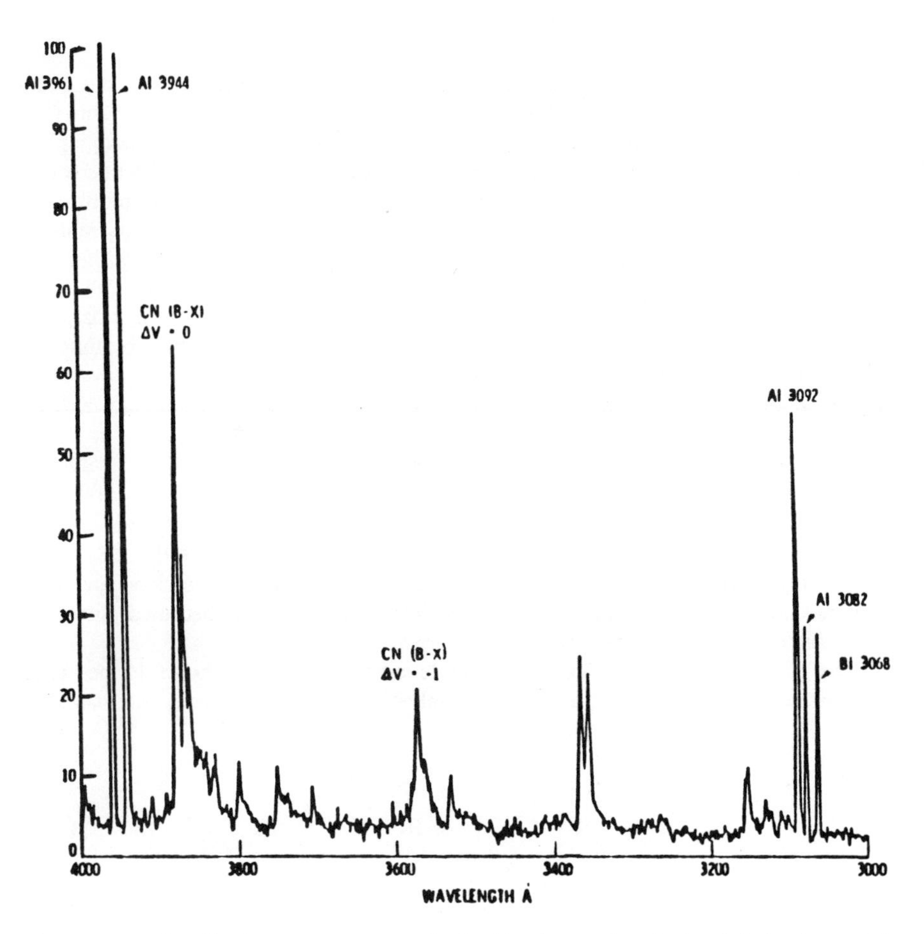

Figure 2-5 ■ Chemiluminescence spectrum obtained in the reaction of trimethyl aluminum with active nitrogen. From D. G. Sutton, K. R. Westberg and J. R. Melzer, *Anal. Chem.* **51**, 1399 (1979).

observed.[7] Examples include active nitrogen reactions with oxygen-containing, phosphorus-containing and sulfur-containing compounds, which produce electronically excited NO, PN and SN, respectively. Electronically excited OH is also observed in reactions of oxygenated compounds.[8] One drawback to selective detection of oxygenated compounds is that O_2 itself reacts with active nitrogen to produce NO*. Rice and co-workers have reported spectra from inorganic compounds containing B, Cl, Br and I when reacted with a flow of active nitrogen.[9] This provides a possible method for selective detection of many different elements by monitoring various wavelengths. This degree of selectivity, which can be combined with its universal hydrocarbon response, gives the active nitrogen detector a significant advantage over the flame ionization detector, which can only be used in a "universal" mode.

Ozone-Induced Chemiluminescence

One of the most widely used chemiluminescence reagents is ozone (O_3). It has been shown that nearly all gas phase reactions of ozone are chemiluminescent.[10] This fact has been capitalized on for the detection of organic compounds in gas chromatography, as well as in other fields of analysis (e.g., atmospheric analysis). The chemiluminescence seen in ozone reactions can be attributed to the weak O—O (24.9 kcal/mol) bond strength in ozone which is broken to form the strong O=O bond (119 kcal/mol) in molecular oxygen and strong C=O bonds (~ 172 kcal/mol) in reactions with many organic compounds.

One of the earliest CL reactions used as a detection method in gas chromatography was that of ozone with olefins.[11] Ozone reacts rapidly with olefins, resulting in chemiluminescence from a variety of excited species. The emitters which have been identified for the O_3/ethylene reaction include $CH_2{=}O(^1A_2)$, vibrationally excited $OH(X^2\Pi, v \leq 9)$ and electronically excited $OH(A^2\Sigma^+)$.[10] $CH_2{=}O^*$ emits at wavelengths of 350 nm and greater, whereas vibrationally excited OH emits at the characteristic Meinal bands in the red and infrared region of the spectrum (700 to 1100 nm). Electronically excited OH can be observed at 306 nm, corresponding to the (0,0) band. When substituted olefins are used, phosphorescence from glyoxal, $(CHO)_2(^3A_u \rightarrow {}^1A_g)$ and methyl glyoxal, $(CH_3CO)_2(^3A_u \rightarrow {}^1A_g)$ are observed. These emissions are both centered near 520 nm and produce the most intense features in substituted alkenes.[12] Some typical CL spectra for various olefins are shown in Fig. 2-6.

The reaction mechanism even for simple olefins has been shown to be quite complicated, but it is thought that a five-membered ozonide is the likely intermediate:

$$O_3 + R_1R_2C{=}CR_3R_4 \longrightarrow R_1{-}\overset{|}{C}(R_2){-}\overset{|}{C}(R_4){-}R_3 \text{ (ring: C–O–O–O–C)} \qquad (2\text{-}6)$$

Several pathways are then possible, some of which result in excited-state products. The overall scheme for *cis*-2-butene is shown in Fig. 2-7. Unimolecular decomposition (Criegee splitting) was initially thought to be the main pathway involved. However, more recent evidence has shown that α-hydrogen abstraction occurs up to 2.5 times

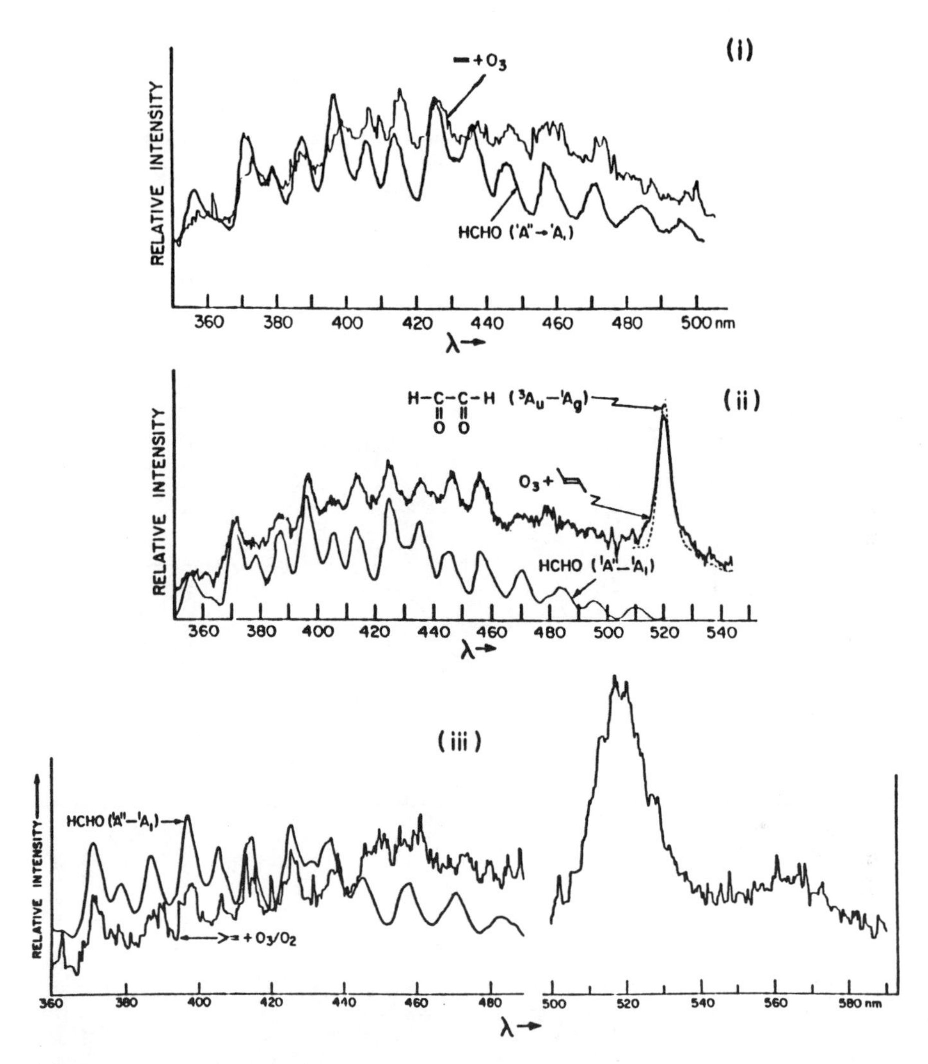

Figure 2-6 ■ Chemiluminescent emission spectra in the visible region for the reaction of 2% O_3 in O_2 with (i) ethylene, (ii) *trans*-2-butene and (iii) isobutene. Reference spectra of $H_2C{=}O\,(^1A_2 \rightarrow {}^1A_1)$ are given in (i)–(iii) and a reference spectrum of $(CHO)_2(^3A_u \rightarrow {}^1A_g)$ is given in (ii). From B. J. Finlayson, J. N. Pitts, Jr. and R. Atkinson, *J. Am. Chem. Soc.* **96**, 5356 (1974).

faster than this Criegee fragmentation.[12] Therefore, it plays a major role in determining the primary emitters. As shown in Fig. 2-7, this pathway leads to the formation of triplet-excited glyoxal and methylglyoxal, in agreement with the emissions from these species being rather strong (see also Fig. 2-6). A general conclusion in O_3/alkene reactions is that when at least one of the carbon atoms on the double bond is bonded to a hydrogen and an alkyl group, emission from triplet glyoxal is observed. If at least

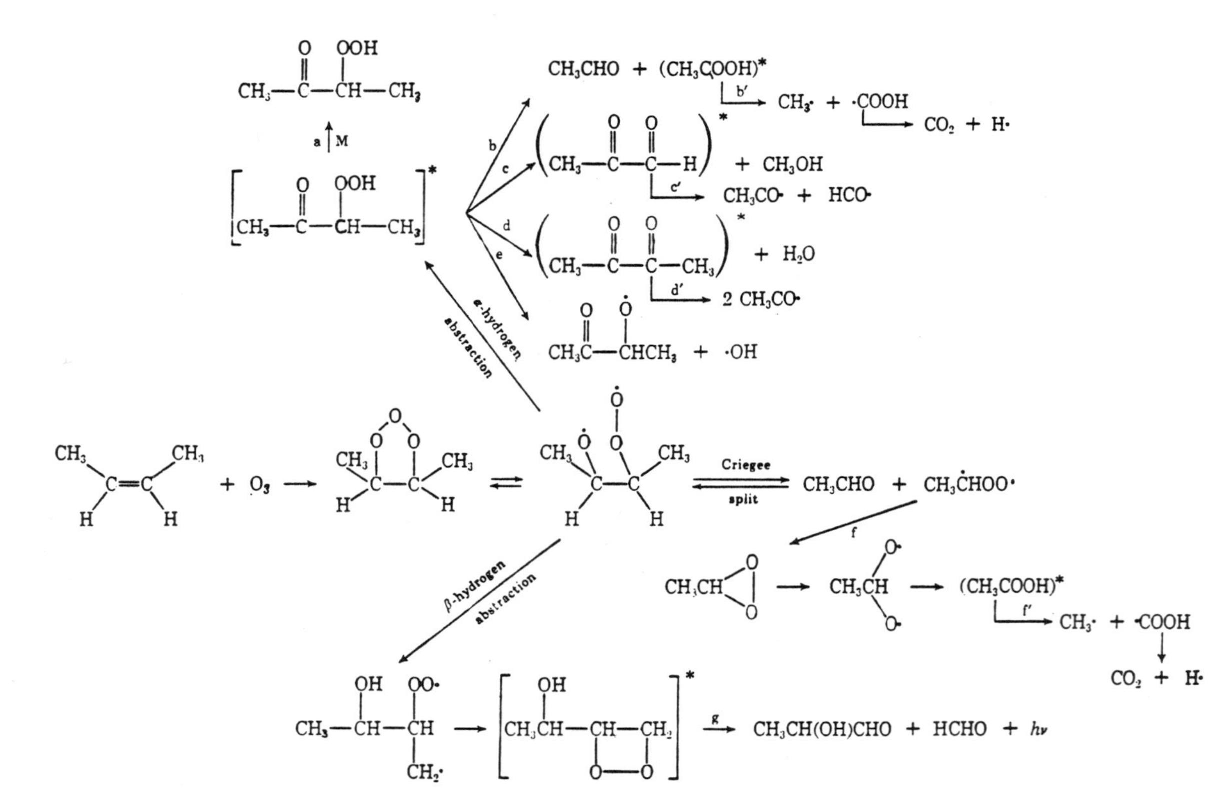

Figure 2-7 ■ Pathways for the reaction of *cis*-2-butene with ozone. From B. J. Finlayson, J. N. Pitts, Jr. and R. Atkinson, *J. Am. Chem. Soc.* **96**, 5356 (1974).

one carbon atom on the double bond is connected to two alkyl groups, methylglyoxal emission is evident.[10]

The channel involving β-hydrogen abstraction has also been shown to be about 8 times slower than that of α-hydrogen abstraction.[12] As can be seen, this channel leads to the formation of $CH_2 = O^*$, which shows much weaker intensity in the chemiluminescence spectra in Fig. 2-6. It should be noted that for terminal alkenes, CH_2O_2 can be formed by Criegee decomposition. This can react further with excess ozone:

$$CH_2O_2 + O_3 \rightarrow CH_2{=}O + 2O_2 \tag{2-7}$$

This reaction is exothermic by 110 kcal/mol, which easily supplies the 80.5 kcal/mol necessary to populate $CH_2 = O(^1A_2)$ and results in another possible channel for the formation of excited-state formaldehyde.[10]

The other major emitter in ozone/hydrocarbon systems is the hydroxyl radical (OH). These can be formed in both vibrationally excited and electronically excited states. $OH(X^2\Pi, v \leq 9)$ occurs when almost any hydrogen-containing compound reacts with ozone.[10] This emission strongly implies that hydrogen atoms are formed in some channels of the reaction. Both the channels involving α-hydrogen abstraction and Criegee splitting have the possibility of producing atomic hydrogen. Herron and Huie have reported that approximately 9% of the total reaction produces H atoms by pathways shown in Fig. 2-7.[13] These hydrogen atoms can then react with excess ozone as

$$H + O_3 \rightarrow OH(X^2\Pi) + O_2 \tag{2-8}$$

for which $\Delta H^0 = -77$ kcal/mol. However, this reaction is not exothermic enough to explain the presence of $OH(A^2\Sigma^+)$, which requires 93.3 kcal/mol of excitation energy. The necessary energy may be produced through energy pooling of vibrationally excited OH or possibly through the reaction

$$CH + O_2 \rightarrow CO + OH^* \tag{2-9}$$

This reaction can provide 159 kcal/mol, but the source of CH radicals is unknown, and the reaction involves a four-center intermediate. However, emission from CH radicals has been observed in reactions of ozone with aromatics, allene and acetylene.[10]

Although olefins react most readily with ozone, by increasing the temperature saturated hydrocarbons and aromatic hydrocarbons exhibit chemiluminescence. This gives a relatively simple method for controlling selectivity when this reaction is used as a detector in GC.[11] As seen in Fig. 2-8, at a temperature of 100°C, only olefins were observed to react. However, at a temperature of 250°C, a chromatogram which is comparable to that obtained with a commercial FID is observed. It is found that at a temperature of 150°C, aromatic compounds begin to show a CL response. Further increase in the temperature to 200°C results in a response for saturated hydrocarbons. Aromatic compounds most likely react by a similar mechanism to that of alkenes. However, due to their aromatic stability, the initial addition of ozone to the double bond is less energetically favorable. Therefore, higher energy (temperature) is needed in order to initiate the reaction. The exact mechanism of reaction for saturated hydrocarbons is unknown and is relatively slow, since not only high temperatures are needed, but CL intensity is much lower.

Although in general CL intensities from each emitter are low for these ozone reactions, emission from the entire visible range can be collected, increasing the net

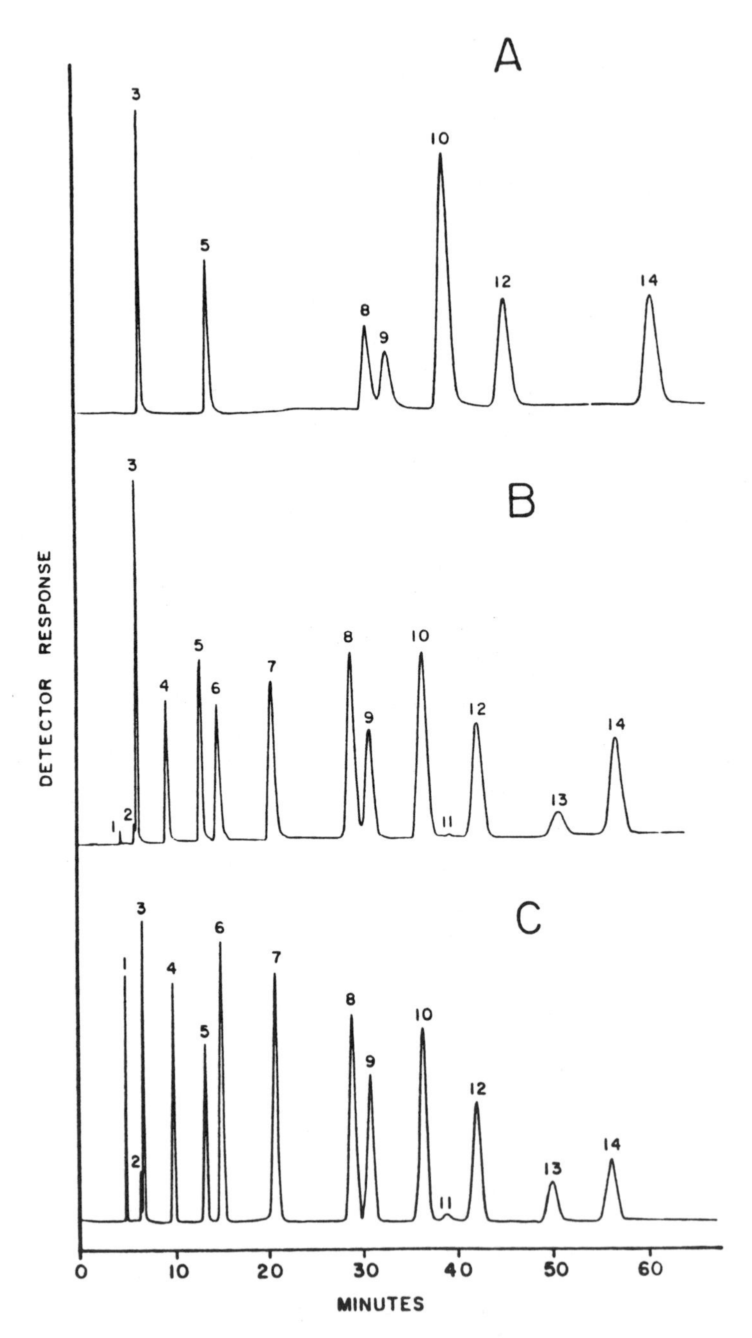

Figure 2-8 ■ Hydrocarbon analysis using ozone chemiluminescence detection at (A) 100°C, (B) 250°C and (C) a chromatogram using a commercial FID. Peaks: 1 = methane, 2 = ethane, 3 = ethylene, 4 = propane, 5 = propene, 6 = isobutane, 7 = *n*-butane, 8 = 1-butene, 9 = isobutene, 10 = *trans*-2-butene, 11 = isopentane, 12 = *cis*-2-butene, 13 = *n*-pentane, 14 = 1,3-butadiene. From W. Bruening and F. J. M. Concha, *J. Chromatogr.* **112**, 253 (1975).

yield of photons and thereby increasing the sensitivity. Detection limits are reported in the nanogram range for most compounds studied.[11] However, by collecting light from the entire spectral range, any spectral selectivity for particular compounds is lost. The reaction has been observed to be first order in both alkene and ozone concentrations[12]; therefore, as long as pseudo-first-order conditions in the detector exist ($[O_3] \gg$ [alkene]), the response of the detector is linear.

Ozone CL reactions can also be used to selectively detect nitrogen-containing compounds, oxygenated compounds and reduced sulfur compounds by various schemes. The selective detection of both nitrogen and oxygenated compounds is based on the CL reaction

$$NO + O_3 \rightarrow NO_2^* + O_2 \tag{2-10}$$

This reaction has become one of the most studied and useful CL reactions to date. Detectors based on this reaction have become extremely important and useful in recent years and are discussed at length in the next chapter.

Reduced sulfur compounds play a major role in the distribution of sulfur within the atmosphere. These compounds generally become oxidized to SO_2 and then further to H_2SO_4. However, the exact amount of sulfuric acid contributed by the oxidation of reduced sulfur compounds is a subject of debate. To understand the role played by reduced sulfur compounds, a method for the determination of sulfur compounds selectively at trace levels is necessary. Other fields in which selective detection of sulfur containing compounds is important include the analysis of foods and beverages, pharmaceuticals, pesticides and fuels. Ozone has been shown to react selectively with reduced sulfur compounds to produce electronically excited $SO_2(^3B_1)$.[14] The emission from SO_2^* occurs in the range 300 to 400 nm (see Fig. 2-9), and by the use of an appropriate filter, a detection scheme selective for sulfur compounds over other compound classes has been described.[15] The only interference noted is from olefin reactions, which produce emission from formaldehyde near 354 nm.

The proposed mechanism involves only three steps for most organosulfur compounds.[10] Using methanethiol as an example, these are

$$CH_3SH + O_3 \rightarrow CH_3O_2 + HSO \tag{2-11}$$

$$HSO + O_3 \rightarrow HO + SO + O_2 \tag{2-12}$$

$$SO + O_3 \rightarrow SO_2^* + O_2 \tag{2-13}$$

The final step of this reaction has $\Delta H^0 = -106$ kcal/mol, which provides the necessary energy for electronic excitation of SO_2. H_2S has also been found to respond, but is thought to produce HSO* from the reaction of ozone with SH radicals.[10] HSO emission occurs in the wavelength range 360 to 380 nm and is also monitored in the detection scheme. The direct production of SH radicals from a reaction of H_2S and O_3 is endothermic by 15 kcal/mol. SH radicals are thought to be produced by the short reaction sequence

$$H_2S + O_3 \rightarrow HSO + HO_2 \tag{2-14}$$

$$HSO + O_3 \rightarrow SH + 2O_2 \tag{2-15}$$

where both steps are exothermic. There is evidence that a single collision reaction between H_2S and ozone results in SO_2^*.[16] This reaction is exothermic by 158 kcal/mol so that it easily provides the necessary energy, but requires a significant amount of

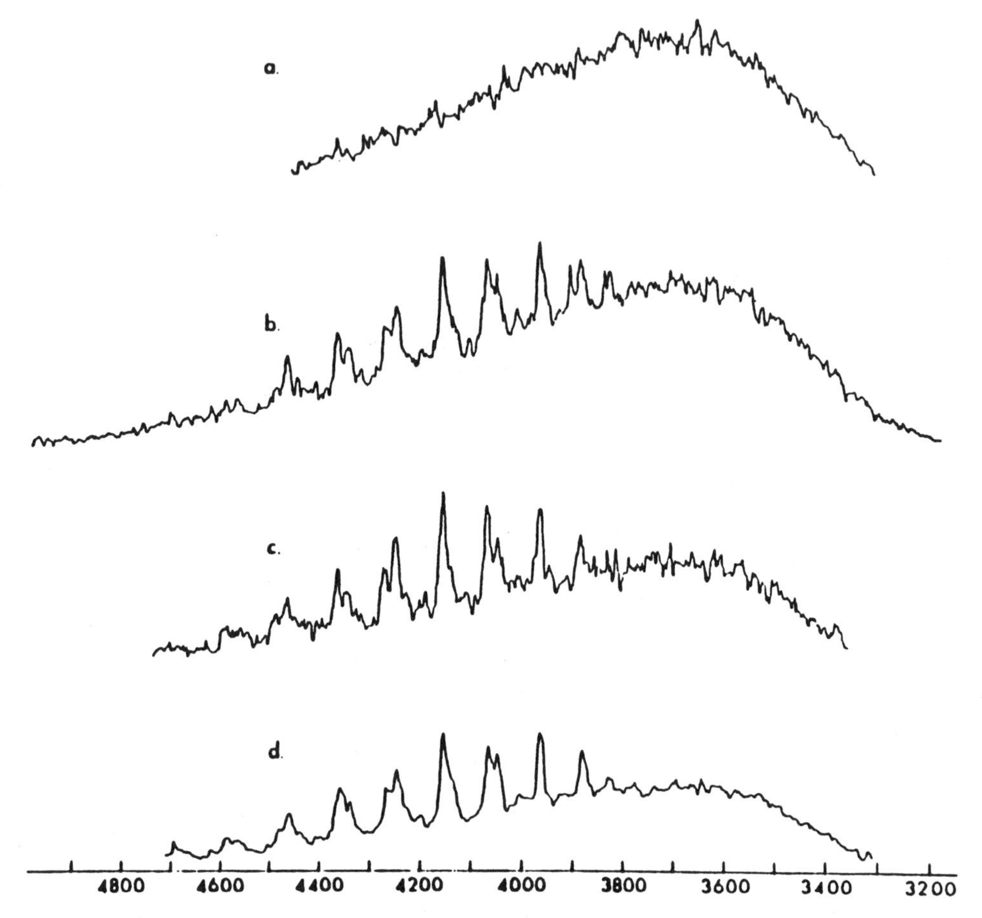

Figure 2-9 ■ Chemiluminescence spectra of $SO_2(^3B_1)$ from the reaction of O_3 with CH_3SH. (a) 9 mtorr O_3, 3 mtorr CH_3SH; (b) 15 mtorr O_3, 7 mtorr CH_3SH; (c) 20 mtorr O_3, 7 mtorr CH_3SH; (d) 30 mtorr O_3, 15 mtorr CH_3SH; 620 mtorr of He in all cases. From R. J. Glinski, J. A. Sedarski and D. A. Dixon, *J. Phys. Chem.* **85**, 2440 (1981).

molecular rearrangement. The single-collision reaction was postulated to occur as

$$O_3 + H_2S \longrightarrow \underset{\mathbf{1}}{[O_3 \cdots SH_2]} \longrightarrow \underset{\mathbf{2}}{[\cdot O{-}O{-}SH_2{-}O\cdot]} \longrightarrow \underset{\mathbf{3}}{[\cdot O{-}O{-}SH_2{-}O\cdot]} \longrightarrow O{=}S{=}O + H_2O \qquad (2\text{-}16)$$

By the use of appropriate filters, so that emission from wavelengths above 400 nm is blocked, fairly selective detection for sulfur compounds can be accomplished. As stated earlier, only olefins, which produce a large amount of emission from formaldehyde, tend to give interferences. The method generally gives detection limits in the parts-per-billion range, with the greatest response for methanethiol.[15] In general, the response decreases in the order $CH_3SH > CH_3SCH_3 > H_2S >$ thiophenes. However, one drawback is that calibration curves tend to be curved and not linear. Calibration curves are highly reproducible, however.[15]

Sulfur Chemiluminescence Detector (SCD)

The reaction of molecular fluorine with organosulfur compounds is the basis of a new sulfur-selective detector[17] recently introduced by Sievers Research, Inc. Molecular fluorine is generated on-line by an electrical discharge of the nontoxic, nonflammable gas SF_6. Any atomic fluorine produced in the discharge is allowed to recombine before entering the chemiluminescence cell where F_2 is mixed with the GC effluent at a pressure of ≈ 1 torr. To protect the vacuum pump oil, a trap containing activated charcoal and soda lime is installed after the detection cell to remove excess fluorine and HF formed in the reaction. The overall schematic is shown in Fig. 2-10.

It has been found that reduced sulfur compounds react rapidly with F_2 to produce several chemiluminescent species. This detector is highly sensitive to organosulfur compounds and shows as much as 7 orders of magnitude of selectivity over many compound classes.[17,18] This offers a major advantage for analysis of low concentrations of sulfur compounds in complex matrices.

Elementary chemical reactions of F_2 are almost entirely unknown at present. Although F atom reactions have been studied at length, the analogous F_2 reactions have gone relatively unnoticed. The lack of kinetic and mechanistic studies of F_2 is somewhat curious considering that most of the exothermicity generated in fluorine chain reactions are contained within steps involving molecular fluorine. The F_2 bond strength is only 37.5 kcal/mol, so that formation of H—F (136 kcal/mol) and C—F (116 kcal/mol) bonds by reactions of radicals with F_2 are extremely exothermic and often produce excited-state species.

The spectrum of the emission signal for the reaction of F_2 with ethanethiol is shown in Fig. 2-11. Progressions of highly excited vibrational overtones of HF can be seen. In the GC detector the $\Delta v = 4$ overtone is monitored between 660 and 750 nm by use of an appropriate bandpass filter. Excitation of HF to levels up to $v = 6$ is common with most thiols, sulfides and disulfides. Under certain conditions, $HF(v = 8)$ can also be seen in the emission spectrum. It is interesting to note that reactions of F atoms with both hydrocarbons and organosulfur compounds have only resulted in vibrationally excited levels of HF up to $v = 4$.[19] One possible explanation of the greater excitation obtained with F_2 is that the mechanism results in H atoms as intermediates. These can then react by

$$H + F_2 \rightarrow HF^{\dagger}(v \leq 9) + F \tag{2-17}$$

However, at present there is no direct experimental evidence that hydrogen atoms are produced.

Several other possible reaction schemes have been formulated to explain the high degree of vibrational excitation in HF. The earliest mechanism postulated involves the addition of F_2 to the parent sulfur-containing molecule.[17] Sulfur atoms exhibit the

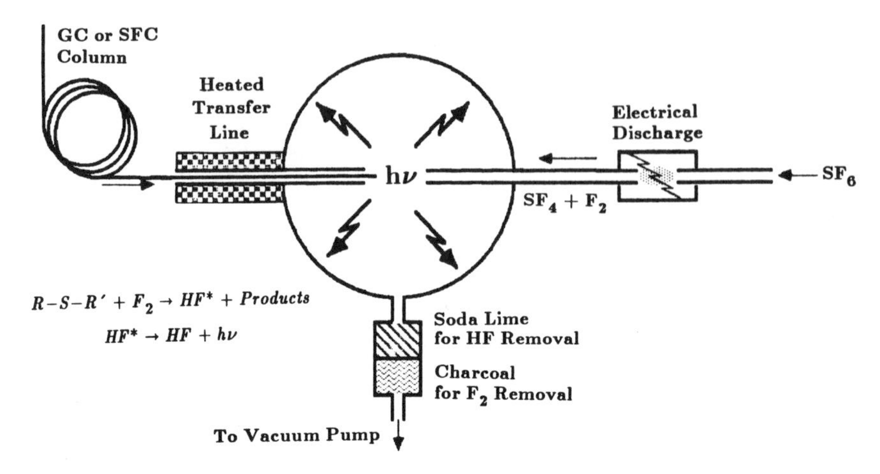

Figure 2-10 ■ Schematic diagram of the Sulfur Chemiluminescence Detector. Courtesy of Sievers Research, Inc.

ability to expand their octet (i.e., become oxidized) when coordinated to highly electronegative atoms such as fluorine by using its empty *d* orbitals. Simple hydrocarbons and oxygen-containing compounds do not have this ability and as such are not expected to respond. This addition reaction could proceed by

$$R_3-C(R_2)(H)-S-R_1 + F_2 \rightarrow \left[R_3-C(R_2)(H)-S(F)-R_1 \cdots F \right] \rightarrow R_3C(R_2)=S(F)-R_1 + HF^{\ddagger} \qquad (2\text{-}18)$$

in which HF is eliminated to form a carbon–sulfur double bond. The energy released upon forming this double bond should be sufficient to produce HF($v \leq 6$). The intermediate in this metathesis reaction could be a charge transfer complex, considering the high electron affinity of F_2 and the low ionization potential of reduced sulfur compounds. Alternatively, the reaction could occur in two concerted steps,

$$R_1R_2-CH-S-R_3 + F_2 \longrightarrow R_1R_2-CH-\underset{\bullet}{\overset{F}{\overset{|}{S}}}-R_3 + F \qquad (2\text{-}19)$$

$$R_1R_2-CH-\underset{\bullet}{\overset{F}{\overset{|}{S}}}-R_3 + F \longrightarrow HF^{\dagger} + R_1R_2-C=\overset{F}{\overset{|}{S}}-R_3 \qquad (2\text{-}20)$$

Either mechanism can explain the fact that the chemiluminescence obtained for

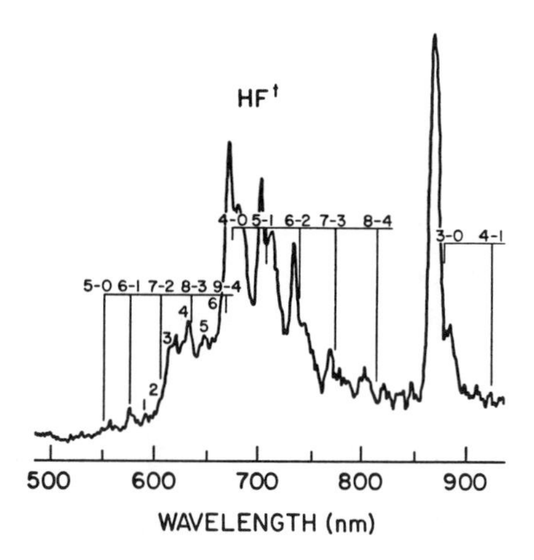

Figure 2-11 ■ Vibrational overtone emission from HF in the reaction of ethanethiol with F_2. From R. J. Glinski, E. A. Mishalanie and J. W. Birks, *J. Photochem.* **37**, 217 (1987).

reaction of F_2 with diphenyl sulfide and di-tert-butyl sulfide is weak in that neither compound contains a β hydrogen. This mechanism can also explain the fact that highly oxidized sulfur compounds do not respond in the detector.

Other emitters have also been found in spectroscopic studies of the F_2/sulfur system. Most prominent is that of triplet-excited thioformaldehyde, $CH_2S(A^3\Sigma^+)$, which is formed in the reactions of $(CH_3)_2S$ and CH_3SH with F_2.[20] The most intense CL emission from this molecule is the (0,0) band located at 670 to 700 nm (see Fig. 2-12.) This emission lies in the same spectral region as the $\Delta v = 4$ overtones of HF, so

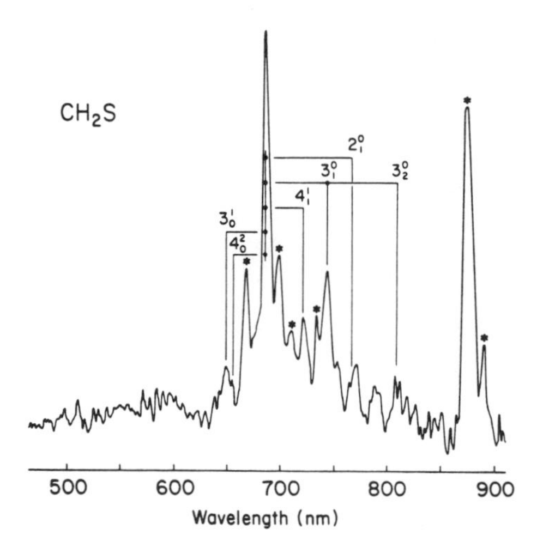

Figure 2-12 ■ Chemiluminescence from triplet-state thioformaldehyde formed in the reaction of methanethiol with F_2. The asterisk (∗) indicates emission bands due to HF as in Fig. 2-11. From R. J. Glinski, J. N. Getty and J. W. Birks, *Chem. Phys. Lett.* **117**, 359 (1985).

that its emission is also monitored in the GC detection scheme using the bandpass filter selected for observing HF vibrational overtone emission. The relative intensity of these two emitters (CH_2S^* and $HF^{\dagger}$) depends on the relative concentrations of F_2 and the organosulfur compound present. When $[F_2] \gg$ [S-cpd], as is the usual operating procedure, HF emissions dominate the spectrum. However, as the sulfur compound concentration is increased relative to the fluorine concentration, HF emission declines and thioformaldehyde emissions begin to dominate.[21]

The mechanism for the formation of CH_2S^* has been proposed to proceed through the intermediate radical CH_3S for both sulfides and thiols.[28] The reaction

$$CH_3SH + F_2 \rightarrow CH_3S + HF + F \qquad (2\text{-}21)$$

is exothermic by about 8 kcal/mol. Further abstraction reactions by F or F_2 could result in the formation of thioformaldehyde by

$$CH_3S + F \rightarrow HF + CH_2S^* \qquad (2\text{-}22)$$

and/or

$$CH_3S + F_2 \rightarrow HF + CH_2S^* + F \qquad (2\text{-}23)$$

These reactions are exothermic by 94 and 56 kcal/mol, respectively, and both provide enough energy to excite thioformaldehyde to its triplet state. These reactions also produce HF, which could be vibrationally excited, but this mechanism taken alone cannot explain the decrease in HF emission as thioformaldehyde emission increases.

Another emitter, shown in Fig. 2-13, which has been identified in these systems is that of $HCF(A^1\Sigma \rightarrow X^1\Sigma)$.[20] The mechanism of its formation is unknown, but the reaction

$$CH + F_2 \rightarrow HCF + F \qquad (2\text{-}24)$$

is exothermic by 78 kcal/mol and releases sufficient energy to create HCF*. HCF emission is only seen when F_2 is in large excess, which probably can lead to the large amount of fragmentation of the parent organosulfur molecules necessary for the production of CH.

Yet another emitter, shown in Fig. 2-14, formed in the reaction of CS_2 with F_2, has been identified as the molecule SF_2.[22] It should also be noted that the emitter SF_2 is

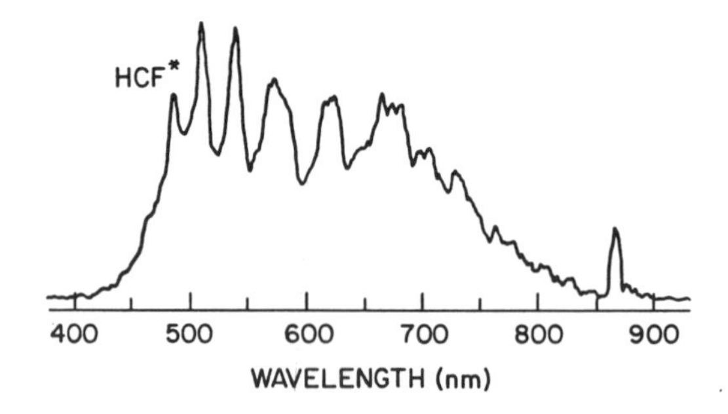

Figure 2-13 ■ HCF emission observed in the reaction of dimethyl sulfide with F_2. From R. J. Glinski, E. A. Mishalanie and J. W. Birks, *J. Photochem.* **37**, 217 (1987).

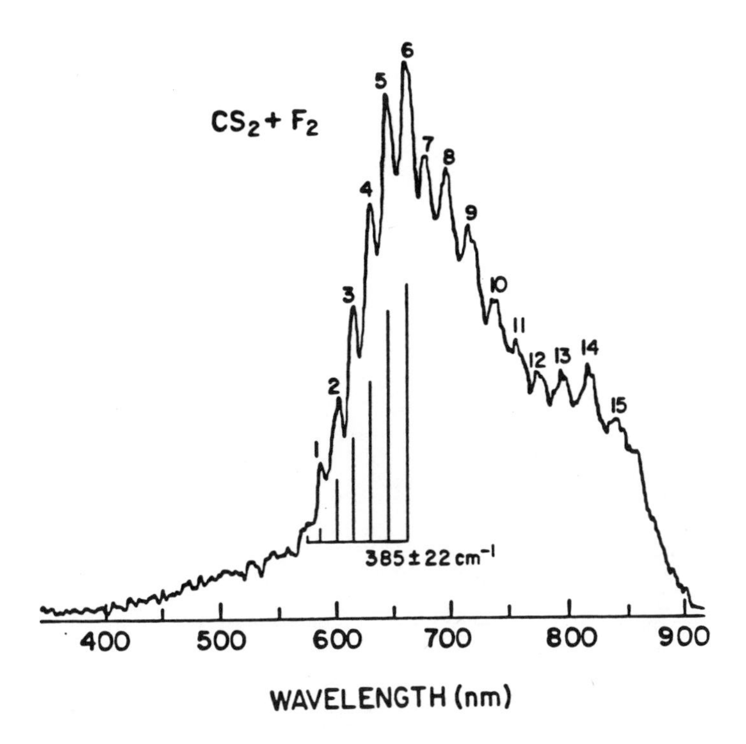

Figure 2-14 ■ SF_2 emission observed in the reaction of CS_2 with F_2 in excess. From R. J. Glinski, E. A. Mishalanie and J. W. Birks, *J. Photochem.* **37**, 217 (1987). See ref. 22 for reinterpretation of this spectrum.

observed in the reaction of many organosulfur compounds with F_2 in the HPLC version of the detector[18,22]; although, at present, no potential mechanisms have been postulated.

The F_2/sulfur detector shows an excellent response to a wide variety of thiols, sulfides and disulfides. Thiophenes also respond; however, their response is much smaller due to a slower reaction rate of these ring compounds with F_2. Detection limits for most thiols and sulfides are in the low picogram range, while thiophenes typically are an order of magnitude poorer. Reported detection limits for many reduced sulfur compounds are given in Table 2-1. The selectivity exhibited over interfering compounds ranges up to 7 orders of magnitude for saturated hydrocarbons and most oxygen-containing compounds.[24] Interfering compounds, which in some applications may serve as the analytes, include olefins and other compounds with weak C—H bonds such as toluene and xylene (benzylic hydrogens), halocarbons and aldehydes. The selectivity ratios of some of these interfering compounds, calculated relative to 1-hexanethiol, are given in Table 2-2. Another key aspect of this detector is its linear response. The detector has been found to be linear over 3 or more orders of magnitude. A working curve for dimethyl sulfide is given in Fig. 2-15. Linearity of the sulfur chemiluminescence detector is a major advantage over other sulfur selective detectors,

Table 2-1 ■ Detection Limits Reported for the Sulfur Chemiluminescence Detector™

Compound	GC (pg)	HPLC (pg)
Ethanethiol	13	130
1-butanethiol	35	180
1-hexanethiol	84	560
1-octanethiol	63	2100
1,3-propanedithiol	8	—
2-methyl-2-propanethiol	—	100
o-aminobenzenethiol	—	1800
Dimethyl sulfide	16	50
Ethyl sulfide	46	70
Allyl sulfide	5	50
n-butyl sulfide	180	430
Sec-butyl sulfide	27	—
Tert-butyl sulfide	—	1800
Phenyl sulfide	—	350
Dimethyl disulfide	—	60
n-butyl disulfide	17	220
Isoamyl disulfide	200	—
Dimethyl sulfoxide	94	—
Dimethyl selenide	—	1200

Source: From J. K. Nelson, R. H. Getty and J. W. Birks, *Anal. Chem.* **55**, 1767 (1983) and E. A. Mishalanie and J. W. Birks, *Anal. Chem.* **58**, 918 (1986). Detection limits obtained using a CH_3CN/H_2O mobile phase.

Table 2-2 ■ Selectivity Ratios for the Sulfur Chemiluminescence Detector™ Using 1-Hexanethiol as the Reference Compound

Compound	Selectivity Ratio
Water	$> 10^7$
Methanol	$> 10^7$
Acetonitrile	$> 10^7$
Acetone	$> 10^7$
Hexane	$> 10^7$
Ethyl acetate	$> 10^7$
Carbon tetrachloride	$> 10^7$
Tetrahydrofuran	3×10^2
Isobutanol	3×10^5
3-chloropropene	3×10^3
Methylene chloride	4×10^4
Toluene	1×10^2
m-xylene	2×10^1
Chlorobenzene	4×10^6
Octane	6×10^4

Source: J. K. Nelson, R. H. Getty and J. W. Birks, *Anal. Chem.* **55**, 1767 (1983).

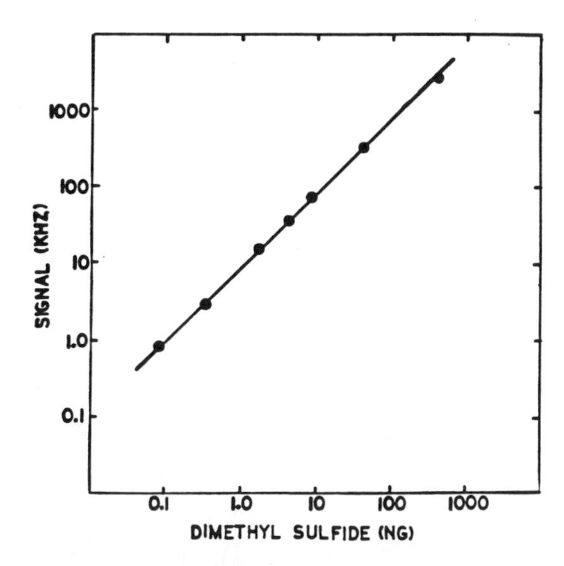

Figure 2-15 ■ Working curve for dimethyl sulfide using the Sulfur Chemiluminescence Detector demonstrating linearity over 4 orders of magnitude (100 pg to 1 μg). The slope of the least squares line is 0.98 ± 0.03. From J. K. Nelson, Ph.D. Thesis, University of Colorado, Boulder, Colorado, 1984, p. 63.

most notably the flame photometric detector (FPD). The FPD exhibits an approximately squared dependence on the concentration of the sulfur compound being detected. The flame also provides a quite unstable background in the FPD, and response is highly dependent on the unstable flame conditions. A major drawback of the FPD is quenching of the response by coeluting hydrocarbons. None of these problems are encountered in the F_2/sulfur CL detector.

The fluorine-induced chemiluminescence detector has also been shown to respond to organoselenium compounds as well.[18] As selenium has the same outer shell electron configuration as sulfur, it is expected to undergo similar reactions. The chemiluminescence emission has been assigned to excited-state triplet selenoformaldehyde, HCF (where F_2 concentration is high) and an unknown emitter.[23] The chemiluminescence spectrum seen for organoselenium compounds is shown in Fig. 2-16. Virtually no HF($\Delta v = 4$) emission was observed. The mechanism of this reaction is assumed to occur along similar pathways as those involving organosulfurs, although detailed studies have not been undertaken. Detection limits of organoselenium compounds were reported in the low nanogram range; however, the emission of selenoformaldehyde lies at the edge of the workable range of the PMT used where the photon efficiency is quite low.

The SCD has recently been adapted for use as a detector for microbore HPLC as well.[18] The high selectivity against common HPLC solvents such as water, methanol and acetonitrile allow the effluent from the HPLC column to be vaporized and

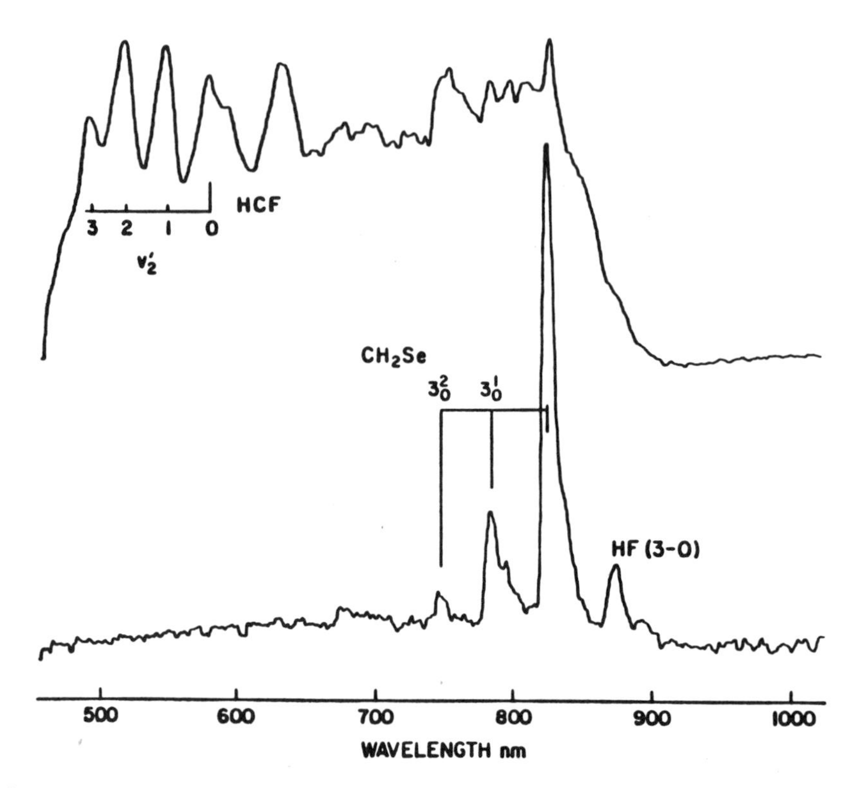

Figure 2-16 ■ Chemiluminescence spectrum obtained for the reaction of dimethyl diselenide (DMDSe) with 10% F_2 in He. Upper trace: 10 mtorr of DMDSe and 330 mtorr of F_2/He. Lower trace: 60 mtorr of DMDSe and 270 mtorr of F_2 He. From R. J. Glinski, E. A. Mishalanie and J. W. Birks, *J. Am. Chem. Soc.* **108**, 531 (1986).

subsequently reacted with fluorine. The CL emitters are the same; however, the intensity of their emissions is affected by large amounts of vaporized methanol/water or acetonitrile/water. The solvent has been shown to take part in the overall reaction, as different emission spectra are seen for different solvent mixtures, but the exact role is undefined. Detection limits in the HPLC system, also given in Table 2-1, are generally in the high picogram to low nanogram range.

A recent application of the SCD is the determination of trace levels of organosulfur compounds in the headspace of beer. Sulfur compounds at the low part-per-billion level and below are responsible for much of the flavor and fragrance of beer. Nearly all sulfur compounds may be determined at levels well below the taste threshold by cryogenic focusing of 3 cc of the beer headspace onto a GC capillary column. Example chromatograms comparing three commercial beers are provided in Fig. 2-17. The compounds identified in all three beers are ethanethiol, dimethyl sulfide and diethyl disulfide. Detection limits for the determination of sulfur compounds commonly found in beer are compared with the taste threshold in Table 2-3. Advantages over the flame

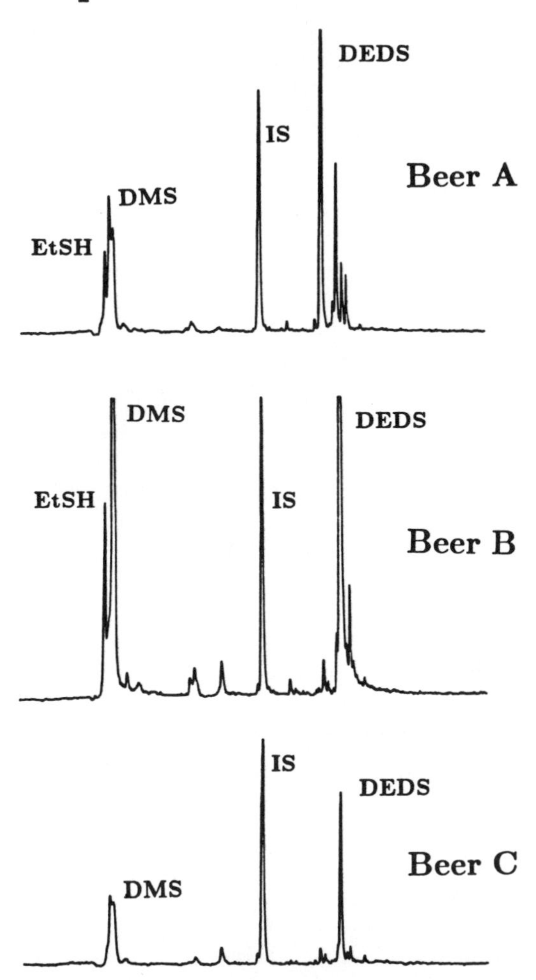

Figure 2-17 ■ Chromatograms obtained for the headspace analysis of three popular beers using the Sulfur Chemiluminescence Detector. Ethanethiol (EtSH), dimethyl sulfide (DMS), and diethyl disulfide (DEDS) are identified. IS is an internal standard added to the beer (diisopropyl sulfide). Courtesy of Sievers Research, Inc.

photometric detector for this application include:

1. Approximately 2 orders of magnitude greater sensitivity.
2. Linearity of response allowing for quantification by standard addition.
3. Insensitivity to coelution of water and ethanol with ethanethiol and dimethyl sulfide.
4. Greater day-to-day reproducibility.

Table 2-3 ■ Detection Limits in Beer by Headspace Analysis Using the Sulfur Chemiluminescence Detector™

Compound	Detection Limit (parts per trillion)	Taste Threshold (parts per trillion)
Dimethyl sulfide	100	30,000
Dimethyl disulfide	75	7,000
Dimethyl trisulfide	300	100
Diethyl disulfide	10	3,000
Ethanethiol	100	3,000

Source: Courtesy of Sievers Research, Inc.

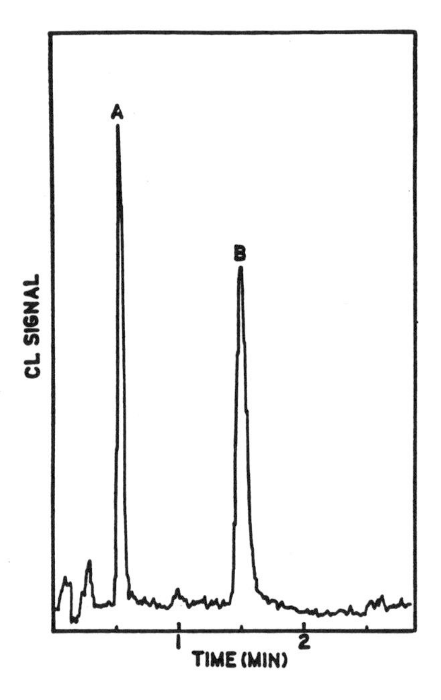

Figure 2-18 ■ Chromatogram of 1 cc of natural gas obtained using the Sulfur Chemiluminescence Detector. Peaks: (A) dimethyl sulfide, (B) *tert*-butyl mercaptan. From J. K. Nelson, Ph.D. Thesis, University of Colorado, Boulder, Colorado, 1984, p. 96.

The detection of odorants in natural gas provides another example of the sensitivity and selectivity of the SCD (Fig. 2-18). An example of the application of this detector to HPLC is provided in Fig. 2-19, which includes chromatograms of a commercial preparation of the insecticide malathion.

Atomic Fluorine as a Chemiluminescence Reagent

When fluorine atoms are used in place of F_2 molecules, the reaction proceeds much more rapidly and the response is nearly universal, similar to that already described for active nitrogen. Fluorine atoms can abstract hydrogen from hydrocarbons at near

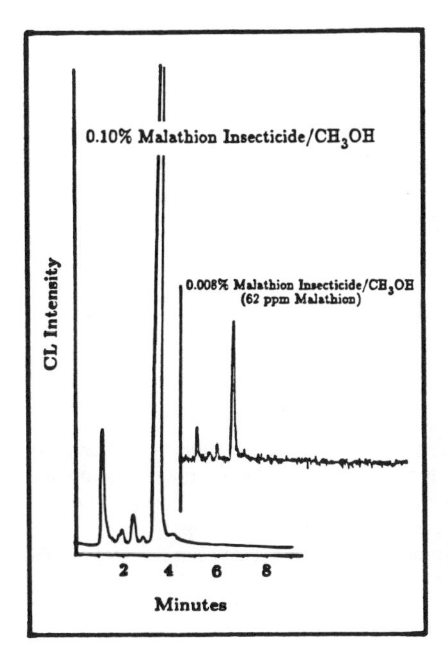

Figure 2-19 ■ Chromatogram of Malathion Insect Control obtained in microbore HPLC using the Sulfur Chemiluminescence Detector. Mobile phase: 75% methanol/water at 100 μl min^{-1}. 1-μl sample. From E. A. Mishalanie and J. W. Birks, *Anal. Chem.* **58**, 918 (1986).

collisional rates with virtually no activation energy. These reactions are highly exothermic, since they form strong H—F (136 kcal/mol) and C—F (116 kcal/mol) bonds, while breaking much weaker C—H (99 kcal/mol) and C—C bonds (88 kcal/mol). Upon initial abstraction, the C—H bond strength in alkyl radicals is dramatically decreased, thereby increasing the exothermicity of subsequent abstractions.

Fluorine atoms can be created in a variety of ways. Microwave discharges of F_2, CF_4 and SF_6 have long been used as atomic fluorine sources. The source used by Spurlin and Yeung is the multiphoton dissociation of SF_6 by a CO_2 infrared laser focused into the center of the detection cell.[24] The generation of F atoms in the center of the cell allows for virtually complete reaction of the F atoms with the analyte molecules without interference from wall loss, which has a near unity efficiency for F atoms. The instrumental diagram is shown in Fig. 2-20. This process entails

$$SF_6 + nh\nu \rightarrow SF_5 + F \qquad (2\text{-}25)$$

$$SF_5 + nh\nu \rightarrow SF_4 + F \qquad (2\text{-}26)$$

These fluorine atoms can then undergo abstraction reactions with hydrocarbons

$$F + R - H \rightarrow HF^{\dagger} + R\cdot \qquad (2\text{-}27)$$

The reaction is exothermic by about 37 kcal/mol for most hydrocarbons, which is sufficient to populate vibrational levels up to $v = 3$ of HF. The (3,0) vibrational overtone band may be detected at 880 nm[25] for most species containing a C—H bond and many other compounds containing hydrogen. To be selective to hydrocarbons, the emissions from electronically excited $C_2(d^3\Pi_g)$ at 470 nm and $CH(A^2\Delta)$ at 431 nm are

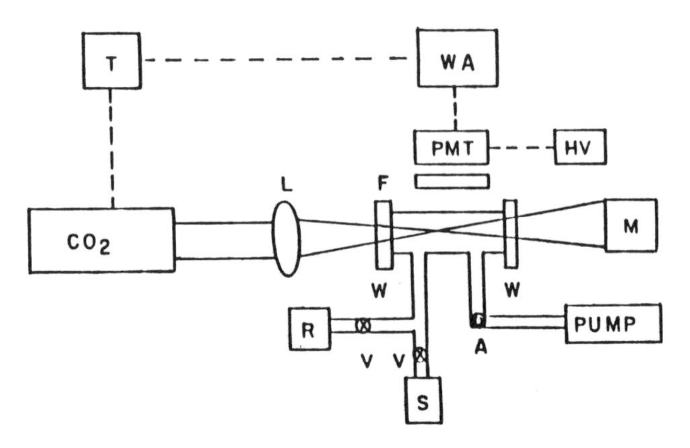

Figure 2-20 ■ Instrumental diagram for the multiphoton infrared dissociation of SF_6 and the subsequent detection of chemiluminescence generated by F-atom reactions with hydrocarbons. T = trigger, CO_2 = pulsed CO_2 laser, L = lens, F = interference filters, W = NaCl windows, PMT = photomultiplier tube, WA = wave form analyzer, HV = power supply, M = power meter, V = leak valves, A = aperture, R = reagent gas of SF_6 and S = analyte source (GC column). From S. R. Spurlin and E. S. Yeung, *Anal. Chem.* **54**, 318 (1982).

monitored in the detector. The exact mechanisms which produce these species are unknown; however, these emissions are common in fluorine/hydrocarbon flames.[19] The final step in the chain reaction must be exothermic by 66 kcal/mol to form excited state CH and 61 kcal/mol in the case of C_2.

The response of the detector shows two concentration regions where linearity is observed. Below 50 ppm, it is thought that quenching and competition between emission and radical reactions is unimportant, and the curve reflects a simple first-order process. At higher concentrations, a high radical concentration begins to compete with the emission process by quenching reactions. This decreases the probability of emission for an excited-state molecule; however, more excited species are formed that more than compensates for quenching.[24] Therefore, the curve remains linear, although the slope is reduced.

As in the case of the active nitrogen detector, selectivity for certain compounds can be achieved by looking at different emitters. Iodinated compounds have been shown to react through short chain reactions to produce excited state IF.[27, 28] The emission of $IF(B^3\Pi_0^+ \rightarrow X^3\Sigma^+)$ occurs between 450 and 800 nm (Fig. 2-21). By using an appropriate bandpass filter centered at 580 nm, where the strongest emissions occur, selectivity for iodinated compounds can be obtained.[29] The reaction is thought to occur through a two-step mechanism

$$F + CH_3I \rightarrow CH_3IF \tag{2-28}$$

$$F + CH_3IF \rightarrow CH_3F + IF^* \tag{2-29}$$

Reaction (2-28) is exothermic by 26 kcal/mol, and the intermediate CH_3IF has been identified.[30] Reaction (2-29) is exothermic by 94 kcal/mol, which is more than enough energy to produce excited-state IF.

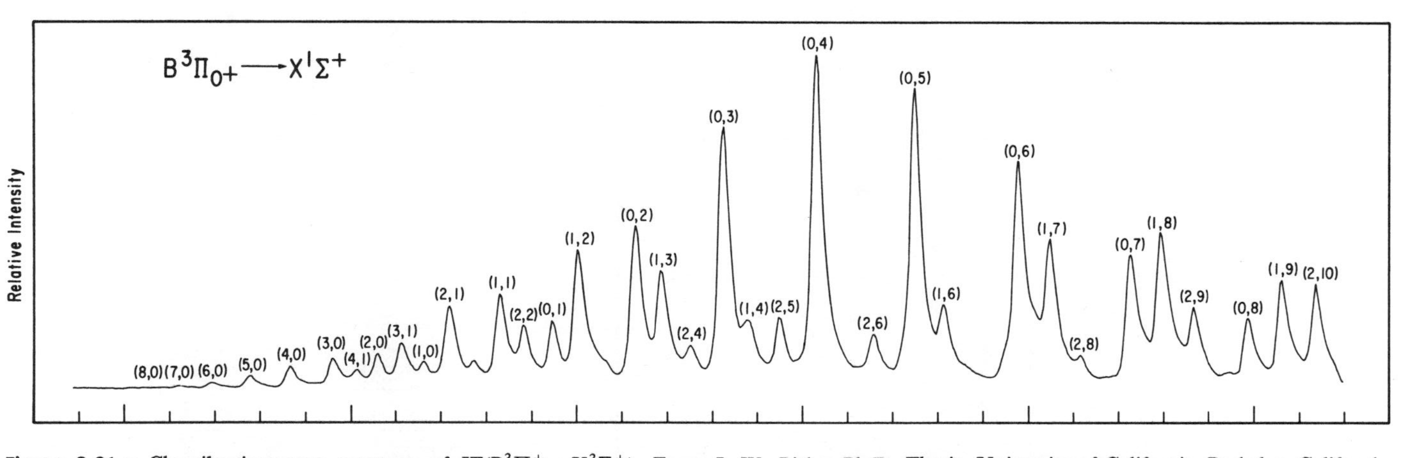

Figure 2-21 ■ Chemiluminescence spectrum of IF($B^3\Pi_0^+ \rightarrow X^3\Sigma^+$). From J. W. Birks, Ph.D. Thesis, University of California, Berkeley, California, 1974.

The detector is highly selective to iodinated compounds over a variety of compounds containing C, H, O, N, Cl and Br atoms; however, the system as reported was not optimized and detection limits were $\approx 1\ \mu g$.[29]

Sodium Vapor Detector

Sodium atoms have also been used as a chemiluminescence reagent for gas chromatography.[31] The detector responds to halocarbons containing more than one halogen atom. A schematic of the detector is shown in Fig. 2-22. A plug of Na metal is heated up to 400°C and the resulting sodium vapor mixed with the GC effluent in a detection chamber maintained at low pressure. When halogenated hydrocarbons are present in the detector cell, the following reactions can occur:

$$R_1XY + Na \rightarrow \cdot R_1X + NaY \tag{2-30}$$

$$\cdot R_1X + Na \rightarrow R_2 + NaX^{\dagger} \tag{2-31}$$

Here R_1 is an alkyl substituent, X and Y are halogen atoms and R_2 is a stable alkene or some other hydrocarbon product. The formation of a strong, ionically bonded, salt molecule and (where possible) formation of a carbon–carbon double bond in the final

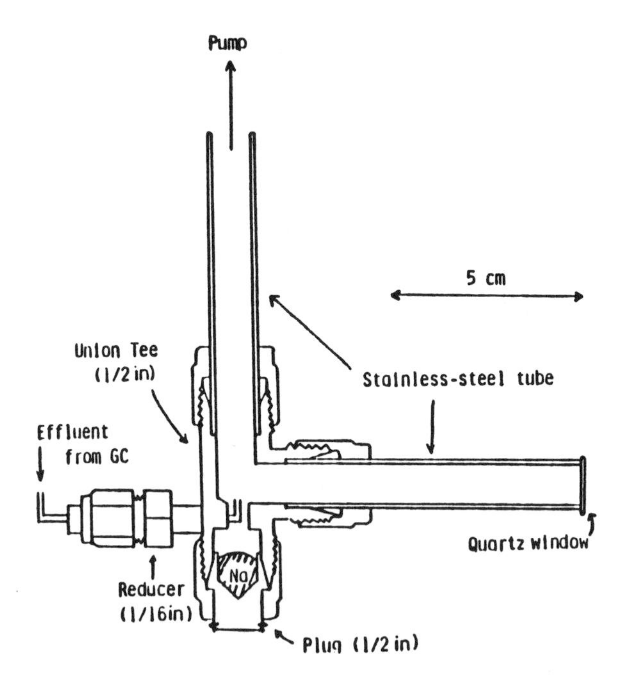

Figure 2-22 ■ Chemiluminescence cell for the detection of polyhalogenated hydrocarbons by sodium-atom chemiluminescence. From M. Yamada, A. Ishiwada, T. Hobo, S. Suzuki and S. Araki, *J. Chromatogr.* **238**, 347 (1982).

product results in the release of a large amount of energy to NaX as vibrational energy. Although a detailed analysis of this system has not been undertaken, a minimum of 48.5 kcal/mol of vibrational energy must be contained in the $NaX^{\dagger}$ molecule. This is the energy required to excite the sodium *D* lines by the following vibrational-to-electronic energy transfer:

$$NaX^{\dagger} + Na(3^2S) \rightarrow NaX + Na^*(3^2P) \quad (2\text{-}32)$$

The excited state Na atom can then fluoresce at the characteristic *D* line (589 nm), which is monitored by means of a photomultiplier tube and filter system.[31]

The detector shows a large dependence on both temperature and pressure. A higher temperature results in more Na atoms in the vapor phase. This, along with the increased energy within the system, results in a faster reaction rate and enhances CL intensity. High pressure has a detrimental effect on response of the detector. This is explained by the collisional deactivation of the $NaX^{\dagger}$ molecules. The response to chloro- and bromosubstituted hydrocarbons is quite sensitive, with detection limits ranging from 4.6×10^{-11} g for $CHBr_3$ to 7.9×10^{-14} g for $Cl_2C{=}CCl_2$.[31] The response to fluorosubstituted compounds is much smaller. This is generally attributed to the much stronger bond strength of C—F bonds (116 kcal/mol) as compared to C—Cl (79 kcal/mol) and C—Br (66 kcal/mol) bonds. The extra energy needed to break the stronger C—F bond decreases the exothermicity of the reaction, so that excited state products are not formed.

A weak CL emission results from certain nitrogen- or oxygen-containing compounds and monohalogenated species. The CL intensities are very low from these species, and the reactions involved are unknown.

Chlorine Dioxide Detector

Another detector which responds to two sulfur gases is based on chemiluminescent reactions with chlorine dioxide, OClO.[32] This novel detection scheme provides an extremely selective method of continuous monitoring of H_2S. The overall reaction occurring within the detector is thought to be

$$2H_2S + ClO_2 \rightarrow S_2 + \tfrac{1}{2}Cl_2 + 2H_2O \quad (2\text{-}33)$$

The elementary steps of this reaction are unknown at present. However, it is thought that the reaction produces S atoms which recombine to form S_2 in its triplet excited state, as in the flame photometric detector. $S_2(B^3\Sigma_u^- \rightarrow X^3\Sigma_g^-)$ emission is monitored between 250 and 400 nm (see Fig. 2-23). The detector exhibits a linear response, a result which is not obvious considering that the formation of S_2 is second order in S atoms. Linearity can be explained by the fact that a large excess of OClO is present in the detector. This can completely convert every H_2S molecule within the detection cell to S atoms. The only possible reaction of these S atoms is recombination, again assumed to be 100% efficient. Therefore, the number of S_2 molecules produced in the reaction is exactly one-half of the number of initial H_2S molecules present, resulting in a linear calibration curve.[32] The detector is quite sensitive to H_2S, with a detection limit of 3 ppb. However, as a GC detector it is limited by the fact that it is actually too selective, responding to only two compounds, H_2S and CH_3SH.

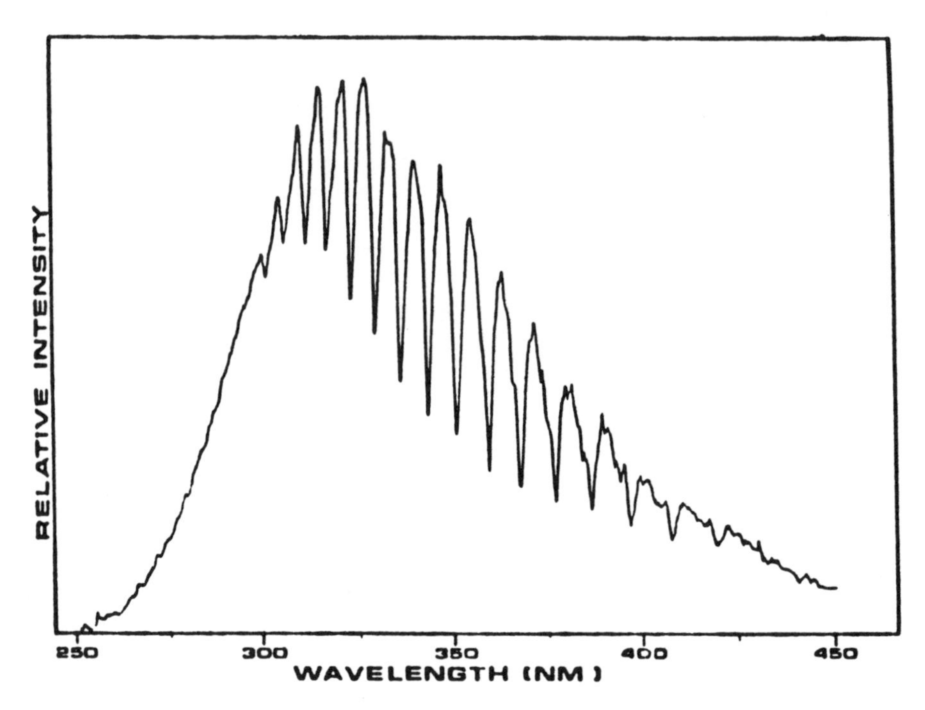

Figure 2-23 ▪ Chemiluminescence spectrum of S_2 observed in the reaction of ClO_2 with H_2S. From S. R. Spurlin and E. S. Yeung, *Anal. Chem.* **54**, 318 (1982).

$SO + O_3$ Detector

A "universal" sulfur-selective detector displaying a linear response and high sensitivity has been developed very recently.[33] In this detector the GC effluent is burned in a hydrogen-rich flame and the flame gases sampled into a chemiluminescence cell by a quartz or ceramic capillary tube where the flame effluent is mixed with ozone. A fraction of the sulfur is converted to SO in the flame. In the reaction cell SO reacts with O_3 to produce SO_2 in an excited electronic state which subsequently emits light:

$$SO + O_3 \rightarrow SO_2^* + O_2 \tag{2-34}$$

$$SO_2^* \rightarrow SO_2 + h\nu \tag{2-35}$$

Emission occurs in the wavelength region 260 to 480 nm, and as many as four electronic states of SO_2 may be involved.[34]

The $SO + O_3$ detector has the advantage over the fluorine-induced chemiluminescence detector described above of not responding to alkenes and other compounds having weak C—H bonds. This detector currently is being developed commercially by Sievers Research, Inc. as their Model 350 Sulfur Chemiluminescence Detector (SCD).

References

1. Sutton, D. G., Westberg, K. R., and Melzer, J. R. *Anal. Chem.* **51**, 1399 (1979).
2. Wright, A. N., and Winckler, C. A. *Active Nitrogen*, Academic, New York, 1968.
3. Safrany, D. R. *Prog. React. Kin.* **6**, 1 (1972).
4. Campbell, I. M., and Thrush, B. A. *Proc. Chem. Soc. (London) Ser. A* **309**, 410 (1964).
5. Wright, A. N., and Winckler, C. A. *Can. J. Chem.* **40**, 129 (1962).
6. Rice, G. W., Richards, J. J., D'Silva, A. P., and Fassel, V. A. *Anal. Chem.* **53**, 1519 (1981).
7. Clay, J. T., and Niemczyk, T. M. *Spectroscopy* **2**, 36 (1987).
8. Melzer, J. E., and Sutton, D. G. *Appl. Spectr.* **34**, 434 (1980).
9. Rice, G. W., D'Silva, A. P., and Fassel, V. A. *Appl. Spectr.* **38**, 149 (1984).
10. Toby, S. *Chem. Rev.* **84**, 277 (1984).
11. Bruening, W., and Concha, F. J. M. *J. Chromatogr.* **112**, 253 (1975).
12. Finlayson, B. J., Pitts, J. N., Jr., and Atkinson, R. *J. Am. Chem. Soc.* **96**, 5356 (1974).
13. Herron, J. T., and Huie, Z. E. *J. Am. Chem. Soc.* **99**, 5430 (1977).
14. Glinski, R. J., Sedarski, J. A., and Dixon, D. A. *J. Phys. Chem.* **85**, 2440 (1981).
15. Kelly, T. J., Gaffney, J. S., Phillips, M. F., and Tanner, R. L. *Anal. Chem.* **55**, 138 (1983).
16. Glinski, R. J., Sedarski, J. A., and Dixon, D. A. *J. Am. Chem. Soc.* **104**, 1126 (1982).
17. Nelson, J. K., Getty, R. H., and Birks, J. W. *Anal. Chem.* **55**, 1767 (1983).
18. Mishalanie, E. A., and Birks, J. W. *Anal. Chem.* **58**, 918 (1986).
19. Duewer, W. H., and Setser, D. W. *J. Chem. Phys.* **58**, 2310 (1973).
20. Glinski, R. J., Mishalanie, E. A., and Birks, J. W. *J. Photochem.* **37**, 217 (1987).
21. Glinski, R. J., Getty, J. N., and Birks, J. W. *Chem. Phys. Lett.* **117**, 359 (1985).
22. Glinski, R. J., and Taylor, C. D., *Chem. Phys. Lett.* **155**, 511 (1989).
23. Glinski, R. J., Mishalanie, E. A., and Birks, J. W. *J. Am. Chem. Soc.* **108**, 531 (1986).
24. Spurlin, S. R., and Yeung, E. S. *Anal. Chem.* **57**, 1223 (1985).
25. Mann, D. E., Thrush, B. A., Lide, D. R., Jr., Ball, J. J., and Acquista, N. *J. Chem. Phys.* **34**, 420 (1961).
26. Patel, R. I., Stewart, G. W., Castleton, K., Gole, J. L., and Lombardi, J. R. *Chem. Phys.* **57**, 461 (1980).
27. Estler, R. C., Lubman, D., and Zare, R. N. *Faraday Disc. Chem. Soc. (London)* **62**, 317 (1977).
28. Braynis, H. S., and Whitehead, J. C. *J. Chem. Soc. (London) Faraday Trans. 2* **79**, 1113 (1983).
29. Getty, R. H., and Birks, J. W. *Anal. Lett.* **12**, 469 (1979).
30. Farrar, J. M., and Lee, Y. T. *J. Am. Chem. Soc.* **96**, 7570 (1974).
31. Yamada, M., Ishiwada, A., Hobo, T., Suzuki, S., and Araki, S. *J. Chromatogr.* **238**, 347 (1982).
32. Spurlin, S. R., and Yeung, E. S. *Anal. Chem.* **54**, 318 (1982).
33. Benner, R. L., and Stedman, D. H., unpublished results.
34. Kenner, R. D., and Ogryzlo, E. A., In: J. G. Burr, Ed., *Chemiluminescence and Bioluminescence*, Marcel Dekker, New York, 1985, p. 139.

3

Detectors Based on the NO + O_3 Chemiluminescent Reaction

Randall L. Shearer and Robert E. Sievers

Department of Chemistry and Biochemistry and Cooperative Institute for Research in Environmental Sciences (CIRES), University of Colorado, Boulder, Colorado 80309

Introduction

Chemiluminescent reactions occurring in the gas phase have been successfully used in making many analytical measurements. Detectors based on chemiluminescent reactions are generally very sensitive, and, since few compounds undergo chemiluminescent reactions, these detectors are inherently selective. The measurement of nitric oxide by its chemiluminescent reaction with ozone[1] constitutes the basis for selective gas chromatography detectors. The utilization of this reaction in detection schemes is attractive owing to its linearity and sensitivity of response with little interference from other species. Sensitivity for nitric oxide at parts-per-trillion levels (v/v) and linearity over 6 orders of magnitude can be achieved by measuring the light emission from the NO + O_3 chemiluminescent reaction.[2–8]

Chromatographic detectors based on the NO + O_3 reaction rely upon the accurate and precise measurement of nitric oxide pulses which may be produced at low concentration levels. Instruments used to measure NO in turn rely upon the detection of photons produced by NO + O_3 chemiluminescence (CL). The photons produced are the result of the combination of reactions (3.1) and (3.3) among the reactions simultaneously occurring [(3-1) to (3-4)][9]:

$$NO + O_3 \xrightarrow{k_1} NO_2^* + O_2 \qquad (3\text{-}1)$$

$$NO + O_3 \xrightarrow{k_2} NO_2 + O_2 \qquad (3\text{-}2)$$

$$NO_2^* \xrightarrow{k_3} NO_2 + h\nu \qquad (3\text{-}3)$$

$$NO_2^* + M \xrightarrow{k_4} NO_2 + M \qquad (3\text{-}4)$$

Experimental measurements of the branching ratio for the formation of excited-state nitrogen dioxide (NO_2^*) range from 0.06[10] to 0.35[11] at room temperature. The light emission from NO_2^* is broad and centers around 1200 nm.[12] Typically, a red-transmitting filter is used to allow transmission of a significant fraction of the NO_2^* emission to the photomultiplier or photodiode sensor, while suppressing CL signals from other species that emit in a different region of the spectrum. The shape of the plot of chemiluminescence intensity of the NO + O_3 reaction vs. wavelength is independent of reaction pressure or the concentrations of the reacting species.[9] Of course, the absolute intensity of the chemiluminescence at a particular wavelength is directly dependent on these as well as other operating conditions. The emission of a photon from excited-state NO_2^* is quenched by collision with other molecules or a wall as shown in Eq. (3-4). This is minimized by operating the nitric oxide/ozone reaction chamber under vacuum, which is typically maintained at approximately 1 to 100 torr by a vacuum pump.

An equation for the observable photon flux ($h\nu_R$) from the nitric oxide/ozone reaction is given by[2]

$$h\nu_R = \left[\frac{k_1}{k_1 + k_2}\right]\left[\frac{k_3}{k_4}\right]\rho\mu_{NO} \tag{3-5}$$

where k_1 through k_4 are the rate constants for Eqs. (3-1) through (3-4). Also, ρ is the volumetric pumping rate of the vacuum pump and μ_{NO} is the mixing ratio of nitric oxide. Equation (3-5) is the result of a steady-state analysis of the rate equations for the important species in Eqs. (3-1) to (3-4). Nitric oxide and ozone concentrations were specified in this treatment and all other parameters were held constant in the treatment, and it was assumed that the nitric oxide and ozone were well mixed in the reaction chamber and that ozone was present in large excess compared to the concentration of nitric oxide. Some approximations regarding the experimental parameters were also made. The observable photon flux is proportional to μ_{NO} since k_1, k_2, k_3, k_4, and ρ are constants.

When used as a detector for gas chromatography, the cell must not contribute significantly to band broadening. This is an additional constraint that is not addressed in the design of some instruments made for the measurement of nitric oxide in ambient air, in which much larger throughput volume flow rates are used. In the derivation leading to Eq. (3-5), it was assumed that the flow rate of the sample approximates the total gas flow rate; for a chromatographic CL detector this is not necessarily so. Taking this into account, the equation for observable photon flux becomes

$$h\nu_R = \left[\frac{k_1}{k_1 + k_2}\right]\left[\frac{k_3}{k_4}\right]\left[\frac{Q_{sample}}{Q_{net}}\right]\rho\mu_{NO} \tag{3-6}$$

where Q_{sample} is the sample flow rate and Q_{net} is the total flow rate of both the sample and the ozone stream into the reaction chamber. The observable photon flux is proportional to μ_{NO} if Q_{sample} and Q_{net} are held constant. This is achieved experimentally by using restrictors or flow controllers to maintain constant flow. The chemiluminescence actually measured is the product of the observable photon flux and the collection efficiency of the photon measuring system. In practice, because of the use of a red-transmitting filter and because photomultiplier tubes generally possess poor response characteristics in the near infrared, only a small fraction of the observable photon flux is measured. Nevertheless, very sensitive measurement of NO is possible despite these limitations, and even greater sensitivity may be achieved with technologi-

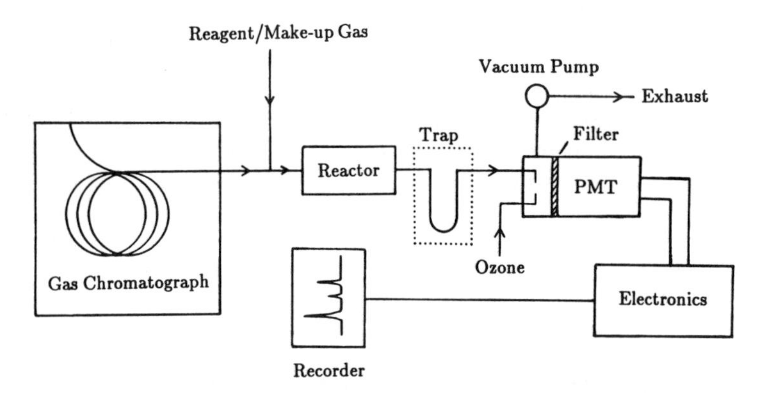

Figure 3-1 ■ Simplified schematic drawing of a general detection system for gas chromatography based on the $NO + O_3$ reaction. Column effluent may be mixed with a reagent gas or it may pass directly into a reactor in which nitric oxide is generated by an analyte. After the reactor, if necessary, a cold trap can be used to trap nonvolatile species before they pass into the CL reaction chamber. In the CL reaction chamber, the effluent is mixed with an excess of O_3 and light emitted from the $NO + O_3$ reaction, after being transmitted through a red-filter, is detected by a photomultiplier tube (PMT). The $NO + O_3$ reaction chamber is maintained at reduced pressure by a vacuum pump.

cal advances in light measurement. In addition, the branching ratio for forming NO_2^* can be increased by operating the $NO + O_3$ reaction chamber at higher temperatures, but this requires that the photomultiplier tube be well insulated to guard against thermal noise, so generally the reaction chamber is operated at room temperature.

Successful detection of analytes based on the $NO + O_3$ chemiluminescent reaction requires that nitric oxide be produced either directly or indirectly by an analyte to be detected. The formation of nitric oxide by an analyte is necessarily limited to a post-column chromatographic process if the measurement of individual components of a mixture is desired. Little or no chromatographic separation may be necessary, however, if a mixture contains only one or a few compounds that can react to produce nitric oxide, or if only the sum total response from a class of reacting compounds is desired. Of course, the separation efficiency required is inherently reduced because the detection process is selective. In any case, nitric oxide produced by an analyte appears as a surrogate pulse derived from the original analyte. Both real-time and chromatographic detection is accomplished for only those compounds or compound classes that react rapidly to produce nitric oxide.

A schematic drawing of a general chromatographic detection system based on the $NO + O_3$ reaction is shown in Fig. 3-1. The effluent from a chromatograph may be mixed with a reagent gas or a makeup gas before entering a reactor in some detection schemes. The column effluent then flows to a small reaction chamber. The reactor may simply consist of a pyrolyzer or some other catalytic reaction zone in which nitric oxide is produced from the compounds that are to be detected. A trap placed between the reactor and the NO detector may be used to condense nonvolatile species, while NO is allowed to pass. Downstream, in the NO detector, nitric oxide is detected, usually by its CL reaction with O_3 in a reduced-pressure reaction chamber, and light from this reaction is detected by a photomultiplier tube (PMT). A red-transmitting filter placed between the CL reaction chamber and the PMT is used to suppress CL signals that occur in a different region of the spectrum, from reaction of ozone with other species, such as olefins, carbon monoxide and some sulfur compounds.[13-20] Finally, the resulting chromatogram is simply the continuously measured nitric oxide signal from a NO detector.

There are currently three gas chromatographic detectors based on the $NO + O_3$ chemiluminescent reaction: the Thermal Energy Analyzer (TEA™), a nitrogen-compound-selective detector and the Redox Chemiluminescence Detector (RCD™). All three of these detectors make use of some post-column reaction in which nitric oxide is generated, and this is measured after mixing with an excess of ozone. In the Thermal Energy Analyzer, the utilization of the fact that *N*-nitroso compounds readily undergo catalyzed pyrolysis to form nitric oxide, has led to the development of a highly selective and sensitive detector of *N*-nitroso compounds.[21-23] A nitrogen-selective detector for GC, based on the $NO + O_3$ CL reaction, has also been developed.[24, 25] In this detector, the effluent of a gas chromatographic column is mixed with oxygen and the resulting mixture is passed through a platinum catalyst chamber, which is heated to $> 800°C$. Only nitrogen-containing compounds that react to from nitric oxide can be detected. Surface-catalyzed redox reactions of nitrogen oxides with reducing compounds may also produce nitric oxide. By contrast, in the Redox Chemiluminescence Detector, the effluent of a gas chromatographic column is mixed with an added reagent gas, typically nitrogen dioxide, and then passed over a heated gold surface.[26, 27] On the gold surface, all analytes (irrespective of whether they contain nitrogen) that can react rapidly with NO_2 can lead to the formation of NO. Each of these gas chromatographic detectors based on the $NO + O_3$ reaction will be described more fully.

Thermal Energy Analyzer

Principles of Operation

The thermal energy analyzer utilizes selective catalytic or pyrolytic decomposition of *N*-nitroso compounds to form nitric oxide. The nitric oxide formed is then measured by its CL reaction with ozone. The TEA is highly selective so it can be used for screening purposes often without separation of a sample being necessary, or it can be used as a detector for either gas or liquid chromatography.

The pyrolysis chamber in TEA is placed serially between a chromatographic column and a NO detector. The chamber may consist of an open tube of quartz or ceramic, an open tube whose walls are coated with a catalytic substrate or simply a tube packed with a catalyst material. In each case, provisions are made so that the temperature of the pyrolysis chamber can be controlled. The temperature of operation of the pyrolysis chamber greatly affects the selectivity of the detector. In *N*-nitroso compounds, the N—NO bond is generally the weakest bond in the molecule and therefore the most easily cleaved. The pyrolysis of *N*-nitroso compounds is

$$(A)(Z)N{-}NO \longrightarrow (A)(Z)N\cdot + \cdot NO \qquad (3\text{-}7)$$

where A and Z represent any organic radicals. The nitrosyl radical (nitric oxide) formed in the preceding reaction is stable at normal pressures and temperatures. Some care must be taken so that NO formed is not lost through any process, such as adsorption on active surfaces. Usually this is not a problem, and cold traps maintained at −150°C and short packed columns of Tenax placed between the pyrolysis chamber and the NO detector have been used to remove potentially interfering organic compounds, while allowing NO to pass unaffected. The organic fragment formed in the preceding reaction may be unstable and further decompose or rearrange to a more stable species, but it is unlikely to form a nitrosyl radical, and therefore it is not likely to be a source of interference.

N-nitroso compounds can undergo simple pyrolytic decomposition to form nitric oxide, as just described, but this requires high temperatures and the pyrolysis may not be completely selective for *N*-nitroso compounds. Under these conditions, for instance, nitro-compounds might also decompose to form NO at sufficiently high temperatures. For this reason, a number of catalysts were investigated as to their effectiveness for cleaving the N—NO bond at lower temperatures.[28-30] Catalysts of WO_3, $W_{20}O_{58}$, acid-passivated stainless steel, nickel oxide and nickel-based alloys were investigated. It was found that a mixture of WO_3 and $W_{20}O_{58}$ was most effective for selective and reproducible cleavage of the N—NO bond.[23] Other investigators recommended that a catalyst consisting of WO_3 adsorbed on the walls of a porous ceramic tube be used because it exhibited a longer catalyst lifetime.[31] Nevertheless, while greater responses for *N*-nitroso compounds at lower temperatures are achieved by using a catalytic pyrolysis chamber, uncatalyzed pyrolysis is attractive because of its simplicity.

Applications

In most laboratories using TEA with GC, packed columns with 10 to 15% of polyethyleneglycol on Chromosorb W as a stationary phase are used because of the polar nature of the solutes. Inferior signal-to-noise ratios of important species can result from column bleeding.[32] Considerable improvement in not only column efficiency but also in reducing the problem of column bleed is achieved through the use of capillary columns. Because lower carrier gas flow rates are employed in capillary GC, greater care must be taken to eliminate unnecessary void volumes in the column/detector interface and cold spots must also be eliminated. One design for a capillary column/TEA interface is shown in the schematic drawing in Fig. 3-2. This interface incorporates not only low dead volume connections but also, because the column is inserted deep into the ceramic pyrolyzer, there are no cold spots in which to lose nonvolatile species. Figure 3-3 shows a TEA-GC chromatogram obtained with a capillary column of an injected sample that contained 1 ng amounts of various nitramines. The identification of the peaks in the chromatogram are: (1) dimethylnitrosamine, (2) methylethylnitrosamine, (3) diethylnitrosamine, (4) methyl-*n*-propylnitrosamine, (5) di-isopropylnitrosamine, (6) di-*n*-propylnitrosamine, (7) methyl-*n*-butylnitrosamine, (8) methyl-*n*-pentylnitrosamine, (9) di-*n*-butylnitrosamine, (10) nitrosopiperidine, (11) nitrosopyrolidine, (12) nitrosomorpholine and (13) methyl-*n*-heptylnitrosamine.

It was found that in the purely pyrolytic mode, the decomposition of *N*-nitroso compounds at 300°C is better than at higher temperatures because the decomposition of the N—NO bond is more selective, even though the absolute response of the

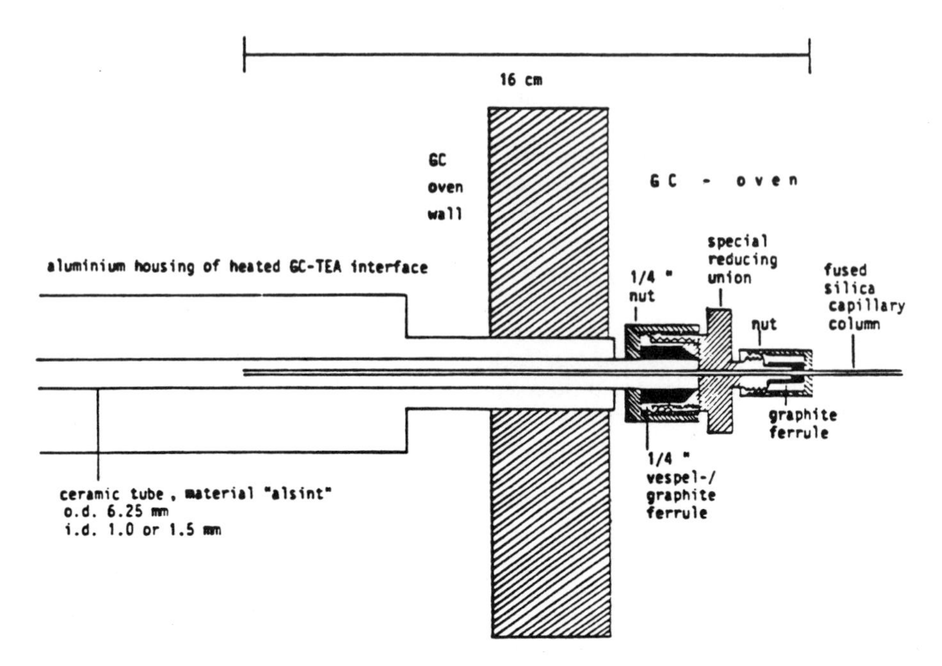

Figure 3-2 ■ Drawing of capillary column/TEA interface with vacuum tight connection between the capillary column and the TEA detector (graph not to scale). Reprinted from H. Borwitsky, *Chromatographia* **22**, 65 (1986), with permission.

detector to nitroso compounds is diminished at lower temperatures.[33] Quantitative conversion of nitrosoamines to NO was found to occur between 300 to 350°C using a simple pyrolysis chamber.[34] The TEA response vs. pyrolytic chamber temperature for a few nitrosoamines is shown in Fig. 3-4. Of a few other compounds tested representing nitroalkanes, nitroaromatics, nitritoalkanes and nitramines, only a nitrite compound appeared to interfere significantly with the detection of nitrosoamines in the temperature range of 300 to 350°C; this is illustrated in Fig. 3-5.

From Fig. 3-5, it is readily seen that the TEA can be used to detect nitro compounds. With the pyrolysis temperature at 800 to 900°C, a detection limit of 0.6 ng for nitroaromatics was observed with a linear dynamic range of over 4 orders of magnitude.[35] Picogram[36] and tens of picogram[37] detection limits were achieved for nitroaromatic compounds when a capillary column was inserted far enough into a ceramic pyrolysis tube so that there were no cold spots between the chromatographic column and the pyrolysis chamber. Thus, it is very important to prevent wall losses of nonvolatile species when very sensitive measurement of these species is desired.

The measurement of nitroaromatic compounds is important in environmental toxicology because they are heavily used in industry, may be formed as waste products in various processes and are present in explosives. A study was undertaken that focused on the analytical measurement of nitrotoluene compounds using TEA because nitrotoluenes are produced in large commercial quantities and have many varied uses.[35] Figure 3-6 shows a GC-TEA chromatogram of a mixture that contained nitrated

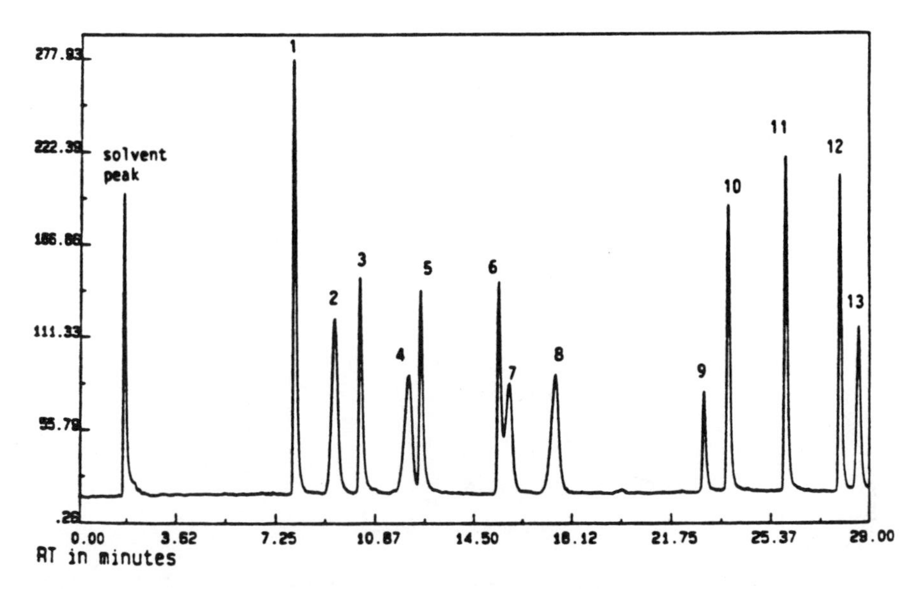

Figure 3-3 ■ TEA capillary GC chromatogram of 13 nitrosamines (identified in text). Column is a 50 m, 0.32 mm i.d. fused silica capillary with a polyethyleneglycol type stationary phase. GC parameters: Injection 1.0 μl spitless, injector temp. 200°C. Carrier gas helium, 1.4 bar. Column temperature 50°C for 1 min, 2.5°C/min to 200°C. Reprinted from H. Borwitsky, Chromatographia **22**, 65 (1986), with permission.

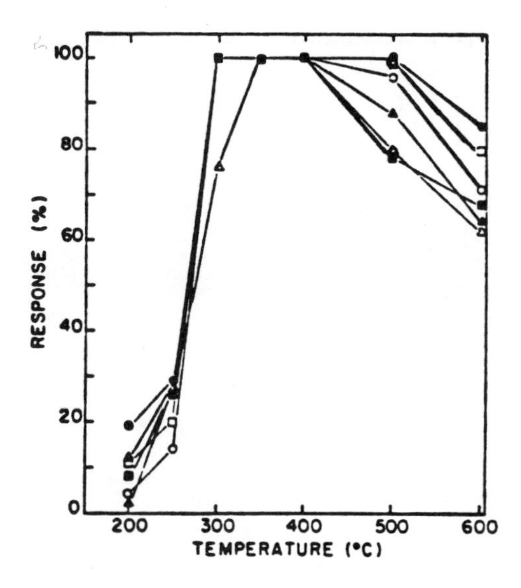

Figure 3-4 ■ TEA response vs. pyrolytic chamber temperature for (□) nitrosodiethylamine, (■) nitrosodipropylamine, (●) nitrosodibutylamine, (△) nitrosopyrrolidine and (▲) nitrosomorpholine. Reprinted from T. J. Hansen, M. C. Archer, and S. R. Tannenbaum, *Anal. Chem.* **51**, 1526 (1979), with permission.

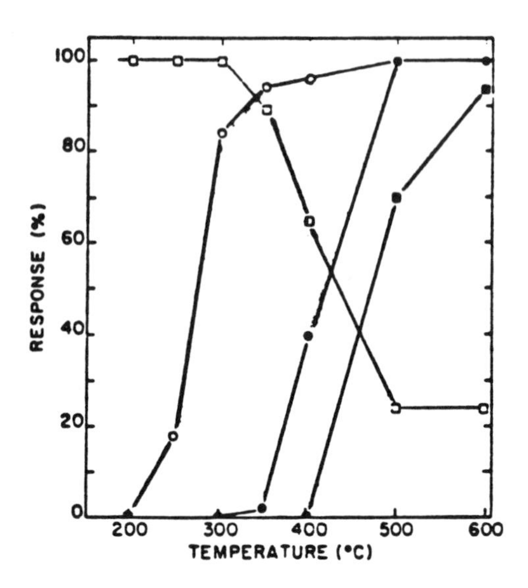

Figure 3-5 ■ TEA response vs. pyrolytic chamber temperature for (●) nitrohexane, (■) nitrotoluene, (□) hexylnitrite and (○) dipropylnitramine. Reprinted from T. J. Hansen, M. C. Archer, and S. R. Tannenbaum, *Anal. Chem.* **51**, 1526 (1979), with permission.

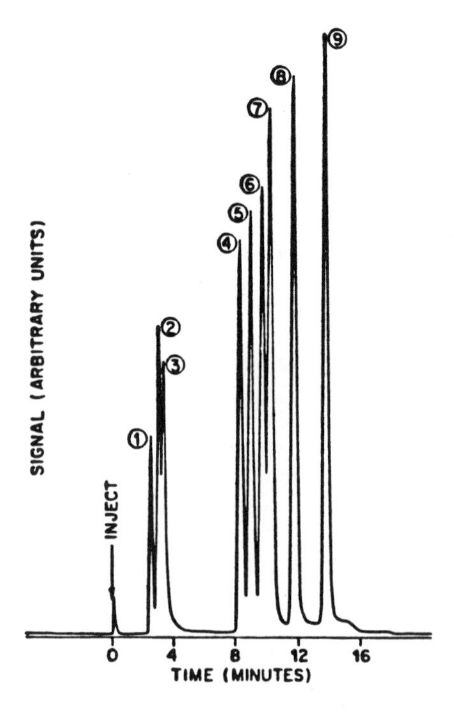

Figure 3-6 ■ GC-TEAchromatogram for a mixture of nitrated toluene derivatives: (1) 2-nitrotoluene, (2) 3-nitrotoluene, (3) 4-nitrotoluene, (5) 2,5-dinitrotoluene, (6) 2,4-dinitrotoluene, (7) 2,3-dinitrotoluene, (8) 3,4-dinitrotoluene and (9) 2,4,6-trinitrotolene. Reprinted from A. L. Lafleur and K. M. Mills, *Anal. Chem.* **53**, 1202 (1981), with permission.

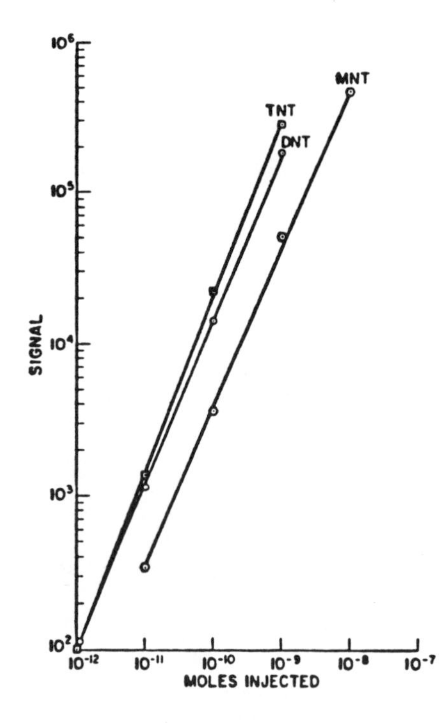

Figure 3-7 ■ Plot of detector response vs. moles injected for GC-TEA detection. Substances tested were 2-nitrotoluene (MNT), 2,4-dinitrotoluene (DNT), and 2,4,6-trinitrotoluene (TNT). Reprinted from A. L. Lafleur and K. M. Mills, *Anal. Chem.* **53**, 1202 (1981), with permission.

toluene derivatives. This chromatogram readily illustrates the determination of nitroaromatic compounds by GC with TEA detection. Figure 3-7 shows a plot of the detector response vs. moles injected for some nitrated toluene derivatives and illustrates the very good linearity that can be achieved with this detector.

Although the Thermo Electron Corporation (Waltham, MA) has marketed the TEA in both nitro and nitroso modes, by far the greatest reported use of the TEA is for the measurement of *N*-nitroso compounds. Particularly in analysis of foodstuffs, cosmetics, tobacco products and in environmental toxicology the measurement of *N*-nitroso compounds, some of which may be carcinogenic, by TEA has become the analytical method of choice. The selective measurement of *N*-nitroso compounds at picogram to subnanogram levels with linearity over greater than 4 orders of magnitude is possible, and, because the TEA is selective, unlike methodologies developed for the measurement of these compounds using other detectors, no sample cleanup or concentration is required in order to satisfy detector requirements. Nevertheless, some cleanup may be necessary to satisfy chromatographic constraints. The inherent selectivity of TEA eliminates the need for elaborate costly and time-consuming cleanup and concentration procedures and reduces possible contamination of samples from such procedures.

Many high molecular weight *N*-nitroso compounds are not amenable to analysis by GC. Some of these compounds are unstable, and, because of their high molecular weight, they are nonvolatile. The TEA has been utilized as a detector for liquid chromatography (LC) in order to address these problems, and LC-TEA is capable of quantitatively detecting nanogram levels of *N*-nitroso compounds.[38, 39]

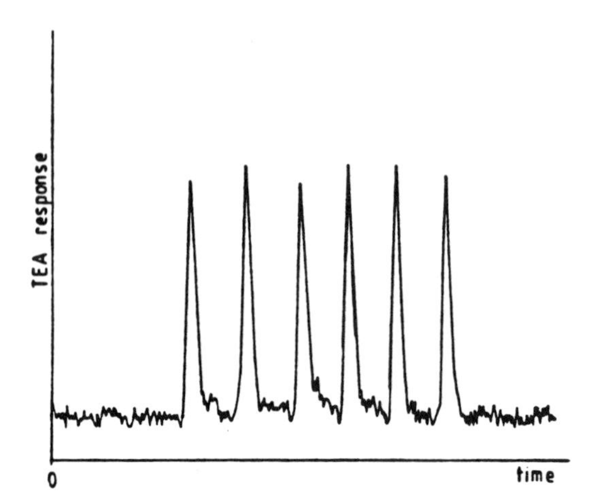

Figure 3-8 ■ Reproducibility of six 4-μl replicate injections of a NDEA sample: Integrator attenuation, 4; solvent system, 2-propanol-*n*-hexane (2.5 : 97.5, v/v); flow rate, 80 μl/min; mean, 28.1 mm peak height; S. D. 1.01; $n = 6$. Reprinted from C. Ruhl and J. Reusch, *J. Chromatogr.* **328**, 362 (1985), with permission.

LC-TEA is very similar in operation to GC-TEA. In LC-TEA, the eluent from a liquid chromatograph is directly introduced into a heated pyrolyzer. A small diameter nozzle and a large pressure drop across it serves to atomize the liquid eluent into a stream of tiny droplets. Cleavage of *N*-nitroso compounds in the heated pyrolyzer occurs just as in GC-TEA. Operation of HPLC-TEA exhibits some limitations in that it cannot be easily used with aqueous or inorganic buffer systems and it cannot be operated continuously for long periods of time.[40] The problem of not being able to operate HPLC-TEA continuously arises from the large amount of solvent that condenses in the cold trap of the TEA, which then must be frequently removed. One approach to solving this problem is to use a secondary vacuum system that continuously clears the cold trap.[41] Another approach is to use microbore LC systems so that large amounts of eluting solvents are not generated.[39] Figure 3-8 illustrates the sensitivity and reproducibility from repetitive injections of 1 ng amounts of *N*-nitrosodiethylamine that can be achieved with microbore high performance LC and TEA detection.

A more recently developed TEA for direct high performance LC and GC analysis of *N*-nitrosamides was described with detection limits in the tens of picograms range.[42] Other papers describing LC-TEA have dealt with the determination of nitroglycerin and its dinitrate metabolites in human plasma,[43] HPLC of volatile nitrosamines,[44] the determination of *N*-nitrosodiethanolamine in cosmetics[45, 46] and characteristics of the TEA in use with HPLC.[47]

The use of GC-TEA for biological analysis has also been reported for the identification and quantification of *N*-nitrosamines in human postmortem organs,[48] for a rapid method for estimating nitrate in biological samples[49] and for the determination of nitroaromatics in biosludges.[50]

TEA-GC has been applied in the cosmetics field as well for the determination of *N*-nitrosodiethanolamine in cosmetics.[51, 52] TEA-GC has also been applied to the analysis of tobacco-specific *N*-nitrosamines[53–55] and HPLC-TEA determination of tobacco-specific nitrosamines has also been reported.[56]

Numerous applications of GC-TEA for beverage and food analysis have also been reported. The application of GC-TEA for the analysis of nitroso compounds in beers, for instance, has been described.[57–59] In foods, the determination of nitrate in dried foods,[60] a purge and trap procedure for determining volatile *N*-nitrosamines in animal feed[61] and a dry column TEA method for determining *N*-nitrosopyrrolidine in fried bacon[62] have been reported using GC-TEA.

The determination of *N*-nitrosodiethylamine in air by GC-TEA[63] and a method for isolation of mononitrated polycyclic aromatic hydrocarbons in particulate matter by LC and determination by GC-TEA[64] have also been described.

Other miscellaneous chromatographic-TEA uses have included trace analysis of explosives at the low picogram level using silica capillary GC with TEA detection,[65] direct determination of *N*-nitrosamines in amines using GC-TEA,[66] a rapid method for determining various nitrosamines in rat urine by GC-TEA,[67] the response of nitramines in the TEA,[68] capillary GC coupled with TEA for the detection of *N*-nitrosamines,[69] the analysis of a model ionic nitrosamine by microbore HPLC using a TEA chemiluminescence detector[70] and selective detection of glycol nitrates by GC-TEA or LC electrochemical detection.[71]

This review has focused on the use of TEA as a chromatography detector, but several other applications are known. Many uses of TEA do not require chromatographic separations. Nevertheless, the publications cited certainly indicate the great importance that TEA has played in analytical measurements of *N*-nitroso and nitro

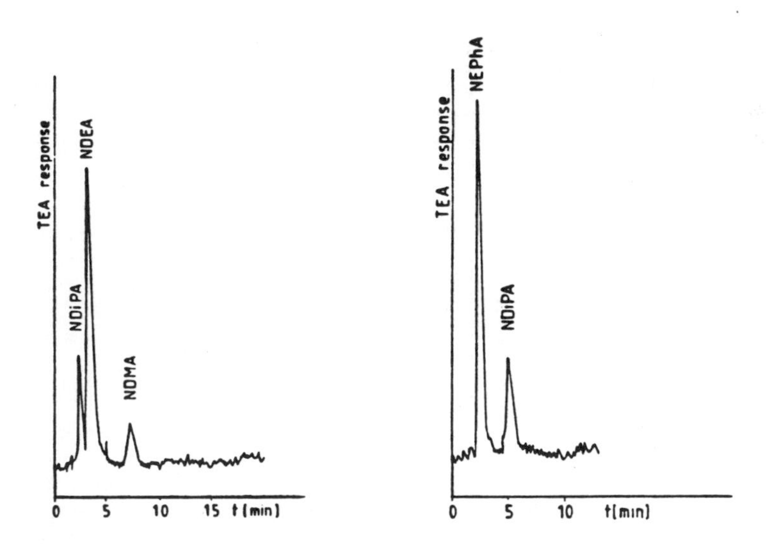

Figure 3-9 ▪ Left: Examination of rubber nipple A for nitrosatable substances (identified as nitroamines). Integrator attenuation, 4; solvent system, 2-propanol-*n*-hexane (2.5 : 97.5), v/v); flow rate, 80 μl/min. NDiPA was added as internal standard for recovery. Right: Examination of rubber nipple B for nitrosatable substances (identified as nitrosamines). Same conditions as chromatogram on left. Reprinted from C. Ruhl and J. Reusch, *J. Chromatogr.* **328**, 362 (1985), with permission.

compounds in wide-ranging fields and applications. For example, Fig. 3-9 illustrates one chromatographic application of TEA in the determination of *N*-nitrosoethylphenylamine (NEPhA), *N*-nitrosodiisopropylamine (NDiPA), *N*-nitrosodiethylamine (NDEA) and *N*-nitrosodimethylamine (NDMA). TEA has also been found to give response to nitramine,[68, 72] and the possibility that response to compounds other than nitrosamine can be observed even with the pyrolyzer at low temperatures must be considered.[73, 74] Of the various TEA-positive compound classes, ways have been devised to distinguish among some of these. Ultraviolet irradiation destroys nitrosamines, nitrosurea, alkylnitrites and nitramine, for example, but not C-nitro compounds;[34] consequently it was reported that a sample need only be irradiated before TEA analysis and only the presence of C-nitro compounds will be detected. Similarly, through chemical means, ascorbic acid and phenol in methanol were found to decompose alkyl nitrite without affecting any other compound,[34] so this too can be used to distinguish between TEA-positive compound classes.

Analysis by TEA has even been extended to other nitrogen-containing compounds, such as amines, by a simple instrumental modification.[75] The modified TEA operates by oxidizing nitrogen-containing compounds, with nitric oxide being generated as one of the oxidation products. The NO formed in this manner is then detected by its CL reaction with ozone as before. A more complete discussion of the modified TEA is given in the following section, which deals with the nitrogen-selective detector, because both are based on the same principle of operation.

Nitrogen Selective Detector

Principles of Operation

The nitric oxide + ozone CL reaction is also the basis for a detector that is selective for oxidizable nitrogen-containing compounds.[24, 25] In order for a compound to be detected, nitric oxide must be formed from its oxidation. Of course, as in TEA, nitroso compounds may give a response by forming NO from simple cleavage without oxidation, but the nitrogen selective detector extends beyond TEA the number of nitrogen-containing species that give a response. The formation of nitric oxide is based on catalytic oxidation of nitrogen-containing species at temperatures of 800 to 1000°C as shown:

$$C_xH_yN_z + \left(x + \frac{1}{4}y + \frac{1}{2}z\right)O_2 \rightarrow z\mathrm{NO} + x\mathrm{CO_2} + \frac{y}{2}\mathrm{H_2O} \qquad (3\text{-}8)$$

The detector is very similar to the TEA except that oxygen is added to the column effluent before it enters a pyrolyzer, but the NO detection is the same. Nitrogen oxides with the nitrogen being in higher oxidation states than in NO may be formed, but in the temperature ranges used in this scheme they thermally decompose to nitric oxide, and, of course, decomposition of NO to N_2 and O_2 is not favored kinetically. Commercial detectors of this type are manufactured by the Thermo Electron Corporation and Antek Instruments, Houston, TX. Generally, detection limits obtained with this detector are on the order of 1×10^{-12} mole for C-bonded nitrogen. The relative response for a compound may be greater if it contains more than one nitrogen atom per molecule. For instance, the ratio of molar response for *N*-nitrosodimethylamine to pyridine is 2:1. Linear response of the detector from C-bonded nitrogen compounds generally extends over at least 4 orders of magnitude.

Applications

Like TEA, the nitrogen-selective detector can be used both as a chromatographic detector and, without sample separations being necessary, as a total nitrogen detector. This detection scheme for determining chemically bound nitrogen by oxidative pyrolysis and CL nitric oxide detection was described by Parks.[76] As a total nitrogen detector, the nitrogen-selective detector was found to be very useful for the determination of total nitrogen in petroleum fractions.[24] A study of the use of this detection scheme for the determination of nitrogen in petroleum fractions demonstrated reliable day-to-day quantitation over a 16 month period. The method also exhibited a number of distinct advantages over other total nitrogen methods, such as microcoulometry, and agreement with the Kjeldahl method was acceptable with an average deviation being $\pm 5\%$. Some of the advantages of CL nitrogen-selective detection compared with microcoulometry were: less setup time, less equilibration and calibration time, no complex and expensive cells to clean, less complicated electronics, more amenable to automation, less lapsed time for analysis and less professional supervision required for routine application. The CL detector was also found to be easily interchanged with a microcoulometric cell to determine nitrogen or sulfur using the same furnace and combustion tube, and by using a splitter on the effluent from a gas chromatograph, simultaneous quantitative detection of nitrogen and sulfur was obtained. One disadvantage of the nitrogen selective CL detector as described by Drushel[24] is that the large internal volume of the combustion tube along with the transfer lines made its use with capillary GC columns impractical.

Other investigators described a CL nitrogen detector based on the same principles of operation but used a quartz tube packed with platinum net[25, 77, 78] rather than a quartz microcombustion tube.[24] In each, however, oxygen is added to the column effluent before it passes through the pyrolyzer. Maximum CL signal intensity was obtained with the pyrolyzer heated to 900°C. With the pyrolyzer at 500°C peak heights were ~ 80% of those at 900°C and exhibited tailing. Through the use of this method of

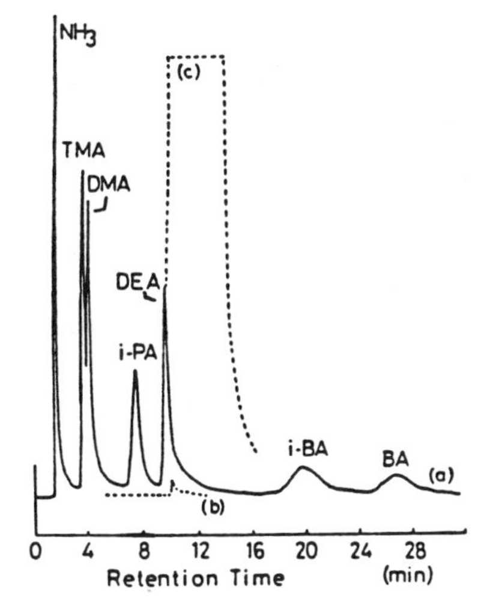

Figure 3-10 ■ Gas chromatogram of amines with GC-CLD. Conditions: 5% squalane plus 2% KOH/Chromosorb 104 (80 to 100 mesh) in a glass column (2 m × 3 mm i.d.). isothermally at 130°C with nitrogen flow rate of 60 ml/min. (a) Mixed standard sample in ethanol solution. (b) Ethanol only. (c) Ethanol with GC-FID. Reprinted from N. Kashihira, K. Makino, K. Kirita, and Y. Watanabe, *J. Chromatogr.* **239**, 617 (1982), with permission.

detection, along with preconcentration of gaseous samples onto Tenax-GC, determination of trimethylamine in ambient air at the ≤ 10 parts per billion by volume (ppbv) level was possible. The on-column limit of detection for trimethylamine was approximately 20 ng. Selective and sensitive measurement of ammonia and amines was demonstrated and there was no interference from microliter injections of organic solvents that do not contain nitrogen. Interestingly, ammonia, acetonitrile and nitromethane exhibited higher sensitivity than amines with this detector by a factor of ~ 1.5. The nitrogen CL detector was also used to determine breakthrough volumes and collection efficiencies of various adsorbents for ammonia and amines.[25] Figure 3-10 shows a nitrogen CL detector chromatogram of a mixed standard sample of amines in ethanol solution, a nitrogen CL detector chromatogram of ethanol only and a flame ionization detector chromatogram of ethanol only. It is readily apparent that DEA (diethylamine) and possibly i-BA (iso-butylamine) could not be determined by FID because they are eluted either under or very close to the solvent peak, but the solvent peak is not sensed by the selective detector and this illustrates an inherent advantage of the CL nitrogen detector. The identities of the other peaks in the chromatogram are TMA (trimethylamine), DMA (dimethylamine), i-PA (iso-propylamine) and BA (butylamine).

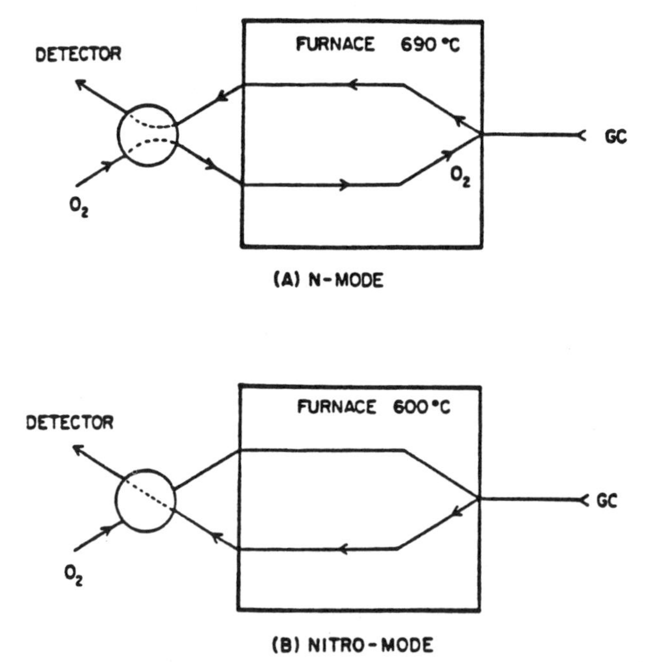

Figure 3-11 ■ Schematic diagram of the GC-TEA™ interface for (A) the nitrogen mode and (B) nitro mode. Reprinted from D. P. Rounbehler, S. J. Bradley, B. C. Challis, D. H. Fine, and E. A. Walker, *Chromatographia* **16**, 354 (1982), with permission.

A modified TEA has also been developed to determine amines and other nitrogen-containing compounds that do not normally give a TEA response.[75] In the modified TEA, a catalyst consisting of an oxide of nickel alloy heated to 650 to 700°C is used to convert the nitrogen in nitrogen-containing compounds into nitric oxide. Again, oxygen is added to a GC column effluent so that organic materials are oxidized to carbon dioxide, water, other products and nitric oxide if the organic material contains oxidizable nitrogen. It is of note that carbon dioxide, water and the other products formed in the oxidation do not interfere with the detection of nitric oxide and a cold trap is therefore not needed. In addition, the modified TEA is equipped with two modes of operation so that it can be operated as a nitroso/nitro specific detector or also in the nitrogen-selective CL mode. A schematic drawing of a modified TEA is shown in Fig. 3-11. In the N mode, a four-way valve is positioned so that oxygen is supplied to the pyrolyzer and the detector responds to all oxidizable nitrogen-containing compounds. By simply changing the position of the four-way valve, the oxygen supply is discontinued and only those compounds that will produce NO by simple pyrolysis (nitroso and nitro compounds) are detected. Figure 3-12 shows a GC

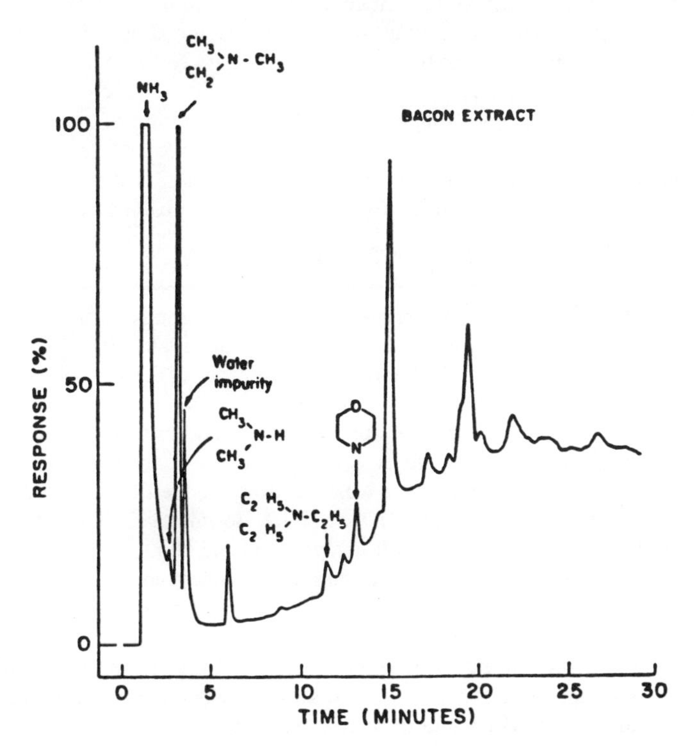

Figure 3-12 ▪ GC-TEA™ nitrogen mode chromatogram of bacon extract showing volatile nitrogen-containing compounds. The GC column was a glass tube, 1.8 m × 2 mm i.d. packed with 60/80 Carbopak B/4% Carbowax 20M/0.8% KOH. Reprinted from D. P. Rounbehler, S. J. Bradley, B. C. Challis, D. H. Fine, and E. A. Walker, *Chromatographia* **16**, 354 (1982), with permission.

application of the detector in the nitrogen mode for the analysis of volatile nitrogen-containing compounds in a bacon extract.

All of the nitrogen-compound selective detectors based on the $NO + O_3$ CL reaction are very selective ($> 10^6$ over normal hydrocarbons) and possess good sensitivity, and they can simplify the determination of nitrogen-containing compounds in a wide variety of complex matrices ranging from petroleum to biological to food and beverage samples.

Redox Chemiluminescence Detection

Principles of Operation

The Redox Chemiluminescence Detector is another selective detector based on the $NO + O_3$ reaction. Like the TEA and the nitrogen-compound-selective CL detector, the RCD is manufactured commercially (Sievers Research, Inc., 1930 Central Ave., Boulder, CO). Unlike TEA and the nitrogen-compound-selective CL detector, in the RCD the responding analytes do not have to contain a nitroso, nitro or oxidizable nitrogen moiety. Instead a nitrogen-containing reagent, such as NO_2 or HNO_3, is reduced to nitric oxide by each analyte and thereby provides a surrogate NO signal proportional to the concentration of that analyte.

The RCD is based on principles similar to those of the operation of a selective CL detector of odd nitrogen oxide species.[6, 8, 79] Odd nitrogen oxide species in the atmosphere can be selectively measured by utilizing the reduction of these compounds by carbon monoxide on a heated surface of gold to form nitric oxide according to[8]

$$XNO_2 + CO \xrightarrow[\Delta]{Au} NO + CO_2 + \text{products} \qquad (3\text{-}9)$$

Quantitative and real-time conversion for several XNO_2 compounds occurs on heated surfaces of gold, where X denotes a radical group, such as alkyl, alkoxy or OH. In this detection scheme, carbon monoxide mixing ratios on the order of 0.2 to 1000 ppmv are used so that the preceding equation is limited only in the XNO_2 species being detected, since CO is in excess. In an inverse manner, an excess of an odd nitrogen oxide species, such as nitrogen dioxide, can be used to detect any reducing analyte which reacts to produce nitric oxide.[26, 27] This greatly broadens the utility of the detector to measure any compound that is rapidly oxidized by NO_2 or HNO_3, forming NO. The coupling of redox reactions of odd nitrogen oxides with analytes and the $NO + O_3$ chemiluminescent reaction forms the basis of redox chemiluminescence detection.

Redox chemiluminescence measurements utilize a two-step reaction and detection scheme.[26, 27] In the first step, an analyte reacts rapidly with a reagent such as nitrogen dioxide on catalyst surfaces at elevated temperatures according to the first equation below. In the second step, the nitric oxide formed during the oxidation of the analyte is subsequently quantitated by its chemiluminescent reaction with added ozone, according to the second equation below.

$$\text{analyte} + NO_2 \xrightarrow[\Delta]{Au} NO + \text{oxidized analyte} + \text{other products} \qquad (3\text{-}10)$$

$$NO + O_3 \longrightarrow NO_2 + O_2 + h\nu \qquad (3\text{-}11)$$

The first reaction is rate-limiting and consequently, the degree of conversion can be modified through the use of different catalysts, catalyst–bed temperatures or other operating conditions. The extent of conversion depends on how rapidly the analyte is oxidized under given reaction conditions. Under equivalent conditions, various analytes may react with the reagent with differing efficiencies. These differences of reactivity account for the selectivity of the detector.

A typical GC-redox chemiluminescence detection system consists of a purified carrier gas, a gas chromatograph, a nitrogen dioxide metering device, a redox reaction zone (catalyst) and a nitric oxide/ozone chemiluminescence measuring system (reaction chamber, ozone source, vacuum pump, optical filter and photomultiplier tube). The nitrogen dioxide is supplied by one or more permeation tubes. The concentration of the nitrogen dioxide used is typically between 100 and 1000 ppmv.

In order for an analyte to be detected, appreciable amounts of nitric oxide must be formed during the short residence time of the analyte in the catalyst bed. Thus, only those compounds that rapidly react to form nitric oxide are detected. The calculated residence time of an analyte in the catalyst bed is < 1 s. This arises from the very small sizes of the catalyst beds used in RCD. Increasing the dimensions of the catalyst bed may have the effect of increasing the amount of conversion of nitrogen dioxide to nitric oxide with a given analyte, but this increase in production of nitric oxide might be offset by a corresponding increase in band broadening and peak asymmetry. This also would increase the formation of nitric oxide from the thermal decomposition of the reagent, nitrogen doxide, which occurs to a limited extent, even in the absence of a reducing analyte. Of course, the nitric oxide formed in this manner appears as a background signal and contributes to the overall noise in the detection system. Also, the thermal decomposition of nitrogen dioxide increases with increasing temperature of the catalyst, so very high catalyst operating temperatures cannot be tolerated. Rapid reaction of most analytes with NO_2 is observed in catalyst beds containing gold between 300 and 400°C. At these temperatures, the formation of nitric oxide caused by reaction with the analyte is maximized relative to the formation of nitric oxide caused by the thermal decomposition of nitrogen dioxide.

Nitric oxide may be formed either directly from the reaction of the analyte with nitrogen dioxide or from secondary reactions which produce nitric oxide indirectly. For instance, the heated catalyst may initiate the dehydrogenation of an analyte and the hydrogen formed in this manner then can react with nitrogen dioxide to form nitric oxide and water. In either case, the nitric oxide formed, quantitated by its chemiluminescence reaction with ozone, is a surrogate measure of the amount of the original analyte. The RCD responds to reducing analytes with selectivities of 10^3 to 10^6 depending on the particular analyte being measured and the operating conditions of the detector.

Applications

The RCD possesses many characteristics that make it a useful GC detector. The sensitivity of the RCD arises from the sensitivity of nitric oxide measurements. The selectivity of the RCD arises from differences in the rates of redox reactions between an analyte and a reagent gas, coupled with the fact that only a few compounds exhibit chemiluminescence when mixed with ozone, and this light emission is generally much weaker or in a different part of the spectrum than the emission from the nitric oxide/ozone reaction. Some compounds can be sensitively detected while others are not. Alkanes do not yield an appreciable redox chemiluminescence response when the catalyst consists of gold coated on soda-lime glass beads, except at temperatures above 400°C.[27] Other important compounds giving little or no RCD response under these conditions are the fully chlorinated hydrocarbons, water, nitrogen, oxygen, carbon dioxide, and the noble gases. These compounds represent the major constituents of many sample matrices and are important mobile phases in GC, supercritical fluid chromatography and liquid chromatography. None of these compounds interferes significantly with redox chemiluminescence detection and, consequently, trace levels of responding analytes can be detected in these matrices without interferences and they can be used as carriers in chromatographic systems employing redox chemiluminescence detection.

Most organic compounds containing oxygen, nitrogen, sulfur or reactive functional groups can be detected by redox chemiluminescence. Other rapidly oxidized compounds such as ammonia, acetylene, carbon disulfide, carbon monoxide, carbonyl sulfide, ethylene, formaldehyde, hydrogen, hydrogen sulfide, hydrogen peroxide, propylene and sulfur dioxide produce good responses.[26, 27, 80] Since several gases give RCD responses, this makes the technique attractive for the analysis of gases by GC. Table 3-1 shows some of the compounds and compound classes that are sensitively detected by RCD with a catalyst consisting of gold coated on soda-lime glass beads at an operating temperature of 400°C. The detection limits of compounds detected by RCD range from high picogram to low nanogram levels, and linear response is observed over greater than 3 orders of magnitude. A response lower than that expected may be observed at high concentrations of an analyte (>1000 ppmv), because the amount of nitrogen dioxide present in the catalyst zone becomes limiting. In addition, saturation of the photomultiplier tube signal may occur.

The selectivity of RCD can be tailored to some particular analyses.[80] The selectivity of the detector can be altered through the use of one or more operating parameters, such as the concentration or composition of the nitrogen oxide reagent gas, the composition of the catalyst and the temperature of the catalyst. Figures 3-13 and 3-14

Table 3-1 ■ Response of Compounds in RCD

Respond		Do Not Respond
Alcohols	Amines	Water
Aldehydes	Nitro-compounds	Saturated hydrocarbons
Ketones	Ammonia	Most chlorinated hydrocarbons
Carboxylic acids	Carbon monoxide	Oxygen
Aromatic hydrocarbons	Hydrogen peroxide	Nitrogen
Olefins	Sulfur dioxide	Carbon dioxide
Sulfides	Hydrogen sulfide	Helium
Thiols	Carbonyl sulfide	Argon
Phosphates	Carbon disulfide	Nitrous oxide
Phosphites	Hydrogen	Tetrahydrofuran

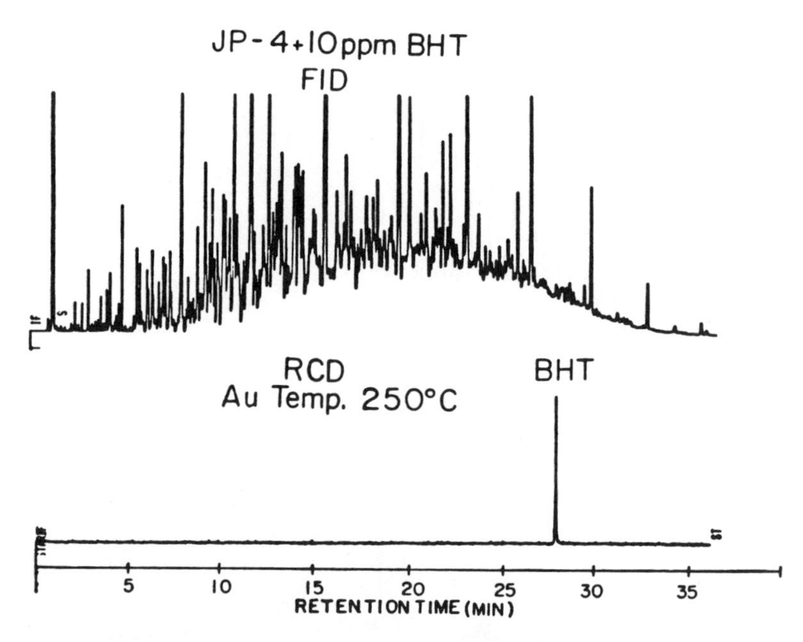

Figure 3-13 ▪ Chromatograms of a JP-4 jet fuel containing 10 ppm of BHT using a flame ionization detector (upper) and a redox chemiluminescence detector (lower). GC conditions: 0.5 μl injection; split 30 : 1. Temperature program: 1 min at 40°C programmed to 300°C at 6°C/min. Column: 25 m × 0.32 mm i.d. Hewlett Packard Ultra #2 capillary column; film thickness 0.52 μm. RCD conditions: gold coated on soda-lime glass beads; catalyst temperature 250°C; 0.5 s integration time. Reprinted from *C and EN*, June 24, 1985, p. 42, ref. 81, with permission.

very elegantly illustrate the tunable selectivity of the RCD. Figures 3-13 and 3-14 show chromatograms of a sample of JP-4 jet fuel containing 10 ppm of the antioxidant BHT (2,6-di-*tert*-butyl-4-methylphenol).[81] In Figure 3-13, a gold catalyst operated at 250°C was used to obtain the RCD chromatogram and in Figure 3-14 the same gold catalyst operated at 350°C was used to obtain the RCD chromatogram. In both figures, a chromatogram of the same sample obtained by a flame ionization detector is shown for comparison. With the gold catalyst operated at 250°C, only the peak corresponding to BHT is observed. With the gold catalyst operated at 350°C, the BHT peak is still observed, but other peaks, corresponding to olefins and aromatics, begin to appear in the RCD chromatogram, but the alkanes are not sensed. This is in stark contrast to the overabundance of the peaks observed with the flame ionization detector. When a more active palladium catalyst is used, many other peaks, including those of the alkanes, are observed even at catalyst temperatures < 300°C.

Complex interactions contribute to the overall effectiveness of catalysts in various chemical transformations. Certain properties of a catalyst will determine whether it is a good catalyst for utilization in redox chemiluminescence detection. Many different materials have been investigated for use as catalysts in RCD.[84] In RCD, catalytic formation of nitric oxide from an analyte is of prime importance. It is important that

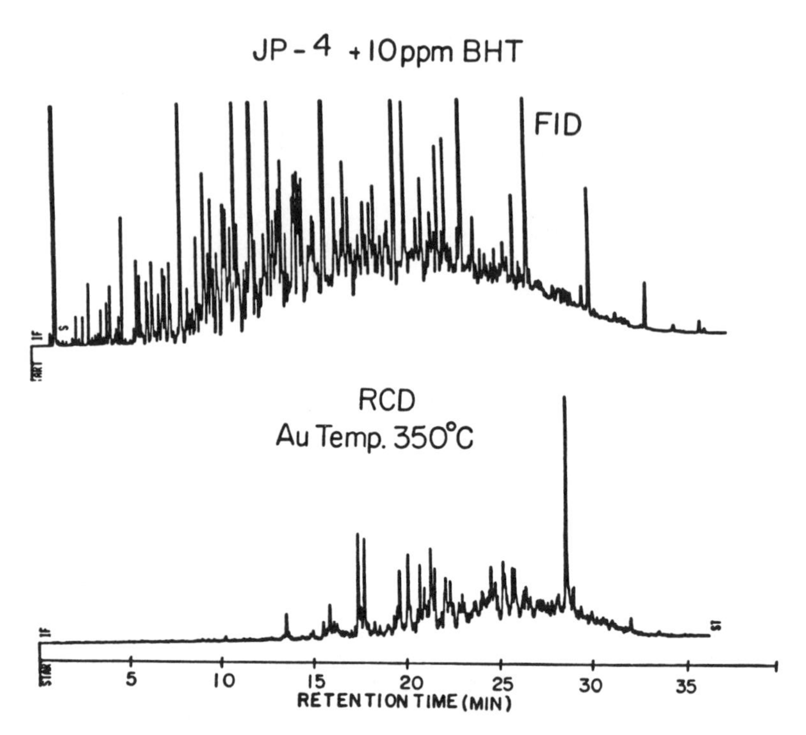

Figure 3-14 ■ Chromatograms of same sample and conditions as in Fig. 3-13 except with RCD catalyst at 350°C. Reprinted from *C and EN*, June 24, 1985, p. 42, with permission.

the catalyst give the desired selectivity and sensitivity, and reactions occurring on the surface must be rapid, so that analyte peaks are narrow. Reproducibility is also very important and, in order to be useful, the catalyst must not produce high background levels of nitric oxide from the thermal decomposition of a reagent such as NO_2. A catalyst might also be ineffective in RCD if it catalyzes the reduction of nitrogen dioxide by an analyte to nitrous oxide or nitrogen. No response for an analyte would be observed from the formation of N_2O or N_2. In RCD, the formation of NO from the thermal decomposition of NO_2 is minimized by keeping times of contact between NO_2 and the hot catalyst surface short. Also, catalyst operating temperatures can be kept as low as possible while still achieving the desired system performance in terms of sensitivity and selectivity. Poisoning may have a deleterious effect on the performance of a catalyst in terms of activity, stability and selectivity. Gold catalysts do not appear to be susceptible to poisoning, while palladium catalysts are much more so, particularly by some sulfur compounds.[80]

RCD responses from many sulfur-containing compounds have been observed with a gold catalyst and no significant poisoning from these compounds was observed.[27, 85] For a palladium catalyst, the composition of the sulfur compound is important in terms of its poisoning effect on the catalyst. Even as little as 100 ng of dimethylsulfide

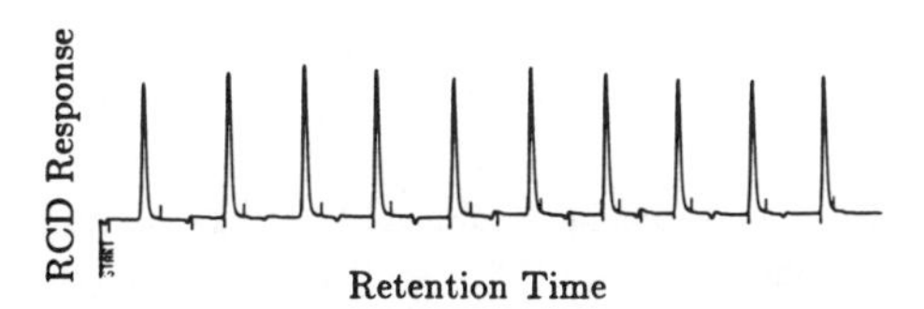

Figure 3-15 ■ RCD response with a palladium catalyst (0.05 g of ~2.5% wt/wt, heated to 400°C) from repetitive injections of 100 ppmv H_2S using a gas-tight syringe. RSD was approximately 7%.

reduced the activity of a palladium catalyst, but many repetitive injections of H_2S and SO_2, while always maintaining an excess of NO_2, did not cause a significant reduction in the activity of this catalyst.[82] Figure 3-15 shows the responses from ten $\frac{1}{2}$-ml injections, made using a gas-tight syringe, of 100 ppmv H_2S (~60 ng) with a Pd catalyst heated to 400°C. The relative standard deviation for these 10 injections was 7%. Over a period of ~4 h, more than 40 injections of H_2S and SO_2 were made and there was no significant reduction in RCD response. Figure 3-16 shows a typical chromatogram obtained with a Pd catalyst at 400°C from a mixture containing nominally 100 ppmv H_2S and 100 ppmv SO_2. This illustrates that a Pd catalyst in RCD can be used to measure these gases, although it would be prudent to use a gold catalyst for samples that may contain unknown sulfur-containing compounds, because gold has not shown a susceptibility to poisoning from any sulfur-containing compounds so far tested.

Sulfur dioxide is known to react stoichiometrically with NO_2 according to the following equation[86]:

$$SO_2 + NO_2 \rightarrow SO_3 + NO \quad (3\text{-}12)$$

This reaction is well known and represents part of the lead chamber process for the manufacture of sulfuric acid. One can write a similar reaction that may occur between H_2S and NO_2:

$$H_2S + 4NO_2 \rightarrow SO_3 + 4NO + H_2O \quad (3\text{-}13)$$

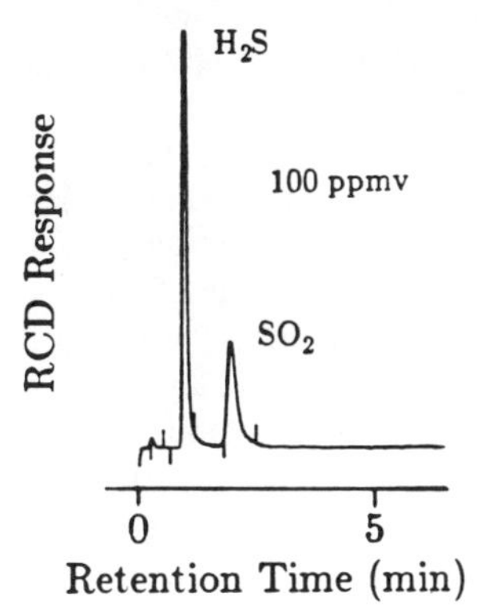

Figure 3-16 ■ Typical RCD chromatogram for a $\frac{1}{2}$-ml injection of a sample mixture containing 100 ppmv H_2S and SO_2 in helium. The catalyst was the same as used for Fig. 3-15.

If these two reactions occur to approximately the same extent in RCD, then one would predict a four fold higher response factor for H_2S compared to the response from SO_2, and within experimental error this is the RCD relative response that is observed. This readily illustrates that one or more molecules of NO may be formed by reaction of a single analyte molecule, depending on the structure of the analyte molecule and RCD operating conditions, such as temperature of the catalyst.

Another example of the use of RCD as a GC detector is in the measurement of carbon monoxide in hydrocarbon matrices.[83] The RCD is particularly sensitive to carbon monoxide, and samples can be rapidly analyzed for CO at subpart per million by volume levels without any sample pretreatment, such as methanation or preconcentration. Rapid reaction of CO with NO_2 to form NO is observed in catalyst beds containing either gold or palladium at temperatures between 300 and 400°C. For the measurement of CO, effective catalyst temperatures are low enough that NO_2 decomposition causes no significant problem. Alkanes do not produce an appreciable redox chemiluminescence response when the catalyst is gold coated on soda-lime glass beads, except at temperatures above 400°C.[27] Therefore, carbon monoxide can be selectively measured in the presence of alkanes when a gold catalyst is used below 400°C. For the application of measuring CO in ethylene, however, a palladium catalyst was used in order to achieve the greatest sensitivity for CO and chromatographic selectivity was used to separate the CO and ethylene. A simple four-port valve was used in order to avoid the potential problem of overloading the detector from ethylene which constituted the major portion of the sample and which gives a RCD response. The implementation of this approach is particularly simple, since ethylene is eluted later than and is easily separated from CO.

The RCD response for carbon monoxide from 1 to 100 ppmv in an ethylene matrix is shown in Figure 3-17. The figure was produced by plotting peak areas from 1 ml sample injections of different concentrations of CO in ethylene. A gas-tight syringe was used to make these sample injections. Good linearity from 0.96 to 96 ppmv was demonstrated. The correlation coefficient of the least squares linear fit of the data points was 0.9991. Relative standard deviations of peak areas from five repetitive injections (using an injection valve with a sample loop) of samples containing 10 and 475 ppmv were 6 and 1%, respectively. Over a period of one week, the relative standard deviations of absolute response were less than 25%. Signal- (peak height) to-noise ratios at 0.18 ppmv CO concentration were between three and four times the peak-to-peak baseline noise. A useful working range for CO concentrations from less than 1 ppmv up to 1000 ppmv was observed. At higher CO concentrations, the CO response was less than expected because of rate limitations resulting perhaps from too little NO_2 available for reaction or possibly stemming from photomultiplier tube saturation.

Sample injections were made every 10 min, demonstrating that this technique for the trace analysis of carbon monoxide in ethylene can be performed repetitively and possibly automated. The injection frequency could be increased by developing a method that perturbs the flow over the catalyst to a lesser extent during the catalyst bypass process. A column backflushing method could be used to shorten analysis times to only twice the time it takes for carbon monoxide to be eluted. This method should also reduce the chance of long term catalyst poisoning if late-eluting sulfur compounds, such as mercaptans or sulfides, were present in the same. With the apparatus used, it is

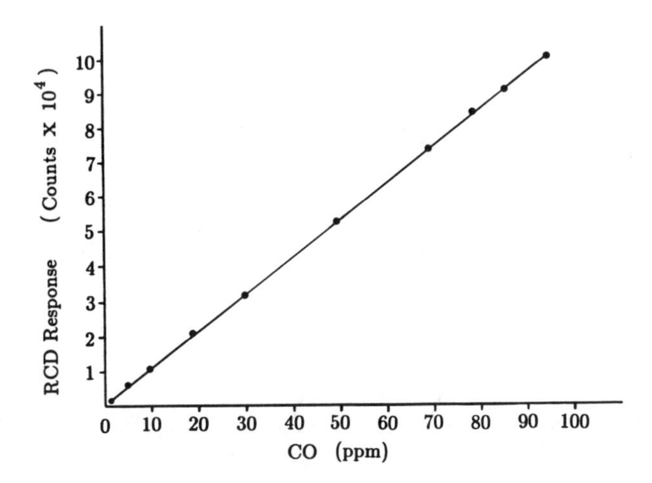

Figure 3-17 ▪ Linearity of RCD response vs. concentration of CO in ethylene with catalyst at 350°C consisting of 2.5 mg of palladium (2.5% by weight) on 230/320 soda-lime glass beads. Reprinted from R. E. Sievers, R. L. Shearer, and R. M. Barkley, *J. Chromatogr.* (1987), with permission.

possible to record simultaneous RCD and thermal conductivity detector chromatograms, since RCD can be serially coupled to the end of a nondestructive detector. In this manner, the RCD can be used to measure certain trace components, while the thermal conductivity detector would be used as a nearly universal detector to measure the major constituents. No appreciable peak broadening in either thermal conductivity detector or RCD chromatograms was observed when the RCD was connected serially in the preceding manner.

RCD is also currently being adapted for use in LC and supercritical fluid chromatography (SFC). For LC-RCD several approaches have been examined in order to carry out the necessary redox reactions to form NO from analytes and several NO precursors have been investigated.[86] While it is possible to vaporize the effluent from a liquid chromatograph and to have the redox reactions occur in the gas phase, a much simpler and less troublesome approach is to add a NO precusor to the eluent from a liquid chromatograph, causing the redox reactions to occur in the liquid phase, and then separate the NO gas from this liquid. This approach is effective when one uses dilute nitric acid as the NO precursor and carries out the redox reactions in a 10 to 20 m long reaction coil consisting of a fused silica capillary tubing with a pressure restrictor placed at its exit.[86] The eluent can be heated well above its boiling point to accelerate redox reactions because the pressure in the fused silica capillary can be maintained up to 2000 psi. A cooling bath downstream from the reaction coil is used to lower the temperature of the eluent before its pressure is reduced. NO is then separated from the eluent and the NO is detected by its CL reaction with O_3. A schematic drawing of a LC-RCD is shown in Figure 3-18.

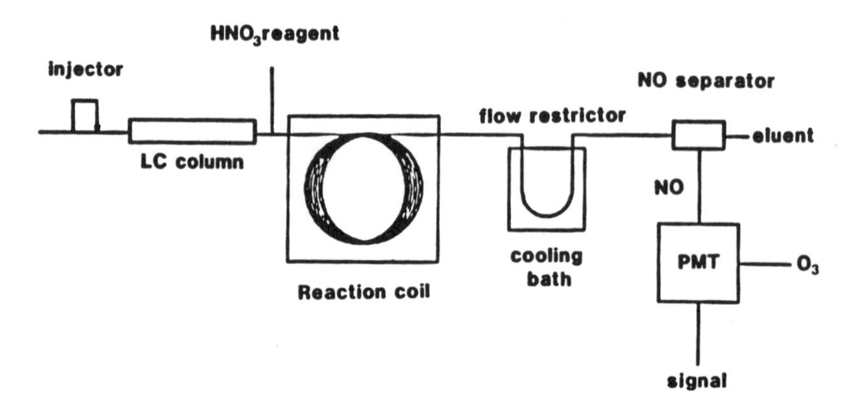

Figure 3-18 ■ Simplified schematic of a LC-RCD device. PMT is photomultiplier tube.

With this design, detection limits for reducing sugars (glucose, mannose, fructose, etc.) and some ions (Fe^{+2}, Sn^{+2}, NO_2^-, SCN^- and $S_2O_3^{-2}$) are on the order of 1 to 5 ng.[86] Other readily oxidizable species, such as ascorbic acid and arachidonic acid, give similar responses. LC-RCD promises to provide new detection possibilities in both LC and ion chromatography.

Work in the field of adapting RCD to SFC is also progressing. The technical problems in adapting SFC to RCD are not as great as LC to RCD and only slight modifications of the GC-RCD have to be made for use in SFC-RCD. The ability to detect caffeine, antioxidants and some other compounds by SFC-RCD has been demonstrated.[87] The use of $YBa_2Cu_3O_7$ as a catalyst in RCD is being explored.[88]

Summary

There are currently three commercially available chromatography detectors based on the NO + O_3 CL reaction. The TEA is based on high-temperature pyrolysis of *N*-nitroso or nitro compounds that produce NO upon decomposition. The nitrogen selective CL detector is based on a similar principle, but oxygen is added to the column effluent before it passes through the pyrolyzer so that nitrogen-containing compounds that can be oxidized to NO are selectively detected. The RCD utilizes post-column redox reactions of analytes with a reagent, such as nitrogen dioxide or nitric acid, which acts as a nitric oxide precursor. Consequently, the detection of analytes by RCD is not limited to compounds that contain nitrogen. All of these detectors are selective, but each has its distinctive place as a detector for important applications in industrial, petrochemical, environmental and other fields, particularly because the use of selective detectors can save valuable time by making lengthy sample cleanup and preparation procedures unnecessary. Modern detection of NO by its CL reaction with O_3 makes measurement at pptv concentration levels with linearity up to 6 orders of magnitude possible. Because this is so attractive, it is very likely that other chromatographic detectors utilizing the same NO detection will be developed. One such scheme may lead to the development of a chemiluminescence oxygen-compound-selective detector, in which oxygen-containing analytes are reduced to their constituent atoms or fragments (such as in a plasma created by microwave or electric discharge) in the presence of a large excess of nitrogen atoms so recombination of atoms would result in the formation of detectable nitric oxide only if an oxygen-containing analyte were present.[82]

References

1. Fontijn, A., Sabadell, A. J., and Ronco, R. J. *Anal. Chem.* **42**, 575 (1970).
2. Ridley, B. A., and Howlett, L. C. *Rev. Sci. Instrum.* **45**, 742 (1974).
3. Winer, A. M., Peters, J. W., Smith, J. P., and Pitts, J. N., Jr. *Environ. Sci. Technol.* **8**, 1118 (1974).
4. Matthews, R. D., Sawyer, R. F., and Schefer, R. W. *Environ. Sci. Technol.* **11**, 1092 (1977).
5. Joshi, S. B., and Bufalini, J. J. *Environ. Sci. Technol.* **12**, 597 (1978).
6. Bollinger, M. J. Doctoral Thesis, University of Colorado, Boulder, CO, 1982.
7. Mehrabzadeh, A. A., O'Brien, R. J., and Hard, T. M. *Anal. Chem.* **55**, 1660 (1983).
8. Bollinger, M. J., Sievers, R. E., Fahey, D. W., and Fehsenfeld, F. C. *Anal. Chem.* **55**, 1980 (1983).
9. Clyne, M. A. A., Thrush, B. A., and Wayne, R. P. *J. Chem. Soc., Faraday Trans.* **60**, 359 (1964).
10. Clough, P. N., and Thrush, B. A. *Trans. Faraday Soc.* **63**, 915 (1967).
11. Birks, J. W., Shoemaker, B., Leck, T. J., and Hinton, D. M. *J. Chem. Phys.* **65**, 5181 (1976).
12. Greaves, J. C., and Garvin, D. J. *Chem. Phys.* **30**, 348 (1959).
13. Toby, S. *Chem. Revs.* **84**, 277.
14. Harrison, H., and Scattergood, D. M. Boeing Scientific Research Laboratories Document D1-82-0480.
15. Campbell, I. M., and Baulch, D. L. *Spec. Per. Repts., Chem. Soc. London*, 1978, 42.
16. Nederbragt, G. W., Vander Horst, A., and Van Duijn, J. *Nature* **206**, 87 (1965).
17. Kelly, T. J., Gaffney, J. S., Phillips, M. F., and Tanner, R. L. *Anal. Chem.* **55**, 135 (1985).
18. Gaffney, J. S., Spandau, D. J., Kelly, T. J., and Tanner, R. L. *J. Chromatogr.* **347**, 121 (1985).
19. Bruening, W., and Concha, F. J. M. *J. Chromatogr.* **112**, 253 (1975).
20. Bruening, W., and Concha, F. J. M. *J. Chromatogr.* **142**, 191 (1977).
21. Fine, D. H., Rufeh, F., Lieb, D., and Roundbehler, D. P. *Anal. Chem.* **47**, 1188 (1975).
22. Fine, D. H., and Roundbehler, D. P. *J. Chromatogr.* **109**, 271 (1975).
23. Fine, D. H., Lieb, D., and Rufeh, F. *J. Chromatogr.* **107**, 351 (1975).
24. Drushel, H. V. *Anal. Chem.* **49**, 932 (1977).
25. Kashihira, N., Makino, K., Kirita, K., and Watanabe, Y. *J. Chromatogr.* **239**, 617 (1982).
26. Nyarady, S. A., and Sievers, R. E. *J. Amer. Chem. Soc.* **107**, 3726 (1985).
27. Nyarady, S. A., Barkley, R. M., and Sievers, R. E. *Anal. Chem.* **57**, 2074 (1985).
28. Fine, D. H., Rufeh, F., and Gunther, B. *Anal. Lett.* **6**, 731 (1973).
29. Fine, D. H., Rufeh, F., and Lieb, D. *Nature (London)* **247**, 309 (1974).
30. Fine, D. H., and Rufeh, F. *N-Nitroso Compounds Analysis and Formation*, International Agency for Research on Cancer, Lyon, 1975.
31. Gough, T. A., Webb, K. S., and Eaton, R. F. *J. Chromatogr.* **137**, 293 (1977).
32. Borwitsky, H. *Chromatographia* **22**, 65 (1986).
33. Parees, D. M., and Prescott, S. R. *J. Chromatogr.* **205**, 429 (1981).
34. Hansen, T. J., Archer, M. C., and Tannenbaum, S. R. *Anal. Chem.* **51**, 1526 (1979).
35. Lafleur, A. L., and Mills, K. M. *Anal. Chem.* **53**, 1202 (1981).
36. Fine, D. H., Yu, W. C. Goff, U., Bender, E., and Reutter, D. *J. Forensic Sci.* **28**, 29 (1983).

37. Douse, J. M. *J. Chromatogr.* **256**, 359 (1983).
38. Oettinger, P. E., Huffman, F., Fine, D. H., and Lieb, D. *Analytical Letters* **8**, 411 (1975).
39. Ruhl, C., and Reusch, J. *J. Chromatogr.* **328**, 362 (1985).
40. Fine, D. H. In Preussmann, R., I. K. O'Neill, G. Eisenbrand, G. Spiegelhalder, and H. Bartsch, eds., *Environmental Carcinogens, Selected Methods of Analysis*, Vol. 6, *N-nitroso compounds*, IARC Scientific Publications No. 45, 1983, p. 443.
41. Widmer, H. M., and Grolimund, K. *Chimia* **35**, 156 (1981).
42. Fine, D. H., Rounbehler, D. P., Yu, W. C., and Goff, E. U. *N-Nitroso Compounds: Occurrence, Biological Effects and Relevance to Human Cancer*, IARC Sci. Publ. No. 57, 1984, p. 121.
43. Woodward, A. J., Lewis, P. A., Aylward, M., Rudman, R., and Maddock, J. *J. Pharm. Sci.* **73**, 1838 (1984).
44. Goff, U. *Environmental Carcinogens, Selected Methods of Analysis*, IARC Sci. Publ. No. 45, 1983, Vol. 6, p. 389.
45. Ho, J. L., Wisneski, H. H., and Yates, R. L., *J. Assoc. Off. Anal. Chem.* **64**, 800 (1981).
46. Takeuchi, M., Mizuishi, K., and Harada, H., *Tokyo-toritsu Eisei Kenkyusho Kenkyu Nempo* **30-1**, 98 (1979).
47. Baker, J. K., and Ma, Cheng-Yu, *Environmental Carcinogens, Selected Methods of Analysis*, Aspects of *N*-Nitroso Compounds, IARC Sci. Publ. No. 19, 1978, p. 19.
48. Cooper, S. F., Lemoyne, C., and Gauvreau, D. *J. Anal. Toxicol.* **11**, 12 (1987).
49. Dull, B. J., and Hotchkiss, J. H. *Food Chem. Toxicol.* **22**, 105 (1984).
50. Phillips, J. H., Coraor, R. J., and Prescott, S. R. *Anal. Chem.* **55**, 889 (1983).
51. Kijima, K., Takarai, T., and Saito, E. *Nippon Koshohin Kagakkaishi* **9**, 32 (1985).
52. Black, D. B., Lawrence, R. C., Lovering, E. G., and Watson, J. R. *J. Assoc. Off. Anal. Chem.* **64**, 1474 (1981).
53. Begutter, H., Klus, H., and Ultsch, I. *J. Chromatogr.* **321**, 475 (1985).
54. Adams, J. D., Brunnemann, K. D., and Hoffmann, D. *J. Chromatogr.* **256**, 347 (1983).
55. Adams, J. D., Brunnemann, K. D., and Hoffman, D. *J. Chromatogr.* **256** 347 (1983).
56. Hecht, S. S., Adams, J. D., and Hoffmann, D. *Environmental Carcinogens, Selected Methods of Analysis*, IARC Sci. Publ. No. 45, 1983, Vol. 6, p. 429.
57. Sen, N. P., Seaman, S., and Brickis, M. *J. Assoc. Anal. Chem.* **65**, 720 (1982).
58. Sen, N. P., and Seaman, S. *J. Assoc. Off. Anal. Chem.* **64**, 933 (1981).
59. Tezuka, T., Sakuma, S., and Katayma, H. *Rep. Res. Lab. Kirin Brew. Co.* **22**, 51 (1979).
60. Ross, H. D., and Hotchkiss, J. H. *J. Assoc. Off. Anal. Chem.* **68**, 41 (1985).
61. Billedeau, S. M., Thompson, H. C., Jr., Hansen, E. G., Jr., and Miller, B. J. *J. Assoc. Off. Anal. Chem.* **67**, 557 (1984).
62. Fiddler, W., Pensabene, J. W., Gates, R. A., and Phillips, J. G. *J. Assoc. Off. Anal. Chem.* **67**, 521 (1984).
63. Matsumura, T., Tanimura, A., Higushi, E., Yamate, N., and Sakata, M. *Nippon Kagaku Kaishi* **10**, 1410 (1979).
64. Tomkins, B. A., Brazell, R. S., Roth, M. E., and Ostrum, V. H. *Anal. Chem.* **56**, 781 (1984).
65. Douse, J. M. F. *J. Chromatogr.* **256**, 359 (1983).
66. Parees, D. M., and Prescott, S. R. *J. Chromatogr.* **205**, 429 (1981).
67. Airoldi, L, Spagone, C., Bonfanti, M., Pantarotto, C., and Fanelli, R. *J. Chromatogr.* **276**, 402 (1983).
68. Walker, E. A., and Castegnaro, M. *J. Chromatogr.* **187**, 229 (1980).
69. Grolimund, K., and Widmer, H. M. *Environmental Carcinogens, Selected Methods of Analysis*, IARC Sci. Publ. No. 45, 1983, Vol. 6, p. 373.
70. Massey, R. C., Crews, C., McWeeny, D. J., and Knowles, M. E. *J. Chromatogr.* **236**, 527 (1982).
71. Tomkins, B. A., Manning, D. L., Griest, W. H., and Jones, R. A. *Anal. Chim. Acta* **169**, 69 (1985).

72. Hotchkiss, J. H., Barbour, J. F., Libbey, L. M., and Scanlan, R. A. *J. Agr. Food Chem.* **26**, 884 (1978).
73. Stephany, R. W., and Schuller, P. L. In B. J. Tinbergen and B. Krol, eds., *Proc. 2nd Int. Symp. Nitrite in Meat Products*, *Zeist*, Pudoc, Wageningen, 1977, p. 249.
74. Gough, T. A., and Webb, K. S. *J. Chromatogr.* **154**, 234 (1978).
75. Rounbehler, D. P., Bradley, S. J., Challis, B. C., Fine, D. H., and Walker, E. A. *Chromatographia* **16**, 354 (1982).
76. Parks, R. E. 27th Pittsburgh Conference on Analytical Chemistry and Applied Spectroscopy, Cleveland, Ohio, March 2, 1976.
77. Kashihira, N., Tanaka, K., Kirita, K., and Watanabe, Y. *Bunseki Kagaku* **29**, 35 (1980).
78. Kashihira, N., Makino, K., Kirita, K., and Watanabe, Y. *Bunseki Kagaku* **31**, 13 (1982).
79. Fahey, D. W., Eubank, C. S., Hubler, G., and Fehsenfeld, F. C. *J. Atmos. Chem.* **3**, 435 (1985).
80. Sievers, R. E., Nyarady, S. A., Shearer, R. L., DeAngelis, J. J., Barkley, R. M., and Hutte, R. S. *J. Chromatogr.* **246**, 395 (1985).
81. *C and EN*, June 24, 1985, p. 42.
82. Shearer, R. L. Doctoral Thesis, University of Colorado, Boulder, CO, 1987.
83. Nyarady, S. A. Doctoral Thesis, University of Colorado, Boulder, CO, 1985.
84. Gray, P., and Yoffe, A. D. *Chem. Reviews* **55**, 1069 (1955).
85. Sievers, R. E., Shearer, R. L., and Barkley, R. M. *J. Chromatogr.* **395**, 9 (1987).
86. Sievers, R. E., DeAngelis, S. S., and Stuber, H. Unpublished results; DeAngelis, J. J., Barkley, R. M., and Sievers, R. E. *J. Chromatogr.* **441**, 125 (1988).
87. Foreman, W. T., Sievers, R. E., and Wenclawiak, B. *Fresenius Z. Anal. Chem.* **330**, 231 (1988).
88. McNamara, E. A., Montzka, S. A., Barkley, R. M., and Sievers, R. E. *J. Chromatogr.* **452**, 75 (1988).

4

Detection Based on Solution-Phase Chemiluminescence Systems

Timothy A. Nieman

Department of Chemistry, University of Illinois, 1209 West California Street,
Urbana, Illinois 61801

Introduction

This chapter is concerned with on-line detection schemes for liquid chromatography (LC) that are based on any of several solution-phase chemiluminescence (CL) systems. The CL systems considered here include luminol, lucigenin, peroxyoxalates, tris(2,2′-bipyridine)ruthenium(II), electrochemiluminescence and the luciferin/luciferase bioluminescence reaction.

All of the CL systems discussed in this chapter would be operated as post-column reactions with the common requirements of a method to deliver and mix the CL reagents with the column effluent and a suitable observation flow cell which allows detection of the CL emission at an appropriate time after the initiation of the reaction (when the column effluent and the CL reagents mix). Figure 4-1 shows a typical setup.

Consideration of the time from mixing to detection is important because the intensity of light emission from a CL reaction is transient rather than constant with time. For a CL reaction, the emission intensity I (in photons emitted per second) is directly proportional to the rate of the chemical reaction dC/dt (in molecules reacting per second),

$$I = \phi_{CL}(dC/dt)$$

where the proportionality constant is the chemiluminescence quantum yield ϕ_{CL} (in photons emitted per molecule reacting). Thus, as the rate of the CL reaction changes with time (due to consumption of reagents), so does the emission intensity change with time. A plot of emission intensity vs. time after reagent mixing would appear as shown in Fig. 4-2. To maximize detector sensitivity, one needs to detect near the time of peak emission intensity. In a flow stream the time from reagent mixing to the observation

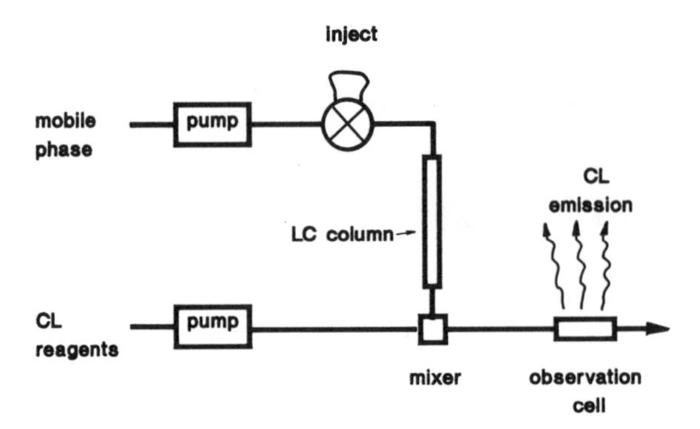

Figure 4-1 ■ General arrangement for chemiluminescence detection in liquid chromatography.

cell is determined by both the volume between these two points and the solution flow rate.

For most CL detection schemes, an eluting analyte causes the CL reaction to proceed at a faster rate than when the analyte is absent. The analyte is detected as an increase in CL emission above a very low background level [Fig. 4-3(a)]. However, in some CL detection schemes, the eluting analyte interferes with or suppresses the CL reaction, in which case the analyte is detected as a valley, or inverted peak, where the CL intensity decreases from a normally high background level [Fig. 4-3(b)].

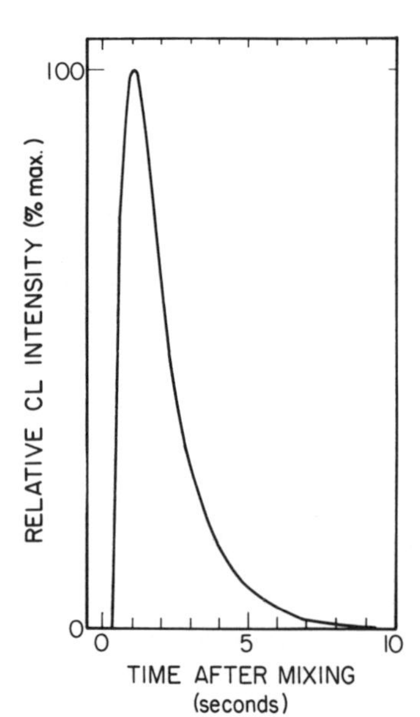

Figure 4-2 ■ Emission intensity vs. time for a chemiluminescence reaction.

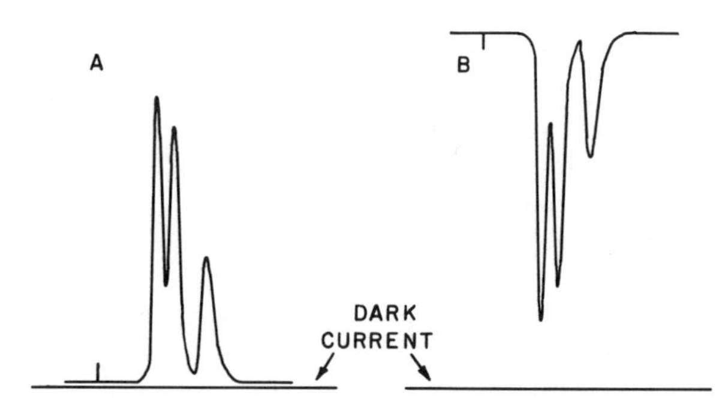

Figure 4-3 ■ Possible chemiluminescence chromatograms. (A) Analyte peaks appear as signal increases from a low level background. (B) Analyte peaks appear as signal decreases from a high level background.

Luminol

Luminol (5-amino-2,3-dihydro-1,4-phthalazinedione) is one of the most commonly used CL reagents. In aqueous alkaline solution, luminol is oxidized to 3-aminophthalate and light. The emission is blue, centered about 425 nm and is from excited state 3-aminophthalate,[1] as shown in Fig. 4-4.

A variety of oxidants are usable, such as permanganate, hypochlorite and iodine.[2] Perhaps the most useful oxidant is hydrogen peroxide. With peroxide as an oxidant, a catalyst is required. Typical catalysts are peroxidase, hemin, transition metal ions (Co^{2+}, Cu^{2+}, Fe^{3+}, etc.) or ferricyanide.[2] Thus, the luminol CL reaction system most often used consists of luminol + H_2O_2 + catalyst + OH^-. Within limits, the CL emission intensity is directly proportional to the concentrations of luminol, H_2O_2 and catalyst. Thus, measurements of CL intensity can be used to quantitate any of these species. As with most CL systems, any of the reaction components can be viewed as the analyte with the others collectively constituting the reagent. Thus, the luminol CL reaction has been used to determine catalysts or species labelled with catalysts, peroxide or species that can be converted into peroxide and luminol or species labelled with luminol.

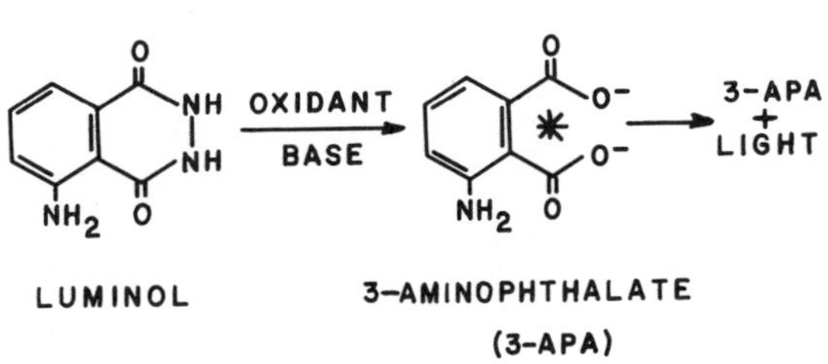

LUMINOL —OXIDANT / BASE→ 3-AMINOPHTHALATE (3-APA)* → 3-APA + LIGHT

Figure 4-4 ■ Luminol chemiluminescence reaction.

Reaction Compatibility with Chromatography

The luminol CL reaction requires an alkaline pH, generally pH 9 to 11, depending upon the catalyst. Compatibility of pH requirements for the separation and the CL detection are a concern. If the separation can be carried out at an alkaline pH, then the separation and detection are directly compatible. Amino acids can be separated on a polymer resin ion exchange column with a $NaOH/Na_2B_4O_7$ mobile phase and subsequently detected with luminol CL (details later).[3] However, generally a post-column pH change is required. This pH change is no problem, because the requisite buffer can be incorporated into the CL reagent solution which is added post-column.

The luminol CL reaction occurs on a short time scale and detection must occur promptly after reagent mixing. With catalysts like Fe^{2+}, Fe^{3+} and Co^{2+}, the CL intensity peaks a second or less after mixing and decays to the background level in just a few more seconds.[3-5] With such a catalyst species it may be necessary to configure the flow system to allow the reagents to come together and mix directly in the observation cell.[6, 7] With catalysts like Cu^{2+}, the CL intensity changes more slowly, reaching a maximum in 2 to 5 s, and not decaying to the background level for 30 s.[3-5]

For the luminol CL reaction as run in aqueous solution, the presence of added organic solvents generally decreases the emission intensity, although the effect is very solvent- and catalyst-dependent. Despite the importance of aqueous/organic solvent systems in liquid chromatography (especially reverse phase), there has been little systematic investigation of the effect of various solvent mixtures on luminol CL. Some limited data are available[4, 8] and are presented in Table 4-1 to show the effect of several 10% organic/90% aqueous solvent systems on the peak CL intensity observed with Co^{2+}, Cu^{2+} and Cr^{3+} catalysts. Peak CL intensity is not the only thing altered; it has been noted that the time scale of emission is also altered by organic solvents.[9] Despite the effect of organic solvents on the metal ion-catalyzed luminol CL reaction, very sensitive detection can be accomplished in mixed aqueous/organic solvent systems. For example, ferricyanide has been used (for luminol detection) in solvent systems containing 15 to 75% methanol,[10, 11] and Co^{2+} has been used (for peroxide detection) in a solvent system containing 70% acetonitrile.[12]

Table 4-1 ▪ Solvent Effects on Luminol Chemiluminescence[a]

	Catalyst		
Organic Solvent	Co^{2+}	Cu^{2+}	Cr^{3+}
Methanol	0.22	0.95	0.08
Ethanol	0.19	0.93	0.23
Acetone	0.94	0.85	0.80
Acetonitrile	0.09	0.98	1.33

[a]The entries are the intensity for a 10% organic /90% aqueous solvent mixture divided by the intensity for a 100% aqueous solution.

Source: Calculated from data in B. M. Strom, M.S. Thesis, Iowa State University, Ames, IA, 1974, and R. Delumyea and A. V. Hartkopf, *Anal. Chem.* **48**, 1402–1405 (1976).

Determinations Involving Catalysts

Direct Detection of Transition Metals. The first use of on-line CL detection in liquid chromatography involved using the luminol reaction for transition metals separated by ion exchange chromatography.[4, 13, 14] Mixtures of Cu^{2+} and Co^{2+} and of Cu^{2+}, Co^{2+} and Ni^{2+} were separated on an anion exchange column as their chloride complexes. Such a separation would normally use HCl in the eluent, but pH compatibility with the luminol CL detection dictated use of LiCl or NaCl instead. It is interesting that since this early work, there have not appeared any other reports using luminol CL to detect transition metals separated by ion exchange.

Detection Based on Suppression of Catalysis. Several reports have appeared in which species that do not directly participate in the luminol reaction are detected by their interactions with other species which are catalysts for the reaction. Picture a detection scheme configured as in Fig. 4-5. Two post-column reagent additions are made. The first addition is a solution of a species which is a catalyst for the luminol CL reaction (e.g., Co^{2+}, Cu^{2+}, Fe^{3+}, Ni^{2+}). The second addition is a mixture of luminol and H_2O_2 in an alkaline buffer. At this point, a constant, relatively intense level of emission will result, and this will be the baseline. Now, suppose an analyte elutes from the LC column and that analyte is a good ligand for the catalyst species. The free (uncomplexed) catalyst concentration is then decreased. Generally, complexed transition metals are much poorer luminol catalysts than the uncomplexed forms. As a result, the decrease in the uncomplexed catalyst concentration (due to the presence of the analyte complexant) decreases the CL emission intensity and a negative peak results. For higher analyte concentrations, the free catalyst concentration (and resulting CL intensity) decreases further and the negative peak becomes bigger. This is shown in Fig. 4-6. The maximum size the peak can be is when essentially all of the catalyst is in the complexed form; higher analyte concentrations yield the same height peak. Although higher analyte concentrations are still detected, the highest analyte concentration that can be quantitated accurately is that which causes the free catalyst concentration, at the center of the band, to drop below the detection limit for that catalyst. As the analyte concentration decreases, the working curve should follow a curve defined by the fraction of free catalyst vs. analyte (ligand) concentration. This curve is nonlinear and is sigmoidal if the analyte concentration axis is logarithmic. The analyte detection limit

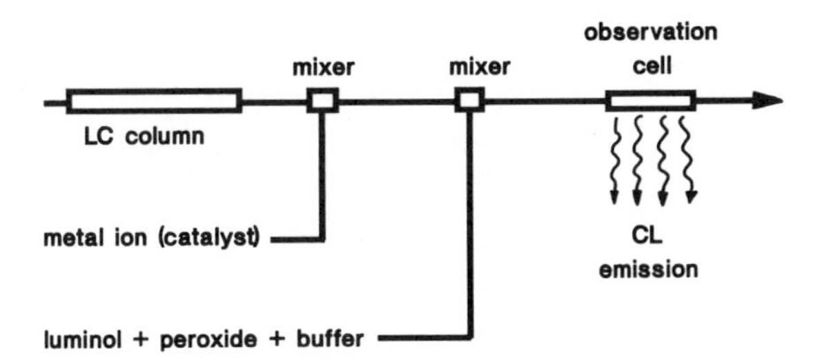

Figure 4-5 ■ Flow system for detection based on suppression of luminol catalysis.

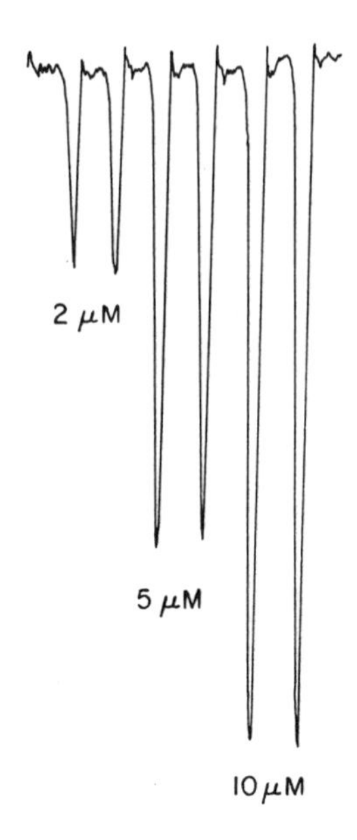

Figure 4-6 ■ Observed signals for sample injections with detection based on suppression of luminol catalysis. From left to right, the samples injected were 2, 5 and 10 μM histidine.

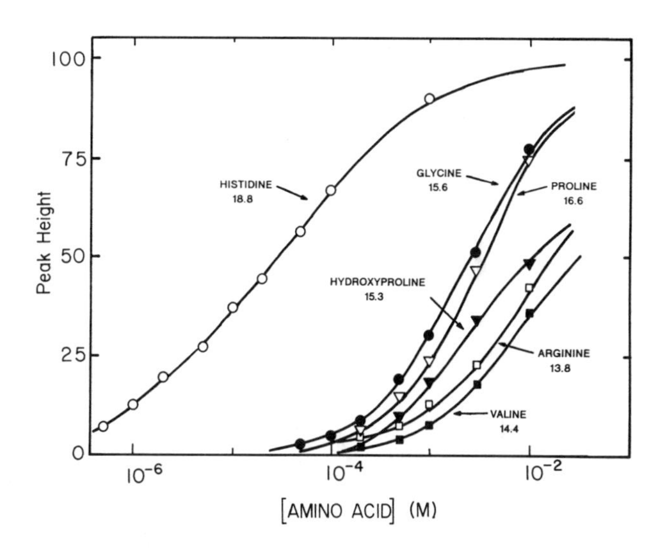

Figure 4-7 ■ Working curves for six amino acids via suppression of Cu^{2+}-catalyzed luminol chemiluminescence. The numbers by the curves are the $\log \beta_2$ values for the Cu^{2+}-amino acid complexes. Reproduced with permission from P. J. Koerner, Jr., and T. A. Nieman, *Mikrochim. Acta* **II**, 79–90 (1987).

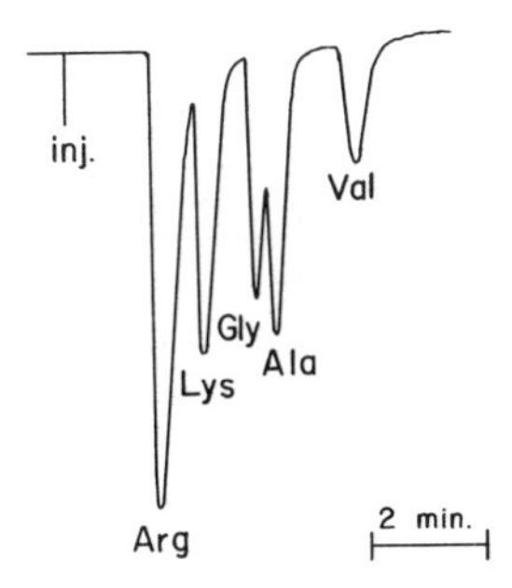

Figure 4-8 ■ Chromatogram of an amino acid mixture detected via suppression of Cu^{2+}-catalyzed luminol chemiluminescence. Arg = arginine, Lys = lysine, Gly = glycine, Ala = alanine and Val = valine; 20 nmol of each injected. Separation on a Dionex HPIC-AS8 anion-exchange column with an aqueous mobile phase containing 23 mM NaOH/6 mM $Na_2B_4O_7 \cdot 10H_2O$. Reproduced with permission from P. J. Koerner, Jr., and T. A. Nieman, *Mikrochim. Acta* **II**, 79–90 (1987).

is then determined by the smallest change in the catalyst concentration which is measurable. The shape of the working curve is predictable based on complexation equilibria, and the relative response to different analytes is predictable based on the relative magnitudes of the complex formation constants.

This general approach has been used in several variations to detect a variety of analytes. Amino acids have been separated via ion exchange and detected based on their complexation with either Co^{2+} or Cu^{2+}.[3, 15] Figure 4-7 presents working curves for several amino acids to show their sigmoidal shape and displacement according to Cu^{2+}-amino acid formation constant. Detection limits range from 1 pmol injected (for histidine with $\log \beta_2 = 18.8$) to 2 nmol (for most other amino acids with $\log \beta_2 = 13$ to 14). Figure 4-8 shows a separation of several amino acids on a polymer anion-exchange column. It is notable that this approach allows detection of both primary and secondary amino acids without derivatization. Most any substance that complexes a transition metal ion catalyst can be detected this way. With Cu^{2+} as a catalyst, catecholamines can be detected down to 1 pmol injected, and have been determined via reverse phase chromatography.[3] Aminoglycoside antibiotics are good Cu^{2+} ligands, and the three components of gentamicin C in commercial gentamicin sulfate have been determined via ion-pair chromatography.[3] By using Cu^{2+} as a catalyst, various proteins are detectable. Thyroglobulin, gamma globulin and ovalbumin have been determined via size exclusion chromatography.[16] Human serum albumin and ovalbumin have been determined via immunoaffinity chromatography.[17] Protein detection limits are generally a few picomoles injected. These applications show a remarkable range of types of separation (ion exchange, reverse phase, ion-pair, size exclusion, immunoaffinity) that are compatible with the same CL detection reaction.

An interesting variation on this theme detects complexing analytes by their ability to remove Cu^{2+} suppression in Fe^{2+}-catalyzed luminol CL.[6, 7] Fe^{2+} is unique among transition metal ions in producing significant CL emission from luminol with simple dissolved O_2 as the oxidant (recall that generally H_2O_2 is required). The presence of Cu^{2+} suppresses CL emission from the luminol–O_2–Fe^{2+} system, possibly by catalyzing the oxidation of Fe^{2+} to Fe^{3+}. To employ this system as an LC detector, the first post-column reagent addition (as in Fig. 4-5) is luminol in an alkaline buffer and the second reagent is an acidic mixture of Fe^{2+} and Cu^{2+}. (In acid, the conversion of Fe^{2+} to Fe^{3+} is very slow so this mixture is stable for several days.) If the Cu^{2+} were not present, significant CL emission would result, but the Cu^{2+} keeps the level of CL emission very low. Now, if an analyte which complexes Cu^{2+} elutes from the column, the free Cu^{2+} concentration decreases; as a result there is lessened suppression of the Fe^{2+}-catalyzed luminol reaction and an increase in CL emission occurs. Thus, positive-going CL peaks result. As the analyte (complexant) concentration increases, the CL signal also increases until it becomes as large as it would be due to the amount of Fe^{2+} present and the total absence of Cu^{2+}. Because the observed signal increases with the analyte concentration, this approach has a fundamental advantage over the earlier approaches in which the signal is inversely proportional to the analyte concentration. Detection limits of 2 to 35 pmol injected were obtained for primary and secondary amino acids without derivitization and amino acid mixtures were determined via anion-exchange chromatography under the same separation conditions as in Fig. 4-8. Once again, other complexants such as amines, organic acids and antibiotics are readily detected based on their ability to complex Cu^{2+}. Phenols, catechols and catecholamines were detected because they directly complex the Fe^{2+} and so create negative peaks.

Determinations Involving Peroxide

If an analyte of interest can be converted into an equivalent amount of hydrogen peroxide, then addition of a luminol-catalyst reagent mixture will allow quantitation of the produced peroxide and thus the original analyte. A photooxygenation chemiluminescence system has been described[12] which functions by using anthraquinone-sensitized photooxygenation. Various aliphatic oxygen-containing analytes (alcohols, ethers, aldehydes and saccharides) were separated on a reverse-phase column using a mobile phase containing anthraquinone-2,6-disulfonate. The column effluent passed through a photochemical reactor where H_2O_2 was produced and subsequently detected via Co^{2+}-catalyzed luminol chemiluminescence. Detection limits for most analytes are in the low microgram range.

Oxidase enzyme-catalyzed reactions have been used to convert various analytes (glucose, sucrose, cholesterol, uric acid, acetylcholine, etc.) to H_2O_2 with the peroxide quantitated via luminol CL to quantitate the original analyte.[18] Most applications reported to date have involved flow injection, not LC, but if one uses an enzyme (or enzyme system) that is class selective rather than compound specific, then application to LC is possible. For example, L-amino acid oxidase (L-AAO) catalyzes the deamination of L-amino acids:

$$\text{L-amino acid} + H_2O + O_2 \xrightarrow{\text{L-AAO}} \text{2-oxo acid} + NH_3 + H_2O_2$$

Immobilized reactors of L-AAO have been used in detection schemes for amino acids separated via reverse phase LC. In one case, the production of NH_4^+ and oxo acid anion was followed conductometrically.[19] In another case, the H_2O_2 produced was monitored fluorometrically following reaction with homovanillic acid (catalyzed by peroxidase).[20] In both cases, the detection limits were 250 to 5000 pmol injected. If instead, one determined the H_2O_2 via luminol CL, a detection limit of 1 to 100 pmol would be anticipated.

As an example, in a very recent report which does involve LC, two consecutive post-column enzyme reactors were used to enable detection of β-glucosides (glucose conjugates generated during metabolism, generally in plants) that were separated on a reverse phase C_{18} column with an acetonitrile/water mobile phase.[21] The column effluent passed first through a β-glucosidase reactor to catalyze hydrolysis of a β-glucoside to glucose plus the corresponding hydroxy aglycon and then through a glucose oxidase reactor to catalyze oxidation of glucose to gluconic acid plus H_2O_2. The H_2O_2 was then quantitated via luminol CL. The use of up to 30% acetonitrile in the buffered mobile phase had very little effect on the efficiency of these enzyme reactors and did not affect the linear range or detection limits for β-glucosides. Chromatographic detection limits were approximately 0.1 μM or 2 pmol injected with a working range of more than three decades.

Recent work with luminol electrogenerated chemiluminescence (ECL) is attractive for use in peroxide determinations.[22, 23] In place of the usual dissolved catalyst species, a positively biased electrode will cause ECL in the presence of luminol and peroxide. The peroxide detection limit is 0.1 μM (3 ppb) or 50 pmol injected. ECL offers several advantages over conventional catalysts; these include control of both the spatial and temporal point at which the CL reaction occurs, and freedom from the need to prepare and deliver catalyst solutions.

Another development potentially applicable to LC involving peroxide detection (but to date reported only with flow injection) is the use of luminol immobilized on small particles and contained in a flow-through reactor placed in the analyte carrier stream.[24, 25] Luminol is leached from the reactor and delivered downstream to where the peroxide and luminol then encounter the catalyst either in dissolved or immobilized form or as an electrode. Thus, a separate solution of luminol (and the associated pump and tubing) is not needed. By using immobilized luminol, peroxide has been detected from 1 μM to 1 mM.

Determinations Involving Species Labelled with Luminol

Because the luminol moiety can be quantitated conveniently and at trace levels, there is interest in use of luminol (or related derivatives) as labels in immunoassay[26] and in pre-column derivatizations for LC.[10, 11, 27] Besides luminol itself, derivatives that are common labels are isoluminol and aminobutylethylisoluminol (ABEI) (see Fig. 4-9). Coupling of one of these compounds to an analyte is accomplished via the primary amine group. The CL quantum efficiency for luminol is 5 to 20 times greater than that for isoluminol (so greater CL emission intensities result). However, substitution at the primary amine group on luminol generally decreases the quantum efficiency, but substitution at the primary amine group of isoluminol often increases the quantum efficiency. ABEI is convenient because the alkyl chain allows coupling with reduced steric hindrance and isolates the isoluminol moiety from the analyte so that the CL quantum yield is relatively independent of analyte.[26]

ABEI has been used as a pre-column labelling reagent for primary and secondary amines and carboxylic acids.[10] The amine labelling reaction made use of *N*,*N*′-disuccinimidyl carbonate (a 4 hr reaction) and the carboxylic acid labelling reaction made use of 2-chloro-1-methylpyridinium iodide and 3,4-dihydro-2H-pyrido[1,2-a]pyrimidin-2-one (a 2 hr reaction). The labelled species were separated by reverse phase on a C_{18} column with a methanol/water mobile phase. The column effluent was mixed first with H_2O_2 and then with alkaline ferricyanide. The detection limit for ABEI-labelled cholic

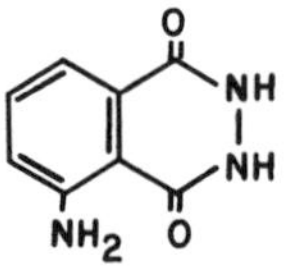

LUMINOL 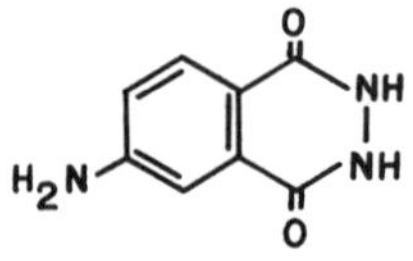 ISOLUMINOL

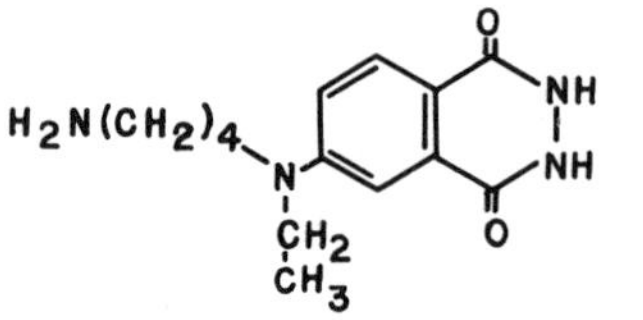

ABEI

Figure 4-9 ■ Structures of luminol, isoluminol and aminobutylethylisoluminol (ABEI).

acid was 20 fmol. Figure 4-10 shows separations of mixtures of carboxylic acids and of bile acids via this approach.

A new CL pre-column labelling reagent for amino acids has recently been reported.[11] The compound is 4-isocyanatophthalhydrazide and is prepared by reaction of thiophosgene with isoluminol; the H_2N— group of isoluminol is replaced by SCN—. The resulting compound couples to amino acids in only 5 to 10 min at room temperature.

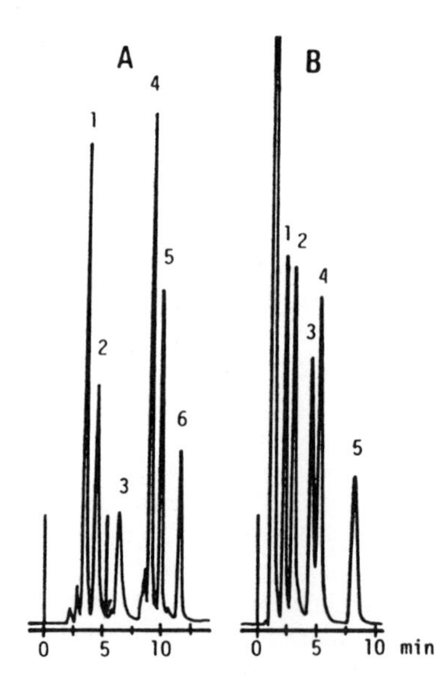

Figure 4-10 ■ Chromatograms of ABEI derivatives of carboxylic acids. (A) Free fatty acids; 1 = acetic, 2 = propionic, 3 = *n*-butyric, 4 = caproic, 5 = caprylic, 6 = capric. Separation on a Waters C_{18} resolve column with a 40 to 70% methanol mobile phase. (B) Bile acids; 1 = ursodeoxycholic, 2 = cholic, 3 = chenodeoxycholic, 4 = deoxycholic, 5 = lithocholic. Separation on a Waters C_{18} resolve column with a 75% methanol mobile phase. Reproduced with permission from T. Kawasaki, M. Meada, and A. Tsuji, *J. Chromatogr.* **328**, 121–126 (1985).

Labelled amino acids were separated by reverse phase on a C_{18} column. The column effluent was mixed first with H_2O_2 and then with alkaline ferricyanide. The detection limits were 10 to 20 fmol.

An indirect ion displacement approach has been reported to yield CL detection for anions.[28] An anionic luminol derivative is prepared by reaction of luminol with succinic anhydride and pyridine; one H— on the amine group of luminol is replaced by $HOOCCH_2CH_3(C{=}O)$—. The anion-exchange resin is then loaded with this reagent and packed into a column. Analyte anions passing through this column displace the luminol derivative which is subsequently quantitated via addition of H_2O_2 and ferricyanide. Detection limits for Cl^-, NO_2^-, NO_3^- and SO_4^{2-} are 1 μM. This detection approach could be applied generally for detection of anions which are inconvenient to detect directly and are difficult to derivatize. The lifetime of the luminol column is limited; only 30 injections of 100 μM Cl^- could be made before noticeable depletion of the luminol column (with corresponding decrease in CL signal) occurred. This detection scheme has been reported to date only for flow injection, but its extension to ion exchange chromatography is apparent (provided one can suitably deal with the problem of depletion of the luminol column by anions in the mobile phase).

Recent work with luminol electrogenerated chemiluminescence (ECL) is beneficially applicable to all these schemes described in the preceding text where the object is to quantitate luminol (or luminol-labelled species).[22, 29] In all the cases described, it was necessary to add both H_2O_2 and catalyst (generally as two separate solutions). By using luminol ECL, a compact detector has been assembled using series dual working electrodes in a thin layer flow cell (Fig. 4-11). The upstream electrode is held at a negative potential and serves to generate H_2O_2. This H_2O_2, plus any luminol in the carrier stream, undergoes an ECL reaction at the downstream electrode which is held at a positive potential, and the resulting light emission is monitored. The detection limit is below 10^{-10} M (50 fmol). At most only a single post-column solution must be added, and that would be a buffer (around pH 10) to change the column effluent from the pH necessary for the separation to the pH needed for the ECL reaction. For separations which can be carried out at alkaline pH, there is no necessity for post-column solution addition of any kind.

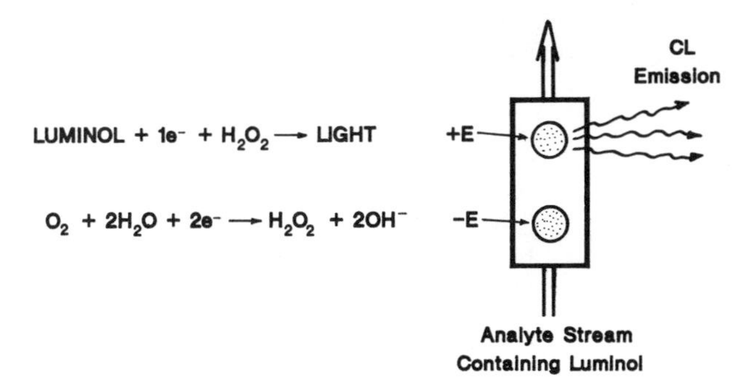

Figure 4-11 ■ Series dual electrode flow cell for generation of peroxide and electrogeneration of luminol chemiluminescence.

LUCIGENIN N-METHYLACRIDONE (NMA)

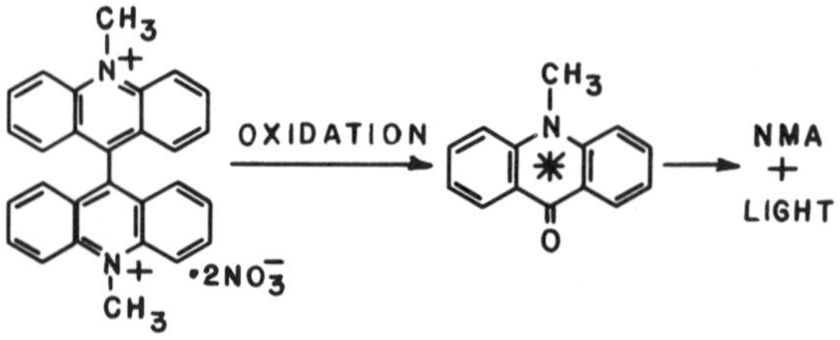

Figure 4-12 ■ Lucigenin chemiluminescence reaction.

Lucigenin

Lucigenin (N,N'-dimethyl-9,9′-diacridinium nitrate) undergoes CL reaction in alkaline solution to yield N-methylacridone and light (Fig. 4-12).[30–34] Several side products also form. The emission is green and due to the primary emitter, excited state N-methylacridone, which emits from about 420 to 500 nm with a maximum at about 440 nm. Because lucigenin absorbs in this region and filters the emission, the exact CL emission spectrum observed is concentration-dependent. The overall reaction is an oxidation, but the mechanism is complicated and probably involves both oxidation and reduction steps. Acridinium esters and acridans (Fig. 4-13) react similarly to lucigenin to also yield N-methylacridone and light; these compounds have found use as labels in immunoassay.[34–36]

Lucigenin can react with either hydrogen peroxide or with a reductant plus oxygen. In the reaction with H_2O_2, the peroxide probably acts both as oxidant and reductant; considerably more light is emitted in the presence of catalysts like Co(II), Fe(III), Fe(II), Cu(II), Cr(III) and Ni(II).[37] In the presence of dissolved oxygen, lucigenin can undergo CL reaction with reductants such as glucose and other reducing sugars, ascorbic acid, uric acid, creatinine and glucuronic acid.[31, 38] The reaction probably begins with an initial reduction followed by oxygen addition (Fig. 4-14). This reaction has been used to sensitively quantitate reductants of clinical interest.[38] It is the application of lucigenin to reductant determinations that has been used as an LC detector.

ACRIDINIUM ESTER ACRIDAN ESTER

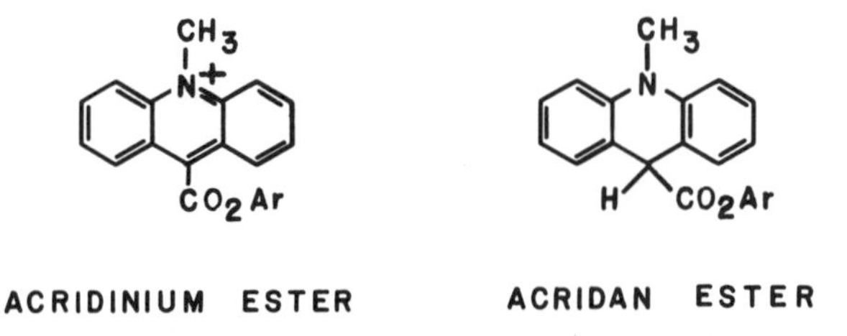

Figure 4-13 ■ Structures of acridinium and acridan phenyl esters.

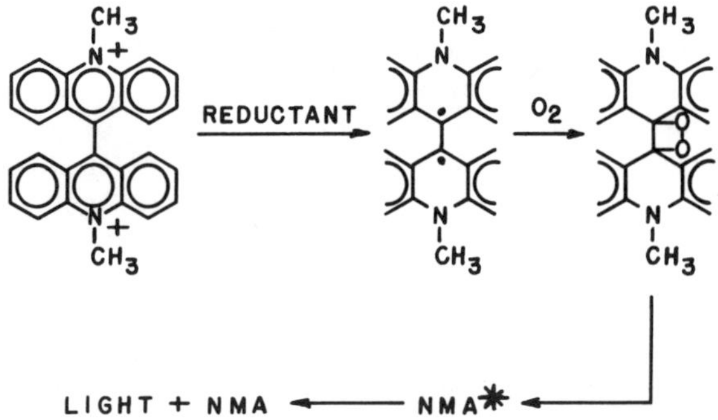

Figure 4-14 ■ Probable features of lucigenin chemiluminescence reaction with reductants and oxygen.

Reaction Compatibility with Chromatography

The pH requirements for the lucigenin CL reaction are more severe than for luminol. The reaction is generally run in 0.1 to 1.0 M KOH or NaOH, so a post-column pH change is necessary. The KOH or NaOH solution and the lucigenin reagent solution must either be added to the column effluent separately or combined into a single solution only shortly before use. The reason is because significant CL background reaction occurs from an aqueous alkaline solution of lucigenin (unless O_2 is excluded), so the mixed solution will degrade over a few hours.

The time scale for lucigenin CL is much longer than for luminol CL. With ascorbic acid the CL emission peaks after only 10 s. However, with glucose (which shows behavior similar to most other reductants) the CL signal takes several minutes to peak and remains relatively constant for several minutes after peaking.[38, 39] Thus, a delay coil inserted between the reagent addition mixer and the observation cell is probably desirable.

An inconvenience with lucigenin CL is that the *N*-methylacridone reaction product is insoluble in water and deposits on the walls of the flow system tubing and on the inside of the detector window. These deposits can be removed by a periodic rinse with nitric acid,[38] but optimum analytical performance dictates that the precipitation needs to be prevented. It has been reported that addition of the surfactant sodium dodecyl sulfate (SDS) to the lucigenin solution (at a concentration below the SDS critical micelle concentration) will prevent *N*-methylacridone from precipitating in a flow system; this level of surfactant will not degrade the CL emission intensity.[40] Some surfactant systems have been reported to yield moderate enhancement of lucigenin CL.[41]

Determinations Involving Reductants

The first use of lucigenin CL in LC detection was to assay mixtures of ascorbic acid and dehydroascorbic acid separated via ion exchange and to demonstrate application to a mixture of glucose, glucuronic acid, creatinine and ascorbic acid similarly separated.[42] Solutions of KOH and lucigenin were added to the post-column effluent. To optimize the detection for ascorbic acid, a delay of 6 to 9 s was inserted between the mixer and the observation cell.

Compounds having an α-hydroxycarbonyl group were found to give intense CL with lucigenin,[43] and this was in agreement with earlier work postulating certain reductants to proceed in this reaction via a 1,2-enediol tautomer intermediate.[44] Based on this observation, lucigenin CL was used to detect several corticosteroids separated by reverse phase chromatography on a C_{18} column with a methanol/water mobile phase.[43] The detection limit was 500 fmol, but response was only linear up to 5 pmol; no pre-column derivatization was need. Phenacyl esters of carboxylic acids give intense CL with lucigenin, so carboxylic acids were derivatized with *p*-nitrophenyl bromide, separated by the same conditions as the corticosteroids and detected by lucigenin CL; the detection limit was also 500 fmol.[43]

Because lucigenin CL is sensitive to certain reductants, it can also be used to quantitate analytes which can be selectively converted to one of those reductants. Glucuronic acid conjugates, or glucuronides, are formed in the liver to detoxify "foreign" molecules (drugs and other potentially toxic organics). A system for determination of conjugated glucuronic acid by means of an immobilized enzyme reactor, an LC separation and lucigenin CL detection has been reported.[45] The method makes use of the enzyme β-glucuronidase to selectively hydorlyze glucuronides to glucuronic acid and the corresponding aglycon. A simplified flow system schematic is shown in Fig. 4-15. To determine total conjugated glucuronic acid, only LC column A is used. Immediately after the immobilized enzyme reactor (IMER), the flow stream contains the reductant glucuronic acid (which was produced in the IMER) plus any other reductants present in the original sample. In a urine or serum sample one would also find appreciable amounts of the reductants ascorbic acid, creatinine, galactose, glucose, lactose, uric acid and glutathione. Because the lucigenin CL reaction is sensitive to all of these species, a method is necessary to differentiate the glucuronic acid from the remainder of the reductants. An anion-exchange column inserted at point A solved the problem by chromatographically resolving the glucuronic acid from the other reduc-

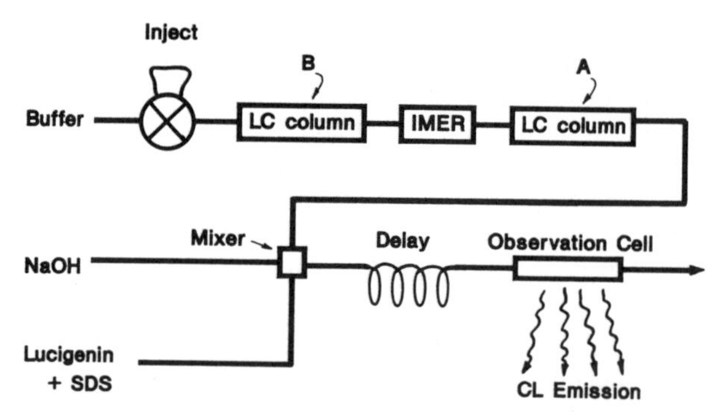

Figure 4-15 ■ Flow system for determination of glucuronides via lucigenin chemiluminescence. The functions of LC columns A and B are discussed in the text.

tants. Lucigenin, SDS surfactant and NaOH were then added to initiate the CL reaction. A 1 min delay coil was inserted between the mixer and the observation cell. For a variety of glucuronides, the detection limits are 5 to 10 μM (500 pmol) and are appreciably lower than achievable with refractive index, UV-visible absorbance or amperometric detection of glucuronic acid. Working curves are linear at least as high as 1 mM. If several glucuronides are each quantitatively converted to glucuronic acid in the IMER, then they will have identical working curves. Although not reported, this approach could be modified to separate and quantitate individual glucuronides by using an LC column at point B instead of point A. Glucuronides separated in time after column B would be converted to peaks of glucuronic acid in the IMER and subsequently quantitated via lucigenin CL.

Determinations Involving Catalysts, Peroxide or Labelled Species

No reports have appeared describing use of lucigenin CL as an LC detector for either peroxide or for transition metal ions that are catalysts. This situation is understandable because luminol CL could be used for these applications and have lower background signal, better detection limits, less severe pH requirements and no solubility problems.

Detection of pre-column labelled species has also not been reported, but one would expect to see activity here in the future. Acridinium phenyl carboxylates are being promoted as labels for immunoassay.[36] Thus, they could also find use as general pre-column derivitizing agents. A noteworthy feature of these substances is that the pH optimum for efficient CL depends on the substituents on the phenyl group. Thus, acridinium esters can be designed to match different pH requirements and have been shown to operate at near neutral pH.[46]

Peroxyoxalates

Peroxyoxalate CL reactions are the hydrogen peroxide oxidations of aryl oxalate esters. These reactions are the most efficient of nonbiological CL reactions with quantum efficiencies in excess of 20 to 30%.

A high energy intermediate is formed by the reaction of the aryl oxalate and hydrogen peroxide:

$$\text{oxalate} + H_2O_2 = \text{intermediate}$$

This intermediate has been postulated to be 1,2-dioxetanedione, but there has been no direct detection of this species; recently, other intermediates have been suggested. This intermediate and a fluorophore then form a charge transfer complex which dissociates to yield excited state fluorophore:

$$\text{intermediate} + \text{fluorophore} = [\text{intermediate}^{\overline{\cdot}}\,\text{fluorophore}^{\dot{+}}] = \text{fluorophore}^*$$

The resulting emission is then from the fluorophore. Unlike luminol and lucigenin, the spectrum of the CL emission is not determined by the consumed CL reagent (here the aryl oxalate) but instead by the fluorophore. The chemical products are ArOH and CO_2.[2, 18, 34, 47, 48] Because electron transfer occurs from fluorophore to intermediate, it is apparent that easily oxidized fluorophores lead to efficient CL production.[49] The mechanism of the peroxyoxalate CL reaction is covered in detail in a separate chapter.

Peroxyoxalate CL can be used to quantitate hydrogen peroxide (as can luminol CL) or fluorophores such as polycyclic aromatic hydrocarbons and dansyl- or fluorescamine-labelled analytes.[49, 50] The most commonly used oxalate esters are bis-(2,4,6-trichlorophenyl)oxalate (TCPO) or bis-(2,4-dinitrophenyl)oxalate (DNPO).

Reaction Compatibility with Chromatography

An advantage of peroxyoxalate CL is that it is useful over a wide pH range which is generally near neutral. TCPO can be used from pH 5 to 9, with the maximum CL intensity occurring at about pH 7.[51, 52] DNPO can be used at more acidic pH, down to pH 3.5.[48, 51] The pH alters the half-life of the reaction. Although DNPO is not as stable as TCPO, the faster kinetics with DNPO lead to a chemical band-narrowing effect which permits use of larger detector cells without band broadening.[53]

A weak base such as triethylamine, tris(hydroxymethyl)aminomethane or imidazole is generally included in the reaction mixture to catalyze the reaction and increase CL intensity.[52]

A major limitation of peroxyoxalate CL is the need for an organic solvent due to solubility, stability and efficiency considerations. Common solvent mixtures have included ethyl acetate/ethanol/water or acetonitrile/tetrahydrofuran/water.[18, 51, 53] Recently, water soluble oxalate esters have been reported, but they are not as efficient or stable as those such as TCPO or DNPO. The peroxyoxalate reaction mechanism probably involves nucleophilic attack of H_2O_2 on the oxalate carbonyl. This means that nucleophilic solvents such as water and methanol can also attack the CL reagent and consume it in non-CL side reactions. For TCPO, water in the solvent mixture decreases the time to reach maximum intensity (in the emission vs. time profile) but does not degrade the maximum intensity until levels of about 40% water are reached. Methanol at amounts as low as 1% can halve the maximum CL intensity.[52]

The CL emission is quenched by easily oxidized species such as bromide, iodide, sulfite, nitrite, substituted ailines and organosulfur compounds. Linear working curves for H_2O_2 are obtained in the presence of these quenchers. But, because the working curve slope is altered by the quencher, one must use standard addition to achieve accurate quantitative results.[54]

Determinations Involving Fluorophores

A large number of publications have appeared using peroxyoxalate CL as a detection mode in HPLC. The system is used generally to detect fluorescent species, so the reagent is a mixture of aryl oxalate and hydrogen peroxide. TCPO and DNPO are the most commonly used oxalates, but recent studies have demonstrated that in addition to those oxalates, others may be more appropriate given the mobile phase pH.[55] Peroxyoxalate detection of fluorophores often results in detection limits 1 to 2 orders of magnitude lower than fluorescence detection. This is because the source instability and source scatter of fluorescence detection are eliminated. Attomole detection limits have been obtained for some compounds.[49] Furthermore, because peroxyoxalate CL is most efficient for easily oxidized fluorophores, this detection mode is more selective than conventional fluorescence and can provide simpler chromatograms for samples with complex matrices.[48]

Detection of Fluorescent Analytes. Polycyclic aromatic hydrocarbons (PAH) and polycyclic aromatic amines such as anthracene, perylene, benzopyrenes, aminoanthracenes, aminopyrenes, etc., can be very sensitively detected with peroxyoxalate CL and have been determined via reverse phase separation with an acetonitrile mobile phase.[49, 56]

Detection of Derivatized Analytes. For analytes that are not natively fluorescent it is necessary to form fluorescent derivatives. This derivatization is generally carried out pre-column. Many of the derivatizing agents commonly used with fluorescence detection are also useful with peroxyoxalate CL detection, so there have been determinations of dansyl-derivatized amino acids,[57–59] aliphatic amines[60] and steroids,[61, 62] *o*-phthalaldehyde derivatives of amines[60] and fluorescamine derivatives of catecholamines.[63] Evaluations of fluorescent derivatives specifically for peroxyoxalate CL have shown 3-aminopyrene,[64] 3-aminofluorene[65] and various amine substituted coumarins[66] are especially sensitive, due to a combination of their fluorescent and redox properties.

Because amino-PAHs are so sensitively detected, a procedure for nitro-PAHs was developed by using on-line post-column zinc reduction to convert the nitro compounds to the corresponding amino compounds.[67]

Tertiary amines have been "derivatized" by extracting them (on-line post-column) with the negatively charged fluorescent counterion 5-dimethylaminonaphthalene-11-sulfonate.[68]

Detection limits for molecules derivatized by these procedures vary, but are generally in the low to subfemtomole range.

Quenched Peroxyoxalate Chemiluminescence. It has been recently demonstrated that easily oxidized nonfluorescers like bromide, iodide, sulfite, nitrite, substituted anilines and organosulfur compounds quench peroxyoxalate CL emission, presumably through destruction of the charge transfer complex (between fluorophore and reaction intermediate).[54, 69] If the CL reagent is a mixture of oxalate, hydrogen peroxide and fluorophore, one can monitor the decrease in baseline emission due to the quencher and thus quantitate the quencher. A mixture of anilines was separated on a column packed with spherical carbon with an acetonitrile/water mobile phase; detection limits were approximately 1 μM.[69]

Determinations Involving Peroxide

Just as with luminol, there has been interest in using post-column reactions to generate hydrogen peroxide from the separated analytes and then use peroxyoxalate CL to quantitate the peroxide.

Acetyl choline and choline were separated by ion-pair chromatography with a pH 5 phthalic acid/triethylamine/sodium octanesulfonate mobile phase. The effluent was then mixed with a pH 8.5 Tris buffer and passed through an IMER column of acetylcholine esterase and choline oxidase to generate hydrogen peroxide. Then TCPO and perylene were added and the resulting CL emission measured. The detection limit was 1 pmol.[70]

L-amino acids were separated on a reverse-phase column using an aqueous pH 7.3 phosphate mobile phase which was directly compatible with the IMER of L-amino acid oxidase which followed. The generated hydrogen peroxide was quantitated by adding an acetonitrile solution of DNPO and using an immobilized fluorophore. Detection limits were 0.3 to 3 μM for all detectable amino acids.[71]

Quinones have been photochemically reacted post-column to yield hydrogen peroxide, which was then quantitated with TCPO.[72, 73]

Immobilized and Solid State Reagent Systems

Considerable recent interest has developed in containing the oxalate and/or fluorophore reagents in a solid state format within flow-through reaction columns. Two instrumental variations are shown in Fig. 4-16. In the single-stream version the injected sample passes through the reagent column so band broadening is possible if voids can form in the column as the CL reagent is consumed. In the split-stream version, such voids are of no concern, but there is the complexity of splitting and recombining the flow stream.

Solid TCPO has been packed into the reagent column.[74] This solid then slowly dissolves into the stream flowing through it. Reactor volumes have been a few hundred microliters and the column lifetime is typically 3 to 8 hr. Acetonitrile/water carrier streams have been used most commonly, but methanol/water streams have also been used. With conventional dissolved TCPO, use of methanolic solvent streams leads to poor reagent stability; the solid-phase reactor allows convenient use of methanol without serious stability problems because of the limited and constant time the TCPO is in solution prior to entering the detector cell.[72] Because the TCPO concentration is determined via dissolution, this concentration is sensitive to flow rate, temperature and solution composition (especially to the amount of water present).[72, 74] For water/acetonitrile solutions, increasing the water content leads to drastic reduction in CL intensities due to the very poor solubility of TCPO in water; use of high acetonitrile concentrations leads to massive dissolution with resulting poor reproducibility and

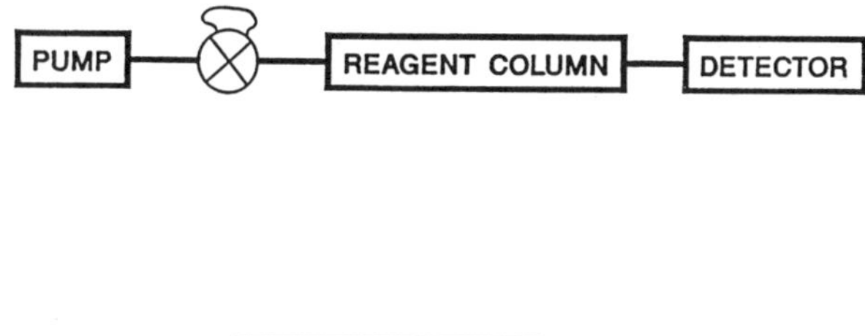

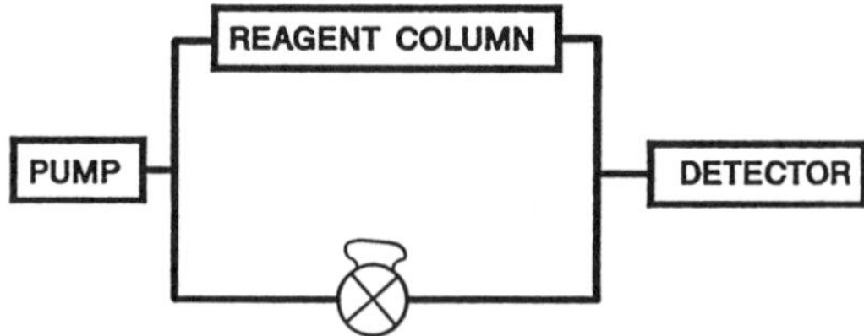

Figure 4-16 ■ Flow system configurations using columns of immobilized or solid state chemiluminescence reagents. Top: Single-stream version. Bottom: split-stream version.

rapid depletion of the reagent column; 20% water/80% acetonitrile seems optimum.[74] If the TCPO is mixed with glass beads prior to packing the reagent column, back pressure and void formation problems are reduced.[75] TCPO reagent columns have been used for detection of hydrogen peroxide with detection limits of approximately 10 nM and linearity for three or more decades of concentration.[72, 75] TCPO columns have been used to determine chromatographically separated fluorophores.[76] In this case, the hydrogen peroxide was placed in the mobile phase and a split-stream approach (separation column and TCPO column in parallel) used so that only the single HPLC pump was necessary. Detection limits by this approach and by the conventional approach (post-column solution additions of TCPO and hydrogen peroxide) were comparable.

The fluorophore 3-aminofluoranthrene has been immobilized by covalent attachment (via a glycidoxypropylsilane) to controlled pore glass. Columns containing this material were placed immediately in front of the detector and used for peroxyoxalate CL determination of hydrogen peroxide.[75] These immobilized fluorophore columns are stable for long term use and show no degradation over one or two months of continuous use.[72, 75]

Immobilized TCPO and fluorophore have been used together in applications involving hydrogen peroxide detection. By placing the TCPO and fluorophore in a single column (with the TCPO upstream), the reaction time between TCPO and peroxide prior to encountering the fluorophore is extremely short and the system becomes relatively insensitive to flow rate. Chromatographically separated quinones have been photochemically reacted post-column to yield hydrogen peroxide which was then detected; single-pump systems were used, both in split-stream[72] and single-stream[73] modes. Glucose has been detected in a system combining in sequence a glucose oxidase IMER, a TCPO reagent column and an immobilized fluorophore column.[77] With a single flow stream of 20% water/80% acetonitrile the detection limit is 800 nM. The detection limit could be improved to 50 nM at the expense of using a second pump; a 100% aqueous stream was used through the IMER and then an additional flow stream containing acetonitrile was added.

The electrochemical behavior of TCPO has been studied.[78] If TCPO is reduced in the presence of O_2 and a fluorophore, then light emission occurs and is from the fluorophore (just as in the conventional reaction between TCPO and H_2O_2). Thus, this electrochemical generation of CL emission can replace the normal requirement for H_2O_2 reagent. Such a system could be used to quantitate fluorophores.

Tris(2,2′-bipyridine)ruthenium(II)

In the ruthenium tris-bipyridine (bpy = bipyridine) system, reaction of $Ru(bpy)_3^{3+}$ and $Ru(bpy)_3^{+}$ yields excited-state $Ru(bpy)_3^{2+}$.[79, 80] Orange emission then arises when this excited state decays to the ground state. $Ru(bpy)_3^{2+}$ is the stable species in solution. One can electrogenerate the $Ru(bpy)_3^{+}$ and $Ru(bpy)_3^{3+}$ from $Ru(bpy)_3^{2+}$ by reduction at about −1.3 V and oxidation at about +1.3 V, respectively. Alternatively, either species can be electrogenerated and the opposite species (and thus light) generated by reaction with the appropriate oxidant or reductant. For example, if $Ru(bpy)^{3+}$ is generated in the presence of oxalate,[79] then in simplified form

$$Ru(bpy)_3^{2+} \rightarrow Ru(bpy)_3^{3+} + e^-$$

$$2Ru(bpy)_3^{3+} + C_2O_4^- \rightarrow Ru(bpy)_3^{2+} + Ru(bpy)_3^{+} + 2CO_2$$

$$Ru(bpy)_3^{+} + Ru(bpy)_3^{3+} \rightarrow Ru(bpy)_3^{2+} + Ru(bpy)_3^{2+*}$$

$$Ru(bpy)_3^{2+*} \rightarrow Ru(bpy)_3^{2+} + \text{light}$$

This chemiluminescent system is noteworthy because the reagent is regenerated and so can be recycled. Recently, it has been reported that aliphatic amines cause CL on mixing with $Ru(bpy)_3^{3+}$.[81]

Reaction Compatibility with Chromatography

The reactions are compatible with reverse-phase LC solvent systems.[79-82] Analytical use of the $Ru(bpy)_3^{3+}$-oxalate reaction has been in aqueous solution, but the reaction has been studied in a variety of solvents, including acetonitrile. With $Ru(bpy)_3^{3+}$-oxalate the CL intensities are constant from pH 4 to 8, and there is no need to remove O_2 to achieve ECL; deoxygenation is reported to improve precision, however. The $Ru(bpy)_3^{3+}$-amine reaction has been done in purely aqueous media and in acetonitrile/water mixtures buffered between pH 4 and 6. Presence of acetonitrile does degrade detection limits somewhat; CL intensities and detection limits are degraded by a factor of 10 when the acetonitrile/water ratio is increased from 10/90 to 55/45. The $Ru(bpy)_3^{+}$-peroxydisulfate reaction has been run in acetonitrile/water mixtures.

In this ECL reaction, light emission occurs over a short time period. In the $Ru(bpy)_3^{3+}$-amine reaction, emission occurred over less than a second, so the post-column reactor needed to be configured to allow the reactants to mix directly in the observation cell.[81]

One point of note is that the $Ru(bpy)_3^{3+}$ [or $Ru(bpy)_3^{+}$] reagent needs to be electrochemically generated shortly before use. For detection of amines, a reservoir of $Ru(bpy)_3^{2+}$ was electrolyzed for 30 to 45 min to convert it to the $Ru(bpy)_3^{3+}$ form which was then added as a post-column reactant.[81, 82]

Detection of Amines

Electrogenerated $Ru(bpy)_3^{3+}$ has been used as a CL reagent for determination of amines. Primary, secondary and tertiary amines are all detected and, in fact, the detection limits improve in that order with tertiary amines detectable at 100-fold lower concentrations than primary amines.[81] It is notable that 1 to 10 pmol detection limits are obtained for tertiary amines because those analytes are difficult to derivatize and detect by other methods. Mixtures of C_1 to C_3 and C_2 to C_6 tertiary *n*-alkyl amines have been separated on a cyano column with acetonitrile/water mobile phases and detected via this CL system.[82]

Determinations Involving Species Labelled with $Ru(bpy)_3^{2+}$

Application of $Ru(bpy)_3^{2+}$ as a CL label has been proposed,[80] but not yet demonstrated. If pre-column derivatizing agents incorporating $Ru(bpy)_3^{2+}$ become available they would prove useful. Recall that the CL reaction is compatible with reverse-phase solvent systems. Detection limits for determination of $Ru(bpy)_3^{2+}$ are very good. By using the $Ru(bpy)_3^{3+}$-oxalate reaction, concentrations from 5×10^{-8} to 1×10^{-5} M can be measured. By using the $Ru(bpy)_3^{+}$-peroxydisulfate reaction, concentrations from 1×10^{-13} to 1×10^{-7} M can be measured.

Electron-Transfer Electrochemiluminescence

Solution phase CL resulting from high energy electron-transfer reactions, often involving aromatic radical ions, has mainly involved electrochemical generation. As a result, this area is traditionally called electrogenerated chemiluminescence or electrochemiluminescence (ECL). In simplest terms, the light-product reaction is

$$R^{\dot{+}} + R^{\dot{-}} \rightarrow R + R^* \rightarrow 2R + \text{light}$$

where the radical cation and radical anion species are produced electrochemically and can be formed from the same initial parent (e.g., both $R^{\dot{+}}$ and $R^{\dot{-}}$ generated from rubrene) or from two different parents (e.g., $R^{\dot{+}}$ from rubrene and $R^{\dot{-}}$ from *p*-benzoquinone).[83, 84] Extensive mechanistic study of numerous chemical systems and environments has occurred, but there has been little application to analytical determinations. For generation and study of such ECL processes there have been two electrochemical approaches. One approach is to use a rapid a.c. potential waveform so that both radical anion and radical cation are generated at the same electrode on alternate cycles. The other approach is to use a d.c. potential to produce radical cations at one electrode and radical anions at the other electrode; in this case the anion and cation need to be transported across the interelectrode gap (by convection or diffusion) to allow the recombination which leads to CL emission.

One study has been directed at using ECL for detection of polycyclic aromatic hydrocarbons in normal phase-solvent systems.[85] The work was done in a thin layer flow cell with close proximity electrodes. One electrode was an optically transparent layer of In_2O_3 doped with Sn and deposited on glass. A d.c. potential was imposed between electrodes. The mobile phase must be aprotic (mixtures of 1,2-dimethoxyethane and *n*-hexane with no electrolyte were used) and free of protic impurities to avoid irreversible side reactions of the radical ions and also corrosion of the electrodes. The mobile phase also must be free from dissolved oxygen because molecular oxygen strongly quenches the fluorescence (and therefore also the ECL) of aromatic hydrocarbons. Because of the extreme potentials applied (10 to 80 V) one must question whether the observed ECL is produced by the normal radical cation–radical anion mechanism; however, the emission spectrum acquired here for rubrene is identical to the emission spectrum of rubrene generated under conventional ECL conditions. The detection limit for rubrene was 10^{-8} M (50 fmol).

A more recent study has been directed at detection in reverse-phase LC.[86–88] Mixtures of polycyclic aromatic hydrocarbons, of pesticides and of phthalate esters were separated on C_{18} columns with acetonitrile/water mobile phases (90/10 to 100/0) containing tetrabutyl ammonium perchlorate electrolyte and detected by ECL. A d.c. potential or very low frequency (0 to 10 Hz) a.c. potential was used. Detection limits for the polycyclic aromatics were 3 to 50 pmol. Curiously, the emission is mainly from the vicinity of the anode even if the anode is located upstream of the cathode by 3 cm. Because of this observation and the extreme conditions used (up to 20 mA and 200 V) there is again question about the mechanism. However, the ECL spectra observed here for perylene and diphenylanthracene match the conventional fluorescence emission spectra for those species.

Bioluminescence

In simplified version, the firefly luciferin/luciferase reaction is[2, 89, 90]

$$\text{luciferin} + \text{ATP} + O_2 \xrightarrow{\text{luciferase}} \text{AMP} + \text{oxyluciferin} + \text{light}$$

where the emission maximum is at 562 nm. The analyte of major importance for this reaction is ATP. Detection limits for ATP range from 10^{-11} to 10^{-14} moles depending upon the purity of the luciferase.

This reaction has been used as the basis for detection of creatine kinase (CK) isoenzymes following anion-exchange separation.[57, 58] The connection comes by using CK to catalyze

$$\text{creatine phosphate} + \text{ADP} \xrightarrow{\text{CK}} \text{creatine} + \text{ATP}$$

thus the ATP produced by action of CK is detected by use of luciferin/luciferase.

For this separation the mobile phase was a linear gradient from 10/90 to 60/40 1.0 M lithium acetate/0.03 M tris, both aqueous at pH 7.4. There was a single post-column addition of a solution containing creatine phosphate, ADP, Mg-acetate, luciferin and luciferase at pH 7.4. Then there was a 2.4 min delay coil prior to the observation cell. In selecting electrolytes for use in the mobile phase one needs to consider the inhibition effects of specific ions on the luciferin/luciferase reaction. For instance, Na^+ is worse than Li^+ and Cl^- is worse than acetate. Also, the bioluminescence signal decreases as the ionic strength increases. This system was demonstrated with separation and detection of the isoenzymes CK-MM, CK-MB and CK-BB in a control sample, and then with a serum sample which had been separately confirmed to contain trace CK-MB.

Conclusions

Because of the separation power of liquid chromatography and the sensitivity of chemiluminescence detection, there has been and will continue to be considerable attention devoted to combining these two techniques. At the risk of being proven inaccurate by time, one can predict research areas that will (or at least should) receive special attention.

One such area is in extension of chemiluminescence detection in liquid chromatography to other analytes, or to detect current analytes more sensitively. Current labelling derivatives can be exploited and others developed. Labels based on luminol (or related compounds), acridan or acridinium esters or $Ru(bpy)_3^{2+}$ either exist or are the subjects of attention. Chemiluminescence detection can be combined with another pre- or post-column reactor which functions to create a CL-detectable form; the glucosidase, glucuronidase and amino acid oxidase enzyme reactors and the photooxygenation reactor described are current examples in this area. "Universal" detection schemes can be explored, as in the ion displacement with luminol.

The second area is in operational improvements. Several possibilities have been demonstrated to avoid the necessity for post-column solution addition by employing chemiluminescence reagents in immobilized or solid state format. These have included the electrochemical generation of luminol and peroxyoxalate chemiluminescence, the electrochemical generation of peroxide and the use of luminol and TCPO immobilized and contained in flow through columns. Exploitation of these techniques and development of similar ones with other chemiluminescence reactions should be undertaken.

References

1. White, E. H., and Roswell, D. F. In J. G. Burr, ed., *Chemi- and Bioluminescence* Dekker, New York, 1985, Chap. 4, pp. 215–244.
2. Seitz, W. R. *CRC Crit. Rev. Anal. Chem.* **13**, 1–58 (1981).
3. Koerner, P. J., Jr., and Nieman, T. A. *Mikrochim. Acta* **II**, 79–90 (1987).
4. Strom, B. M. M.S. Thesis, Iowa State University, Ames, IA, 1974.
5. Nekimken, H. L. Ph.D. Thesis, University of Illinois, Urbana, IL, 1985.
6. Nussbaum, M. A. Ph.D. Thesis, University of Illinois, Urbana, IL, 1987.
7. Nussbaum, M. A., and Nieman, T. A. Unpublished.
8. Delumyea, R., and Hartkopf, A. V. *Anal. Chem.* **48**, 1402–1405 (1976).
9. MacDonald, A. Ph.D. Thesis, University of Illinois, Urbana, IL, 1984.
10. Kawasaki, T., Meada, M., and Tsuji, A. *J. Chromatogr.* **328**, 121–126 (1985).
11. Spurlin, S., and Cooper, M. M. *Anal. Lett.* **19**, 2277–2283 (1986).
12. Gandelman, M. S., and Birks, J. W. *J. Chromatogr.* **242**, 21–31 (1982).
13. Neary, M. P., Seitz, W. R., and Hercules, D. M. *Anal. Lett.* **7**, 583–590 (1974).
14. Seitz, W. R., and Hercules, D. M. In M. J. Cormier, D. M. Hercules, and J. Lee, eds., *Chemiluminescence and Bioluminescence* Plenum, New York, 1974, pp. 427–449.
15. MacDonald, A., and Nieman, T. A. *Anal. Chem.* **57**, 936–940 (1985).
16. Hara, T., Toriyama, M., and Ebuchi, T. *Bull. Chem. Soc. Jpn.* **58**, 109–114 (1985).
17. Hara, T., Toriyama, M., Ebuchi, T., and Imaki, M. *Bull. Chem. Soc. Jpn.* **59**, 2368–2370 (1986).
18. Seitz, W. R. In M. A. DeLuca, ed., *Methods in Enzymology*, Academic, New York, 1978, Vol. LVII, pp. 445–462.
19. Taylor, D. W., and Nieman, T. A. *J. Chromatogr.* **368**, 95–102 (1986).
20. Kiba, N., and Kaneko, M. *J. Chromatogr.* **303**, 396–403 (1984).
21. Koerner, P. J., Jr., and Nieman, T. A. *J. Chromatogr.* **449**, 217–228 (1988).
22. VanDyke, D. A. Ph.D. Thesis, University of Illinois, Urbana, IL, 1986.
23. VanDyke, D. A., and Nieman, T. A. Unpublished.
24. Hool, K., and Nieman, T. A. *Anal. Chem.* **59**, 869–872 (1987).
25. Hool, K., and Nieman, T. A. *Anal. Chem.* **60**, 834–837 (1988).
26. Kricka, L. J., and Carter, T. J. N. In L. J. Kricka and T. J. N. Carter, eds., *Chemical and Biochemical Luminescence*, Dekker, New York, 1982, Chap. 8, pp. 153–178.
27. Sato, M., Yamada, T., and Horikawa, M. *Denki Kagaku oyobi Kogyo Butsuri Kagaku* **51**, 111–112 (1983).
28. Cooper, M. M., and Spurlin, S. R. *Anal. Lett.* **19**, 2221–2230 (1986).
29. VanDyke, D. A., and Nieman, T. A. Unpublished.

30. Maskiewicz, R., Sogah, D., and Bruice, T. C. *J. Am. Chem. Soc.* **101**, 5347–5354, 5355–5364 (1979).
31. Totter, J. R. *Photochem. Photobiol.* **22**, 203–211 (1975).
32. McCapra, F., and Perring, K. D. In J. G. Burr, ed., *Chemi- and Bioluminescence*, Dekker, New York, 1985, Chap. 6, pp. 259–320.
33. Thorpe, G. H. G., Kricka, L. J., and Carter, T. J. N. In L. J. Kricka and T. J. N. Carter, eds., *Clinical and Biochemical Luminescence*, Dekker, New York, 1982, Chap. 3, pp. 21–42.
34. McCapra, F., and Beheshti, I. In K. VanDyke, ed., *Bioluminescence and Chemiluminescence: Instruments and Applications*, CRC Press, Boca Raton, 1984, Vol. I, Chap. 2, pp. 9–42.
35. Weeks, I., and Woodhead, J. S. In L. J. Kricka, P. E. Stanley, G. H. G. Thorpe, and T. P. Whitehead, eds., *Analytical Applications of Bioluminescence and Chemiluminescence*, Academic, New York, 1982, pp. 1985–1988.
36. Weeks, I., Beheshti, I., McCapra, F., Campbell, A. K., and Woodhead, J. S. *Clin. Chem. (Winston-Salem)* **29**, 1474–1479 (1983).
37. Montano, L. A., and Ingle, J. D., Jr. *Anal. Chem.* **51**, 919–926 (1979).
38. Veazey, R. L., and Nieman, T. A. *Anal. Chem.* **51**, 2092–2096 (1979).
39. Veazey, R. L. Ph.D. Thesis, University of Illinois, Urbana, IL, 1980.
40. Klopf, L. L., and Nieman, T. A. *Anal. Chem.* **56**, 1539–1542 (1984).
41. Hinze, W. L., Riehl, T. E., Singh, H. N., and Baba, Y. *Anal. Chem.* **56**, 2180–2191 (1984).
42. Veazey, R. L., and Nieman, T. A. *J. Chromatogr.* **200**, 153–162 (1980).
43. Maeda, M., and Tsuju, A. *J. Chromatogr.* **352**, 213–220 (1986).
44. Veazey, R. L., Nekimken, H. L., and Nieman, T. A. *Talanta* **31**, 603–606 (1984).
45. Klopf, L. L., and Nieman, T. A. *Anal. Chem.* **57**, 46–51 (1985).
46. Carter, T. J. N., and Kricka, L. J. In L. J. Kricka and T. J. N. Carter, eds., *Chemical and Biochemical Luminescence*, Dekker, New York, 1982, Chap. 7, pp. 135–152.
47. Mohan, A. G. In J. G. Burr, ed., *Chemi- and Bioluminescence*, Dekker, New York, 1985, pp. 245–258.
48. Imai, K., and Weinberger, R. *Trends in Anal. Chem.* **4**, 170–175 (1985).
49. Sigvardson, K. W., Kennish, J. M., and Birks, J. W. *Anal. Chem.* **56**, 1096–1102 (1984).
50. Grayeski, M. L. In J. G. Burr, ed., *Chemi- and Bioluminescence*, Dekker, New York, 1985, pp. 469–473.
51. Weinberger, R. *J. Chromatogr.* **314**, 155–165 (1984).
52. Hanaka, N., Givens, R. S., Schowen, R. L., and Kuwana, T. *Anal. Chem.* **60**, 2193–2197 (1988).
53. de Jong, G. J., Lammers, N., Spruit, F. J., Brinkman, U. A. Th., and Frei, R. W. *Chromatographia* **18**, 129–133 (1984).
54. Gooijer, C., VanZoonen, P., Velthorst, N. H., and Frei, R. W. *J. Biolum. Chemilum.*, to be published.
55. Honda, K., Miyaguchi, K., and Imai, K. *Anal. Chim. Acta* **177**, 103–110 (1985).
56. Sigvardson, K. W., and Birks, J. W. *Anal. Chem.* **55**, 432–435 (1985).
57. Mellbin, G. *J. Liq. Chromatogr.* **6**, 1603–1616 (1983).
58. Miyaguchi, K., Honda, K., and Imai, K. *J. Chromatogr.* **303**, 173–176 (1984).
59. Imai, K., Matsunaga, Y., Tsukamoto, Y., and Nishitani, A. *J. Chromatogr.* **400**, 169–176 (1987).
60. Mellbin, G., and Smith, B. E. F. *J. Chromatogr.* **312**, 203–210 (1984).
61. Koziol, T., Grayeski, M. L., and Weinberger, R. *J. Chromatogr.* **317**, 355–366 (1984).
62. Nozaki, O., Ohba, Y., and Imai, K. *Anal. Chim. Acta* **205**, 255–260 (1988).
63. Kobayashi, S. I., Sekina, J., Honda, K., and Imai, K. *Anal. Biochem.* **112**, 99–104 (1981).
64. Honda, K., Miyaguchi, K., and Imai, K. *Anal. Chim. Acta* **177**, 111–120 (1985).
65. Mann, B., and Grayeski, M. L. *J. Chromatogr.* **386**, 149–158 (1987).
66. Grayeski, M. L., and Devasto, J. K. *Anal. Chem.* **59**, 1203–1206 (1987).
67. Sigvardson, K. W., and Birks, J. W. *J. Chromatogr.* **316**, 507–518 (1984).

68. Kwakman, P. J. M., Brinkman, U. A. Th., Frei, R. W., De Jong, G. J., Spruit, F. J., Lammers, G. F. M., and van den Berg, J. H. M. *Chromatographia* **24**, 395–399 (1987).
69. van Zoonen, P., Kamminga, D. A., Gooijer, C., Velthorst, N. H., Frei, R. W., and Gubitz, G. *Anal. Chem.* **58**, 1245–1248 (1986).
70. Honda, K., Miyaguchi, K., Nishino, H., Tanaka, H., Yao, T., and Imai, K. *Anal. Biochem.* **153**, 50–53 (1986).
71. Jansen, H., Brinkman, U. A. Th., and Frei, R. W. *J. Chromatogr.* **440**, 217–223 (1988).
72. Poulsen, J. R., Birks, J. W., van Zoonen, P., Gooijer, C., Velthorst, N. H., and Frei, R. W. *Chromatographia* **21**, 587–595 (1986).
73. Poulsen, J. R., Birks, J. W., Gubitz, G., van Zoonen, P., Gooijer, C., Velthorst, N. H., and Frei, R. W. *J. Chromatogr.* **360**, 371–383 (1986).
74. van Zoonen, P., Kamminga, D. A., Gooijer, C., Velthorst, N. H., and Frei, R. W. *Anal. Chim. Acta* **167**, 249–256 (1985).
75. Gubitz, G., van Zoonen, P., Gooijer, C., Velthorst, N. H., and Frei, R. W. *Anal. Chem.* **57**, 2071–2074 (1985).
76. van Zoonen, P., Kamminga, D. A., Gooijer, C., Velthorst, N. H., and Frei, R. W. *J. Liq. Chromatogr.* **10**, 819–827 (1987).
77. van Zoonen, P., de Herder, I., Gooijer, C., Velthorst, N. H., Frei, R. W., Kuntzberg, E., and Gubitz, G. *Anal. Lett.* **19**, 1949–1961 (1986).
78. Brina, R., and Bard, A. J. *J. Electroanal. Chem.* **238**, 277–295 (1987).
79. Rubinstein, I., Martin, C., and Bard, A. J. *Anal. Chem.* **55**, 1580–1582 (1983).
80. Ege, D., Becker, W. G., and Bard, A. J. *Anal. Chem.* **56**, 2413–2417 (1984).
81. Noffsinger, J. B., and Danielson, N. D. *Anal. Chem.* **59**, 865–868 (1987).
82. Noffsinger, J. B., and Danielson, N. D. *J. Chromatogr.* **387**, 520–524 (1987).
83. Faulkner, L. R. In M. A. DeLuca, ed., *Methods in Enzymology*, Academic, New York, 1978, Vol. LVII, pp. 494–526.
84. Faulkner, L. R., and Glass, R. S. In W. Adam and G. Cilento, eds., *Chemical and Biological Generation of Excited States*, Academic, New York, 1982, Chap. 6, pp. 191–227.
85. Schaper, H. *J. Electroanal. Chem.* **129**, 335–342 (1981).
86. Blatchford, C., and Malcolme-Lawes, D. J. *J. Chromatogr.* **321**, 227–234 (1985).
87. Blatchford, C., Humphreys, E., and Malcolm-Lawes, D. J. *J. Chromatogr.* **329**, 281–284 (1985).
88. Hill, E., Humphreys, E., and Malcolm-Lawes, D. J. *J. Chromatogr.* **370**, 427–437 (1986).
89. Bostick, W. D., Denton, M. S., and Dinsmore, S. R. *Clin. Chem.* (*Winston-Salem*) **26**, 712–717 (1980).
90. Bostick, W. D., Denton, M. S., and Dinsmore, S. R. In K. VanDyke, ed., *Bioluminescence and Chemiluminescence: Instruments and Applications*, CRC, Boca Raton, 1985, Vol. II, Chap. 15, pp. 227–246.

5

The Peroxyoxalate Chemiluminescence Reaction

Richard S. Givens and Richard L. Schowen

The Department of Chemistry and The Center for Bioanalytical Research,
The University of Kansas, Lawrence, Kansas 66045

Introduction

Overview

This chapter is an in-depth analysis of the peroxyoxalate reaction that ultimately produces fluorescence emission from a fluorophore. The primary object of this treatise is to define and evaluate those reaction variables that might influence the application of peroxyoxalate chemiluminescence to chemical analysis at trace levels. For our purposes, the fluorophore serves the role of the analyte. We further have confined the system to a "flowing stream" reaction of the type that one would employ for an HPLC or flow injection chemiluminescence detection.

This chapter, therefore, restricts coverage to the essential parameters and features of the peroxyoxalate chemiluminescence reaction which one would encounter in the preceding applications. No attempt has been made to provide a comprehensive treatment of the analytical applications themselves. Those that are discussed are introduced because of their historical importance or because the application is illustrative of a particular feature or variable of interest.

Historical

While the "flowing stream" reaction is of primary concern here, it is important to note that the forerunner to the current research and applications of peroxyoxalate chemiluminescence was the discovery by Chandross[1] of a bluish-white emission when oxalyl chloride and hydrogen peroxide were reacted in a static solution of dioxane or benzene. He further demonstrated that the reaction successfully sensitized the luminescence from a number of fluorescent compounds, such as 9,10-diphenylanthracene (DPA), anthracene, and *N*-methyl acridone. A hydroperoxy oxalyl chloride intermediate which subsequently decomposed to HCl, CO and O_2 was suggested to be responsible for the luminescent reaction, as well as for the weak emission observed. Chandross[1] also suggested that in the presence of certain fluorescent compounds, the excited triplet of the fluorophore was generated with enough excess energy to cross to the fluorescent singlet state. In the absence of a fluorescer, however, oxalyl chloride serves as the acceptor, emitting the bluish-white light.

The Dioxetanedione Intermediate

Rauhut and his co-workers[2] at Cyanamid subsequently developed the chemiluminescent reactions of aryl oxalate esters and hydrogen peroxide with a variety of fluorophores in quest of a long-lived light source for commercial and military uses. Scheme 1 outlines Rauhut's mechanism[2] for the hypothetical chemical sequence that generates the excited state of the fluorescer. Although these two chemical systems are formally related, the nature of the aryl oxalate chemiluminescent reaction appears to be quite different and to include specific contributions from the phenolic function of the ester (*vide infra*).

An important and attractive feature of the aryl oxalate ester system, particularly when compared with the other chemiluminescent reactions, is its versatility. The extensive work of Rauhut[2] has shown that a wide range of fluorescent compounds will function as the fluorophore. An additional attraction is that among the known nonbiological chemiluminescent reactions, aryl oxalate esters and oxamides are reported to have the highest quantum efficiencies.[3]

The key (and most controversial) feature of the Rauhut mechanism is the role of the dioxetanedione intermediate (5-2), a species that has not been independently detected

$$H_2O_2 + ArO_2CCO_2Ar \xrightarrow[-ArOH]{} \underset{\mathbf{1}}{ArO_2CCO_3H} \tag{5-1}$$

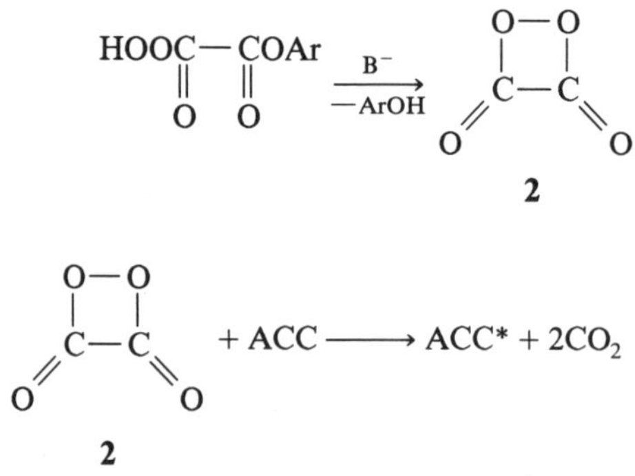

$$\mathrm{HOOC(=O){-}C(=O)OAr} \xrightarrow[-ArOH]{B^-} \underset{\mathbf{2}}{\text{(1,2-dioxetanedione: O–O / O=C–C=O ring)}} \tag{5-2}$$

$$\underset{\mathbf{2}}{\text{(1,2-dioxetanedione)}} + ACC \longrightarrow ACC^* + 2CO_2 \tag{5-3}$$

$$ACC^* \longrightarrow ACC + h\nu_{CL} \tag{5-4}$$

Scheme 1 ■ Rauhut's mechanism[2] for peroxyoxalate chemiluminescence. ACC is the fluorescent acceptor and B^- is a base catalyst.

or characterized (*vide infra*). Nevertheless, the premise upon which most of the studies on improving peroxyoxalate chemiluminescence have been based targets the efficient formation of a dioxetanedione intermediate. For example, the design of new oxalate esters has focused on increasing the electrophilic character at the acyl carbon and on improving the nucleofugal properties of the leaving group in order to enhance the formation of the hydroperoxy half ester **1** and to improve the subsequent closure of the four membered ring. A number of improved oxalate esters were, in fact, discovered by this approach.[4]

The CIEEL Modification

The intermediacy of **2** has had a certain appeal in view of the contemporary work on the chemiluminescent reactions of dioxetanes **3**[5–7] and dioxetanones **4**.[8–11]

3 **4**

These closely related dioxetane containing substrates also chemically excite fluorophores by a well defined chemically initiated electron-exchange luminescence[11] (CIEEL) mechanism initially championed by Schuster.[8] Based on those studies, McCapra[12] expanded Rauhut's mechanism to include the CIEEL process (Scheme 2).

$$ArO_2CCO_2Ar \xrightarrow[-ArOH]{H_2O_2} \underset{\mathbf{1}}{ArO_2CCO_3H} \tag{5-1'}$$

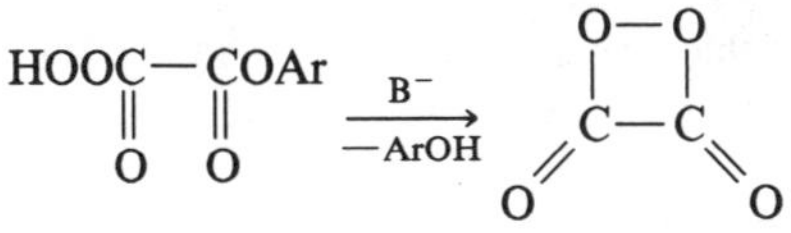

$$HOOC{-}COAr\ (\text{with } C{=}O,\ C{=}O) \xrightarrow[-ArOH]{B^-} \mathbf{2} \tag{5-2'}$$

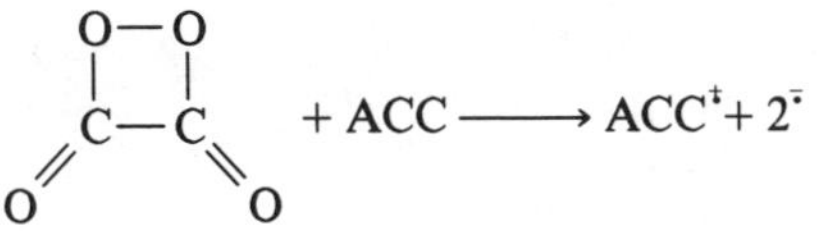

$$\mathbf{2} + ACC \longrightarrow ACC^{\dot{+}} + \mathbf{2}^{\dot{-}} \tag{5-3'}$$

$$\mathbf{2}^{\dot{-}} \longrightarrow CO_2 + CO_2^{\dot{-}} \tag{5-3'a}$$

$$CO_2^{\dot{-}} + ACC^{\dot{+}} \longrightarrow ACC^* + CO_2 \tag{5-3'b}$$

$$ACC^* \longrightarrow ACC + h\nu_{CL} \tag{5-4'}$$

Scheme 2 ■ The Rauhut–McCapra mechanism.[12] ACC is the fluorescent acceptor and B^- is the base catalyst.

Paralleling the mechanism for dioxetanones, the CIEEL mechanism requires a bimolecular oxidation-reduction reaction yielding an ion pair composed of a semidione radical anion and a fluorophore radical cation. Subsequent fragmentation of the semidione yields CO_2 and a reduced carbon dioxide ($CO_2^{\cdot-}$), a stronger reducing agent than the semidione. "Back" electron transfer to the fluorophore radical cation regenerates the acceptor; however, this process generates the fluorophore in its first singlet excited state. The emission that follows is the characteristic fluorescence of the acceptor.

In support of the CIEEL mechanism, McCapra reported that a linear correlation existed between the chemiluminescence intensity and the singlet energy of three anthracene fluorophores, a parallel to the emission intensity dependence on the oxidation potential observed for **4**.[12] This result contrasts, however, with an earlier, more extensive study by Lechtken and Turro,[13] who examined a dozen acceptors with singlet energies ranging between 50 and 105 kcal/mol. They found that no direct correlation existed between the efficiencies (ϕ_{CL}) and the singlet energies (E_S). While these two reports depict inconsistencies concerning the nature of the chemical excitation process, they do further illustrate the wide range of fluorophores amenable to peroxyoxalate activation.

Application of Peroxyoxalate Chemiluminescence to Liquid Chromatography Detectors

The extensive range of energies of excitable fluorophores, the reported high efficiencies, the potential for very high sensitivity and the ready availability of the reagents have attracted researchers' attention to peroxyoxalate chemiluminescence as the energy source for sensitive analytical detectors. Among the applications of the peroxyoxalate chemiluminescent reactions, the detection of analyte from column eluents was first reported by Williams, Huff and Seitz[14] in 1976 for an assay of hydrogen peroxide. The essential feature of their study was the use of a post-column flow reactor connected just prior to the detector. 2,4,6,-trichlorophenyl oxalate (TCPO, 1.3 mM) and perylene (0.2 mM) in ethyl acetate were combined with triethylamine (TEA, 0.35 mM) in methanol and this mixture was mixed with the analyte H_2O_2 as it was eluted from the analytical column and directed into a CL cell where the CL emission from perylene was monitored by a PMT. A limit of detection of 10 nM H_2O_2 with a signal-to-noise (S/N) of 2 and a dynamic range between 10 nM and 1 mM was realized.[14] A limited number of the parameters which affect the CL intensity were also investigated, including the concentration of the fluorescer, the oxalate and the triethylamine catalyst and the effect of the pH and buffer concentration. This first successful application of the peroxyoxalate CL to an automated assay was soon followed by a number of other reported applications to liquid chromatography by Imai (*vide infra*).

The Peroxyoxalate Chemiluminescent Reaction Mechanism

The Intensity/Time Profiles

For over a decade, the Rauhut mechanism (Scheme 1) remained the working hypothesis for the peroxyoxalate reaction. A number of attempts to further characterize, trap or identify the putative dioxetanedione **2** were without success.[15–18] It is now generally

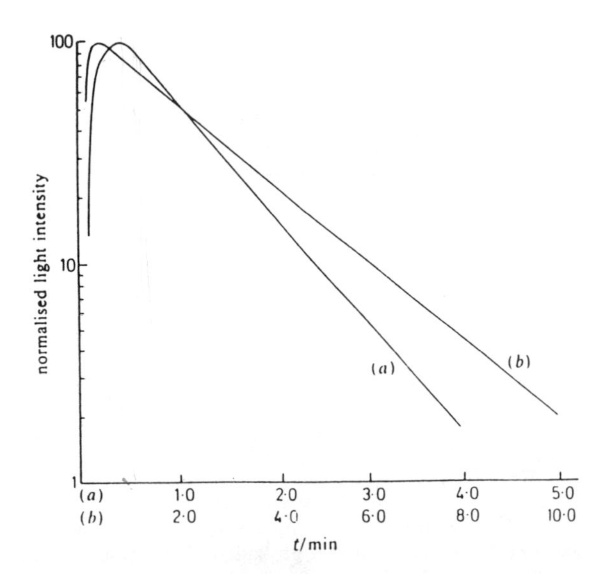

Figure 5-1 ■ Decay of chemiluminescence for reaction mixture containing (a) PCPO (5×10^{-4} M), H_2O_2 (2×10^{-2} M) and DPA (5×10^{-5} M) and (b) PCPO (5×10^{-4} M), H_2O_2 (1.25×10^{-5} M) and DPA (5×10^{-3} M). Reprinted from C. L. R. Catherall, T. F. Palmer and R. B. Cundall, *J. Chem. Soc., Faraday Trans. II* **80**, 823–836 (1984) with the permission of the Chemical Society, London.

conceded that previous reports concerning the identification or detection of **2** were incorrect and that, to date, there exists no independent evidence that **2** is the required, short-lived intermediate in peroxyoxalate chemiluminescent reactions.[18, 19]

In a carefully designed study, Catherall, Palmer and Cundall[19] examined the kinetic parameters and the chemiluminescent quantum efficiencies for the reaction of 2,3,4,5,6-pentachlorophenyl oxalate (PCPO), hydrogen peroxide and sodium salicylate (catalyst) with 9,10-diphenylanthracene (DPA) as the fluorophore. The intensity/time profile from this system rapidly reached a maximum, followed by an exponential decay as illustrated in Fig. 5-1. Because the rise occurred so rapidly, only the decay rate constants (k_f) were determined, and it is from these results that a mechanistic picture was deduced, as illustrated in Scheme 3. This detailed study established first-order dependence for both PCPO and hydrogen peroxide (1 : 1 stoichiometry) and a concentration dependence for the catalyst where a maximum rate was observed between 2 and 5×10^{-5} M sodium salicylate. Increasing the H_2O_2 concentration lowered the chemiluminescent quantum efficiency, however, which asymptotically approached a minimum. The decay rate constants were independent of the DPA concentrations, paralleling the observations of Rauhut *et al.*[2]

Kinetic Modeling

In thinking about intensity/time data in terms of mechanistic models, it is important to keep in mind that the plots are not analogous to the plots of product concentration

$$PCPO + H_2O_2 \longrightarrow X \quad (5\text{-}5)$$

$$PCPO + H_2O_2 \longrightarrow \text{nonchemiluminescent products} \quad (5\text{-}6)$$

$$X \longrightarrow \text{nonchemiluminescence decay} \quad (5\text{-}7)$$

$$X + F \longrightarrow XF\ (\text{complex}) \quad (5\text{-}8)$$

$$XF \longrightarrow F^* + \text{products} \quad (5\text{-}9)$$

$$XF \longrightarrow F + \text{products} \quad (5\text{-}10)$$

$$F^* \longrightarrow F + h\nu \quad (5\text{-}11)$$

$$F^* \longrightarrow F + \Delta \quad (5\text{-}12)$$

Scheme 3 ■ The Palmer–Cundall mechanism for peroxyoxalate chemiluminescence[19] where X is the reactive intermediate formed from hydrogen peroxide and PCPO and is responsible for the chemiexcitation of the fluorophore and F* is the first singlet excited state of 9,10-diphenylanthracene (DPA) as demonstrated independently by comparison of the wavelength-dependent emission spectrum with the fluorescence spectrum of DPA.

vs. time, of the sort one often uses in mechanistic interpretation. The photons may clearly be considered products of the reaction, but intensity/time plots show the rate of production of photons ($d(h\nu)/dt$) rather than the *number* of photons produced. Intensity/time plots are thus rate vs. time plots and it is their integrals which will give yield vs. time, analogous to concentration vs. time. Thus, the intensity at any time $t(I_t)$ is given by the expression

$$I_t = d(h\nu)/dt = k_{11}[F^*] \quad (5\text{-}13)$$

Under conditions of high and therefore constant oxalate concentration, the intensity is proportional to the oxalate, fluorophore and hydrogen peroxide concentration according to the processes outlined in Scheme 3. Kinetic treatment of Eqs. (5-5) to (5-12) assuming steady-state conditions for X, XF and F* leads to the integrated equation

$$I_t = \phi_F \phi_{CT}(k_8[F])/(k_8[F] + k_7)k_5[PCPO]_0[H_2O_2]_0 \exp\{-(k_5 + k_6)[PCPO]_0 t\} \quad (5\text{-}14)$$

where ϕ_F is the fluorescence efficiency of the fluorophore and ϕ_{CT} is the fraction of XF complexes which yield F*. The integration of Eq. (5-14) from $t = 0$ to ∞ provides the expression for the luminescence yield Y given in

$$Y = \phi_F \phi_{CT}[k_8[F]/(k_8[F] + k_7)][k_5/(k_5 + k_6)][H_2O_2]_0 \quad (5\text{-}15)$$

Thus, under the conditions of excess oxalate, the light yield is dependent on the initial hydrogen peroxide concentration. Equation (5-15) predicts that there should be a linear dependence between the total quanta emitted and the hydrogen peroxide

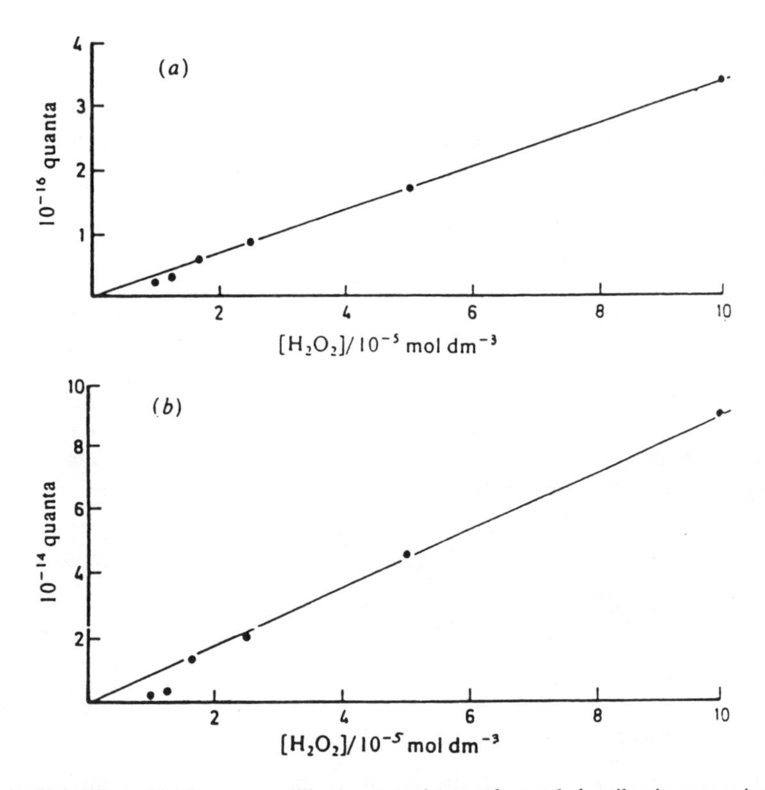

Figure 5-2 ■ Effect of hydrogen peroxide concentration on the total chemiluminescence intermediate of solutions containing (a) 5×10^{-3} M DPA (gradient = 3.4×10^{20} quanta M^{-1}) and (b) 5×10^{-6} M DPA (gradient = 9×10^{18} quanta M^{-1}). Reprinted from C. L. Catherall, T. F. Palmer and R. B. Cundall, *J. Chem. Soc., Faraday Trans. II* **80**, 823–836 (1984) with the permission of the Chemical Society, London.

concentration at constant fluorophore concentration. This was experimentally verified for two different DPA concentrations as illustrated in Fig. 5-2.

The ratio of the slopes for these two DPA concentrations in Fig. 5-2 can be manipulated to obtain the ratio of the two rate constants [Eq. (5-15)] for the decay of the reactive intermediate X, yielding a value of 5.2×10^3 M^{-1} for k_7/k_8. If one assumes, as the authors have, that the bimolecular process (k_8) is diffusion controlled (i.e., 8.8×10^9 M^{-1} s^{-1}), then k_7, the unimolecular decay of the reactive intermediate X, is approximately 1.6×10^6 s^{-1} which translates to a lifetime of ~ 500 ns at 298K.

The quantum efficiency for chemiluminescence (ϕ_{CL}) at a given fluorophore concentration is then determined, since $\phi_{CL} = Y/[H_2O_2]_0$, by

$$\phi_{CL} = Y/[H_2O_2]_0 = \phi_F \phi_{CT} [k_8[F]/(k_8[F] + k_7)][k_5/(k_5 + k_6)] \quad (5\text{-}16)$$

A similar study under pseudo-first-order conditions in hydrogen peroxide revealed that step (5-6) (Scheme 3) is a minor process and that the rate of decay of chemiluminescence is proportional to the initial H_2O_2 concentration. This signals an increase in the importance of side reactions of the intermediate X, possibly involving hydrogen peroxide.

Catherall *et al.*[19] employed this analysis to probe for other properties of the reactive intermediate. Delayed addition of the fluorophore simply resulted in interception of the intensity/time profile without diminution in the intensity observed at the same point on a normal intensity/time plot. The presence of oxygen produced only the expected minimal fluorescence quenching and gave no evidence for diversion of X to other pathways.

The chemiluminescence quantum yields were shown to be proportional to the concentrations of the fluorophore, whereas the decay rates were independent of [F].[19] The authors established the CIEEL mechanistically significant linear relationship between the half wave oxidation potential and the normalized efficiency (ϕ_{CL}/ϕ_F) for four fluorophores (DPA, anthracene, 9,10-dibromoanthracene and 1,2-di-2(5-phenyl-oxazolyl)benzene). From their work, they proposed that the tetrahedral intermediate **5**, a possible precursor to dioxetanedione **2**, is the activator (X in Scheme 3):

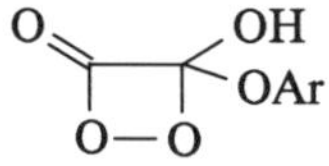

O OH OAr O—O

5 (Ar = C_6Cl_5)

Biphasic Intensity/Time Profiles

In an effort to further explore the important parameters that contribute to the peroxyoxalate reactions, we also have examined a variety of oxalate ester/hydrogen peroxide reactions in diverse environments.[20] The kinetic and mechanistic picture appears to be more complex in media such as ethyl acetate with a minimal water content (for example, with only the water introduced along with the H_2O_2, used as a 30% solution). The kind of experimental data[20] that must be accounted for by any mechanistic proposal are illustrated by Fig. 5-3.

In Fig. 5-3, the time course of the chemiluminescent light intensity is portrayed for three different concentrations of triethylamine as catalyst. Although at high concentrations (e.g., 0.2 mM) the catalyst produces what appears as a rapid burst of light which decays exponentially (curve c) similar to the result of Catherall, Palmer and Cundall[19] (Fig. 5-1), curve fitting of the decay phase of the emission suggests that the decay is not truly exponential. The decay is fitted much more closely by a function which incorporates two or even three exponential components (F. J. Alvarez and B. Matuszewski, unpublished observations). The multiphasic character of the decay is confirmed at lowered catalyst concentrations (curve b, 0.1 mM), where an initial rapid drop is followed by a decay that itself cannot be represented as simply exponential. Finally, at low catalyst concentration (e.g., 0.05 mM, curve a), the emission occurs in two distinct bursts.

This behavior is inconsistent with a mechanism in which the reactants are converted to a single intermediate species which then interacts with the fluorophore to generate light with simultaneous and/or concomitant formation of products. Such a mechanism

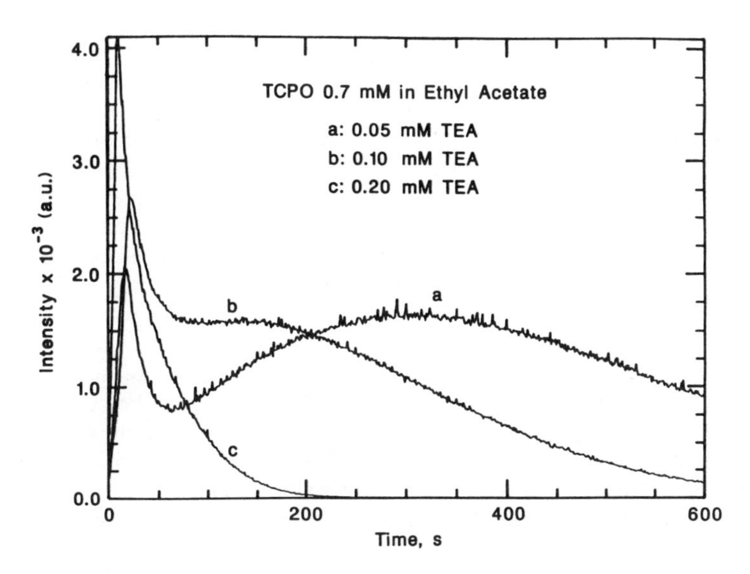

Figure 5-3 ■ Time courses of the chemiluminescence intensity from oxalate-hydrogen peroxide systems in ethyl acetate as solvent. The curves correspond to the three indicated concentrations of the triethylamine (TEA) catalyst. Reprinted from F. J. Alvarez, N. J. Parekh, B. Matuszewski, R. S. Givens, T. Higuchi, and R. L. Schowen, *J. Am. Chem. Soc.* **108**, 6435–6437 (1986) with the permission of the American Chemical Society.

should generate light in an intensity proportional to the concentration of this single intermediate as given by the expression in Scheme 4.

Most simply, the intensity/time curve would then show a single-burst emission characterized by an exponential rise phase and an exponential decay phase (see following text). More complex mechanisms might introduce lag features into the rise or decay, but there is no simple means for such a one-intermediate mechanism to produce more than a single distinct burst of light. Thus, the most venerable of the proposed mechanisms, that in which the reactants are converted to 1,2-dioxetanedione which generates light by interaction with the fluorophore, is excluded by these observations. Indeed, any mechanism, including that of Catherall *et al.*,[19] with only one intermediate that eventually leads to light emission is excluded as a complete accounting of the system.

In fact, mechanisms with only *two* intermediates, both of which lead to light production, are also excluded. Such mechanisms can readily produce multiexponential decay of the light intensity, but they can cannot produce the "valley" in the light intensity. This can be seen qualitatively by the following considerations. Suppose the intermediates occur along parallel pathways of reaction, and we imagine the two bursts to be generated by, first, an intermediate which is both formed and lost rapidly (initial burst) and, second, an intermediate which is formed and lost more slowly (subsequent burst). This mechanism does not serve to reproduce the observed sequence of two bursts, because either the entire supply of reactants will be totally depleted along the first route before the second route begins to produce light or the two light bursts will be additive and produce an apparent single burst. Thus, model calculations according to this mechanism are incapable of producing a pattern of two sequential bursts of light.

Alternatively, therefore, consider the two intermediates to be in series along the same reaction pathway. This model also cannot account for the sequential-burst kinetics because the fall in light intensity from loss of the first intermediate will be counterbalanced by a rise (by no means necessarily an equal rise) in intensity from the coupled increase in the concentration of the second intermediate. Since light production from any two immediately sequential intermediates is necessarily coupled in this way, the result must be a monotonic decay of light intensity. This would not produce the biphasic intensity/time profile.

Differently put, the "valley" between the bursts of emission requires a "storage form" in which material can accumulate, having been lost from the first light-generating species but not yet transformed into the second light-generating species. Such a "waiting room" allows (in principle) for a complete decay of light output from the first intermediate, as material passes from it into the waiting room, before the onset of light output from the second intermediate when material passes out of the waiting room into it.

Kinetic Model

These considerations force a mechanism incorporating at least two light-generating intermediates and at least one "storage" intermediate. The simplest version of such a mechanism is presented in Scheme 4. In Scheme 4, the reactants are converted through the agency of the catalyst to an intermediate X with a pseudo-first-order rate constant k_x. X may suffer three fates: It may proceed to the intermediate Z (rate constant k_z), or it may react to generate products (overall rate constant k_a), either with the production of light (fraction f), or without the production of light (fraction $1-f$).

The intermediate Z is the "storage form" or waiting room and it is thus ascribed no route for light production. In this simplified version of the mechanism, it is given only a single exit route, that to the next intermediate Y (with rate constant k_y). In more complex versions of the mechanism, which would be equally consistent with the data, this intermediate could have a number of fates, including direct formation of products, formation of smaller amounts of light than either X or Y or reversion to X.

The intermediate Y, the second light-generating species, is allowed exit by two routes. The sum of these is characterized by the overall rate constant k_b, the decay branching into pathways either with (fraction g) or without (fraction $1-g$) production of light. Irreversible formation of Y from Z is again a simplification introduced for convenience and not a necessary feature of the model.

In Scheme 4, both light-generating routes, those from X and from Y, are shown as passing through further intermediates X′ and Y′, pictured as the proximal species which interact with the fluorophore to produce the chemiluminescent excited state. The formation of X′ and Y′ from X and Y, respectively, is taken to be slower than their interactions with the fluorophore. This model constitutes one way to account for the observed absence of any dependence of the decay rate constants on the concentration of the fluorophore [diphenylanthracene (DPA) in this system]. For example, Alvarez and Parekh (unpublished experiments), consistent with the previous work of Catherall

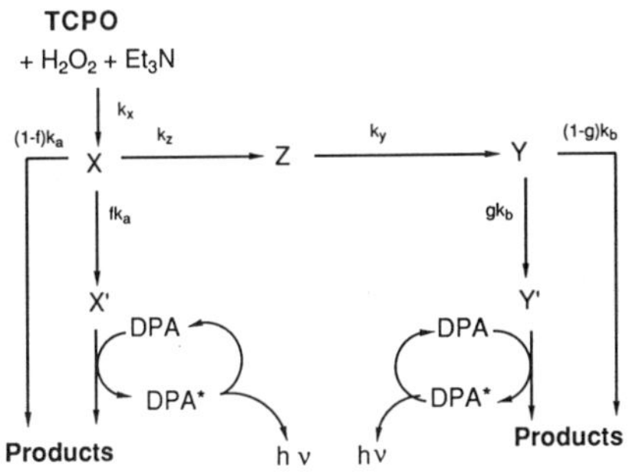

TCPO
+ H_2O_2 + Et_3N

k_x
$(1-f)k_a$
k_z
$X \longrightarrow Z \longrightarrow Y$
k_y
$(1-g)k_b$
fk_a
gk_b
X'
Y'
DPA
DPA*
DPA
DPA*
Products
hν
hν
Products

$$k_T = k_a + k_z$$

$$I = \frac{dh\nu}{dt} = fk_a[X] + gk_b[Y]$$

$$I = OX_0\left\{\frac{fk_ak_x}{(k_T-k_x)}\left(e^{-k_xt} - e^{-k_Tt}\right) + \frac{gk_bk_xk_yk_z}{(k_T-k_x)}\left[\frac{\left(e^{-k_xt} - e^{-k_bt}\right)}{(k_y-k_x)(k_b-k_x)} - \frac{\left(e^{-k_Tt} - e^{-k_bt}\right)}{(k_y-k_T)(k_b-k_T)}\right] - \frac{gk_bk_xk_yk_z}{(k_T-k_y)}\left[\frac{\left(e^{-k_yt} - e^{-k_bt}\right)}{(k_y-k_x)(k_b-k_y)}\right]\right\}$$

Scheme 4 ■ Minimal mechanism for the generation of light in the $TCPO/H_2O_2/TEA$ system. The kinetic expressions derivable from the mechanism are shown at the bottom. $[OX]_0$ means initial concentration of TCPO. Reprinted from F. J. Alvarez, N. J. Parekh, B. Matuszewski, R. S. Givens, T. Higuchi and R. L. Schowen, *J. Am. Chem. Soc.* **108**, 6435–6437 (1986) with permission of the American Chemical Society.

et al.,[19] found that the decay rate constants for the system illustrated in Figs. (5-1) and (5-3) vary by less than 28% when the DPA is varied from 5.0 μM to 5.0 mM. This kinetic order of zero shows that the fluorophore cannot be interacting with X or Y in the steps that chiefly determine the rate of their decay. Thus it is suggested that X and Y are not themselves capable of light-generating interaction with the fluorophore, but that they are actually precursors of the active materials X′ and Y′, which form in the rate-limiting step for decay of X and Y. The intermediates X′ and Y′ must then react rapidly with the fluorophore to generate light. An alternative model would hold that the decay of X and Y does not proceed chiefly by reaction along the light-producing pathways. The light-producing routes could in fact be quite minor, with the main routes of decay being parallel, non-light-producing reactions. This would be consistent with, but not required by, the less than unity (often very low) quantum efficiencies of chemiluminescence generally observed. There is currently no basis for choosing between the serial and parallel models for the kinetic order of zero displayed by DPA in the decay of chemiluminescent intensity.

The mechanistic model of Scheme 4 is described by the following system of differential equations:

$$-d[\text{oxalate}]/dt = k_x[\text{oxalate}] \tag{5-17}$$

$$d[\text{X}]/dt = k_x[\text{oxalate}] - (k_a + k_z)[\text{X}] \tag{5-18}$$

$$d[\text{Z}]/dt = k_z[\text{X}] - k_y[\text{Z}] \tag{5-19}$$

$$d[\text{Y}]/dt = k_y[\text{Z}] - k_b[\text{Y}] \tag{5-20}$$

$$d(h\nu)/dt = fk_a[\text{X}] + gk_b[\text{Y}] = I_t \tag{5-21}$$

where [oxalate] is the TCPO concentration (initial value $[\text{oxalate}]_0$), I_t is the time course of the chemiluminescent light intensity and the k's are pseudo-first-order rate constants which can depend on concentrations of H_2O_2, catalyst and perhaps other variables in ways yet to be established. This system of equations may be solved to generate the expression for I_t shown in Scheme 4, where $k_T = k_a + k_z$.

Ideally, this expression would fit the data shown in Fig. 5-3, for example by the method of least squares, to obtain best-fit values for all the parameters. These are, however, numerous and complex; the equation contains seven adjustable quantities, each of which depends on experimental variables in unspecified ways and in some cases are related to the observables by exponential relationships. The result is that meaningful fits are not readily obtained.

The semiquantitative adequacy of the proposed mechanistic model can, however, be demonstrated with the use of the kinetic formulation of Scheme 4. Figure 5-4 shows three calculated curves to be compared with the experimental curves in Fig. 5-3. The curves in Fig. 5-4 were obtained by an elementary form of kinetic simulation in which the analytical solution to the kinetic system (Scheme 4; $I_t/[\text{oxalate}]_0$) was plotted with selected values of the kinetic parameters. These parameters for each of the three curves a, b and c, expressed in arbitrary units of reciprocal time, are

Curve	f	g	k_x	k_z	k_y	k_a	k_b
a	1.0	1.0	6.0	6.5	0.3	1.0	0.5
b	1.0	1.0	7.0	4.5	0.8	0.5	1.45
c	1.0	1.0	12.0	3.5	2.4	0.5	0.5

These values were chosen quite arbitrarily and it is certain that other sets of parameters would produce very similar simulated curves. Nevertheless, the resemblance between the experimental and calculated curves in general appearance is strong. In fact, the simulation suggests that a model with these general features should be capable of being reduced to quantitative agreement with experiment.

Some features of the simulation are instructive. The experimental variable that produced differences among the three experimental curves was triethylamine (TEA) concentration. The value of [TEA] changes by twofold between curves a and b, and by twofold between curves b and c. Note that in the simulations, no rate constants except for k_y and k_b were changed by more than twofold between adjacent curves, i.e., curves for conditions differing by twofold in [TEA]. The values of k_y and k_b were both

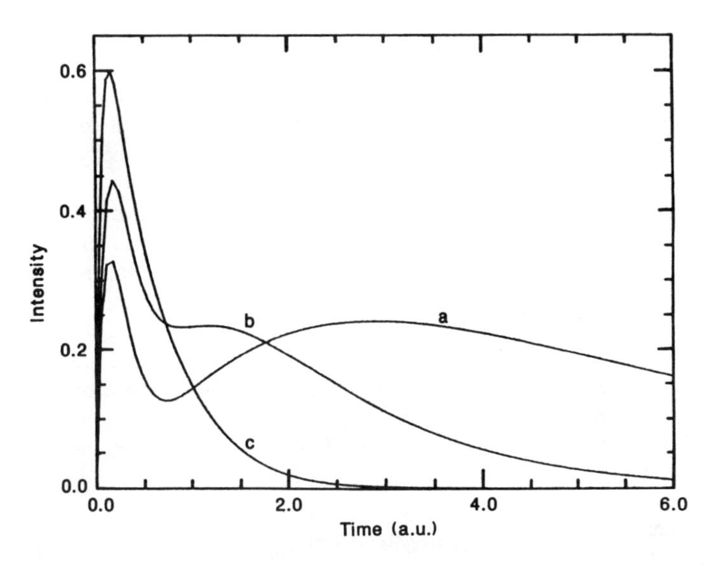

Figure 5-4 ■ Calculated simulations of the curves in Fig. 5-3, with use of the kinetic model of Scheme 3 and the parameters given in the text. Reprinted from F. J. Alvarez, N. J. Parekh, B. Matuszewski, R. S. Givens, T. Higuchi and R. L. Schowen, *J. Am. Chem. Soc.* **108**, 6435–6437 (1986) with the permission of the American Chemical Society.

changed by about threefold between adjacent curves, while the remaining constants were changed by factors of 1.2 to 2.0. All of these changes correspond to positive or negative kinetic orders in TEA that are not far from unity. This is exceedingly realistic, since a number of the chemical processes possibly involved in the overall reaction scheme should be either accelerated by base (e.g., nucleophilic attack at carbonyl by H_2O_2) or inhibited by base (e.g., electron transfer to $-CO-OOH$). Acceleration should generate a positive kinetic order of unity in TEA, inhibition a negative kinetic order of unity.

It may seem surprising that such drastic alterations in the general character of the I_t curves should result from changes in the component rate constants that are only two- to fourfold. That this occurs can be understood by examining ratios of rate constants between curves a and c (fourfold change in [TEA]). Note that the quantity k_z/k_y changes from 22 (curve a) to 1.5 (curve c). When the influx to Z is 22 times faster than efflux, Z accumulates and the "valley" of curve a appears. When influx and efflux become nearly equal, material is transformed essentially from X directly to Y and there is no "valley," merely the monotonic picture of curve c. Also, the quantity k_y/k_b changes from 0.6 (curve a) to 4.8 (curve c). This ratio measures the influx/efflux ratio for Y. When it is smaller than unity, the final intermediate Y disappears rapidly and the curve falls quickly to zero emission (curve a). When the ratio is large, material accumulates in Y and slowly "leaks out" through the chemiluminescent route.

Although the model lacks the capacity at this point to render a quantitative and conceptually satisfying account of the kinetic characteristics of the reaction, it does permit the following conclusions to be drawn:

1. More than a single material formed from the initial reactants must give rise in the end to chemiluminescence, although it need not be true that the intermediates present in highest concentration are those that interact directly with the fluorophore.
2. In addition to the two or more light-generating intermediates, at least one intermediate must be present which leads to no light or much less light than the other intermediates, and which serves as a "storage form" between one or more "early" intermediates and one or more "late" intermediates, both of which produce light.
3. The balance in concentrations, formation rates and decay rates among these, and thus the duration, frequency and amplitude of the emitted pulses of chemiluminescence, can be strongly affected by relatively small alterations in reaction conditions such as catalyst concentration, solvent composition, etc.

Simplified Time-Course Behavior in Mixed Aqueous Solvents: The Pooled Intermediates Model

The data in Fig. 5-5[21] are characteristic of the behavior of this system in the acetonitrile-water mixtures characteristic of reverse-phase HPLC mobile phases and, therefore, the relevant solvents for the application of chemiluminescent methods to analytical applications in chromatographic effluents. The curves are quite reminiscent of curve c in Fig. 5-3. Curve c, although the simplest in appearance of the curves in Fig. 5-3, is actually of some complexity, the best-fit function for the decay side of the curve requiring at least a biexponential function. The same is true in mathematical detail of

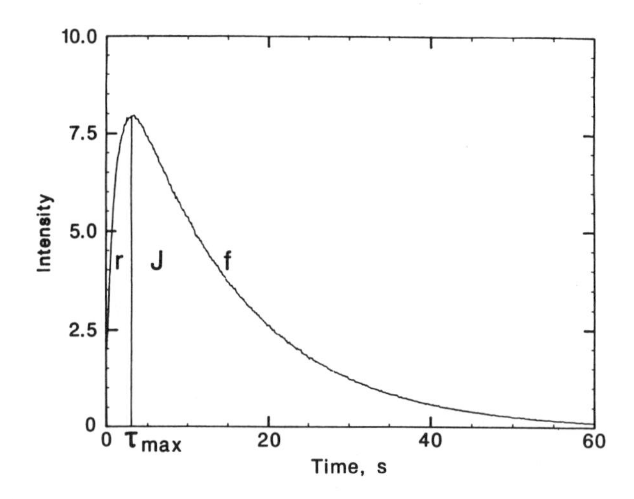

Figure 5-5 ■ Typical time/intensity profile for the same system as in Fig. 5-3, but now in 75% acetonitrile/25% water. See text for definitions of r, f, J and τ_{max}.

the curve shown in Fig. 5-5. Nevertheless, an account of these curves which is adequate for many purposes[22] can be achieved by the procedure outlined in the following text.

The complex mechanism presented in Schemes 1 to 4 can be viewed in vastly simplified form in terms of three pools of substances:

$$A \xrightarrow{r} B \xrightarrow{f} C \tag{5-22}$$

The reactant pool A is converted with pseudo-first-order rate constant r into a pool of intermediate substances B, which are in turn converted with rate constant f into the product pool C. The intermediate pool B is assumed to be rapidly mixed so that its components may be regarded as generating chemiluminescence as if from a single substance (clearly the model may be applied only to circumstances where this is accurate to within acceptable limits, and conversely its success in application to any set of circumstances is presumptive evidence that such a situation prevails). Since B will rise with rate constant r (= rise) and fall with rate constant f (= fall), only a single pulse of light can be described.[22] The model can be represented by the following system of differential equations:

$$-d[A]/dt = r[A] \tag{5-23}$$

$$d[B]/dt = r[A] - f[B] \tag{5-24}$$

$$d[C]/dt = f[B] \tag{5-25}$$

The solution for $[B]$ is

$$[B] = \{[A]_0 r/(f-r)\}\{e^{-rt} - e^{-ft}\} \tag{5-26}$$

This can be quantitatively fitted[22] to curves such as those in Fig. 5-5 with an acceptable precision for many cases in aqueous solution, if $[B]$ is taken as above to be proportional to the chemiluminescence intensity,

$$d(h\nu)/dt = \phi[B] \tag{5-27}$$

Reformulating the equation in terms of I_t yields

$$I_t = \{Mr/(f-r)\}\{e^{-rt} - e^{-ft}\} \tag{5-28}$$

where M is a theoretical maximum level of intensity if the reactants were entirely converted to a chemiluminescence-generating material and r and f are, respectively, the first-order rate constants for the *rise* and *fall* of the burst of chemiluminescence.

A typical application of the *pooled-intermediate* model to the TCPO reaction as a function of the hydrogen peroxide concentration is illustrative. As shown in Table 5-1, the rise (r) and fall (f) rate constants are essentially independent of the hydrogen peroxide concentration as it is increased from 1.25 to 30 mM. It should be noted that the rise rate is about 20 to 30 times faster than the fall rate for the TCPO system under these reaction conditions. Thus, for cases which are appropriate, it becomes possible to describe the luminescent observations in terms of only three parameters. For practical purposes, this may have great advantages and, as we shall see, there are also useful advantages from the viewpoint of kinetic and mechanistic investigations.

A further advantage of this pooled-intermediates model is that it allows determination of approximate values of the parameters with minimal labor and, conversely, allows for prediction of many interesting characteristics of the luminescent process. Consider the following features of the light bursts and their relationships to the

Table 5-1 ■ Effect of the Hydrogen Peroxide Concentration on the Rate Constants and Other Properties for CL Reaction in 75% Aqueous Acetonitrile[a]

$[H_2O_2]$ (mM)	r (s^{-1})	f (10^{-2} s^{-1})	J (cm)	Φ (10^{-4} Einstein/mol)
1.25	—	3.42(0.20)	3.8(0.2)	7.8(0.17)
2.5	0.72(0.04)	3.57(0.15)	5.4(0.1)	9.14(0.29)
5.0	1.04(0.08)	3.16(0.23)	6.9(0.3)	11.83(1.09)
10.0	1.20(0.02)	3.29(0.18)	7.9(0.4)	12.72(0.89)
30.0	2.03(0.18)	3.12(0.47)	8.7(0.8)	13.37(0.74)
60.0	—	4.34(0.61)	9.6(0.4)	11.29(0.37)
120.0	—	6.29(0.76)	9.8(0.8)	8.11(0.59)

[a] [TCPO] = 0.50 mM; [DPA] = 0.25 mM; [imidazole] = 2.5 mM; pH = 7.0 at 25°C. See text for definition of r, f and J.

Source: Orlovií, M., Schowen, R. L., Givens, R. S., Alvarez, F., Matuszewski, B., Parekh, N. *J. Org. Chem.* **54**, (1989) in press.

parameters M, r and f:

intensity at the maximum $$J = M(f/r)^{(f/[r-f])} \quad (5\text{-}29)$$

time of maximum intensity $$\tau_{max} = \{\ln(f/r)\}/(f-r) \quad (5\text{-}30)$$

A simple advantage of the model is that if insufficient data are available to permit a direct determination of the three model parameters from the data, they can be estimated from these relationships. For example, it not uncommonly occurs that the rise time of a burst is too fast to allow direct fitting to obtain a good value of r (as noted in the study of Catherall *et al.*[19] *vide supra*), but the decay is usually slow enough to produce good calculated values of f. It is also the common experience that the time and intensity at the burst maximum J and τ_{max} are well defined. This being so, r can be calculated from τ_{max} and f by solution of the equation for τ_{max} given in the foregoing text. (This is a transcendental equation and cannot be solved in closed form, but it can readily be solved by successive approximations.) Now that J, r and f are known, M can be calculated from Eq. (5-29) for J. Thus, even if the data-collection conditions are quite unfavorable, values of the model parameters can be calculated with reasonable ease.

In addition, the model permits an estimate of the total light yield from the reaction after the collection of only a part of the emitted light. The light yield Y is given by

$$Y = \int_0^{\infty} I_t \, dt = M/f \quad (5\text{-}31)$$

Thus, if enough data can be obtained for good estimates of M and f to be made, then Y can be calculated from this equation. For the TCPO/H_2O_2 system, the quantum efficiency in 75% aqueous acetonitrile is only 10^{-4} Einsteins/mol. The efficiency is moderately dependent on the hydrogen peroxide concentration, as is the emission peak height (Table 5-1).

The model is practically useful, therefore, for analytical chemistry, in that the characteristics of the light burst can be predicted for any set of conditions once it is established how the parameters M, r and f vary with reaction conditions. Ideal analytical conditions for detection purposes can thus be established through the use of calibration results.

The model is also useful for fundamental kinetic studies. The parameters, particularly r and f, can be calculated as a function of experimental conditions. Since r and f, in correctly designed experiments, are pseudo-first-order rate constants, their dependence on the concentrations of reagents and catalysts contains much mechanistic information.

The success of the pooled-intermediates model in treating the data for highly aqueous systems suggests that intermediate species are interconverting in these media with sufficient rapidity to satisfy the requirements of the model. In particular, passage through "dark" intermediates, such as Z in the complex model of Scheme 4, appears to be fast and kinetically insignificant.

Considerations of Peroxyoxalate-Derived Chemiluminescence in HPLC Detector Applications

We begin with the proviso that the most common applications of HPLC to separations have been with reverse-phase columns (RP-HPLC) which generally involve mixed aqueous-organic solvent systems. Post-column addition of the chemiluminescence reagents, specifically the oxalate ester and hydrogen peroxide, to the column eluent

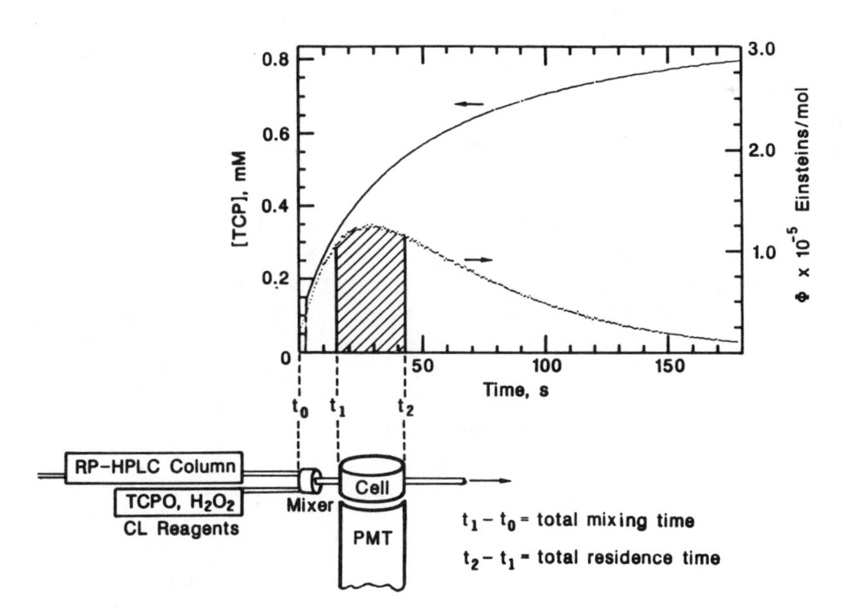

Figure 5-6 ■ Schematic of the fraction of chemiluminescence time profile observed in a flowing stream detector.

containing the fluorophore generates the chemiluminescence during its passage through the detector flow cell. A conventional photomultiplier tube (PMT) for detecting the CL emission is the principal component of a chemiluminescence detector.

The chemical reaction occurs in a continuous flow reactor which is comprised of the mixing chambers, the connecting tubing and the flow cell. The chemical reaction itself parallels the pooled-intermediate model that was applied to the *time* course of the aqueous-organic solvent systems previously outlined.

In order to maximize the sensitivity of this detector, a significant fraction of the total emission must be captured by the PMT, which may be optimized by a number of techniques, e.g., (1) placing polished concentric mirrors behind the cell to more efficiently collect the radiant energy and (2) increasing the residence time of the premixed reagents in the flow cell relative to the pre- and post-cell residence times. The fraction of light captured by the cell is a function of the residence time ($t_2 - t_1$, Fig. 5.6) and can be determined from the ratio of the integrated intensity during passage of the reagents through the cell relative to the total light produced [Eq. 5-32)]. In this model, the fraction of emission captured relative to the total emission generated under a specific set of reaction conditions is derived from the pooled-intermediate model. Finally, the total quanta captured by the detector will be a function of two factors, the quantum efficiency for the chemiluminescent reaction and the capture efficiency of the detector. This is formalized in expression (5-33):

$$\frac{Y_{cap}}{Y_{tot}} = \frac{\int_{t_1}^{t_2} I_t \, dt}{\int_0^{\infty} I_t \, dt} = \frac{Mr/(f-r) \int_{t_1}^{t_2} (e^{-rt} - e^{-ft}) \, dt}{Mr/(f-r) \int_0^{\infty} (e^{-rt} - e^{-ft}) \, dt}$$

$$= \frac{r(e^{-ft_1} - e^{-ft_2}) - f(e^{-rt_1} - e^{-rt_2})}{r-f} \qquad (5\text{-}32)$$

$$\Phi_{cap} = \frac{Y_{cap}}{Y_{tot}} \cdot \Phi_{CL} \qquad (5\text{-}33)$$

The application of the peroxyoxalate method for an HPLC detection system was first reported by Kobayashi and Imai[23] in 1980. The basic design was that of a post-column flow reactor into which were pumped solutions of 5 mM TCPO in ethyl acetate and 10 mM H_2O_2 in acetone. The resulting reaction mixture was then combined with the HPLC column eluent in a second mixing chamber and fed directly into a commercial HPLC-fluorescence detector with the source turned off. The column eluent contained 0.05 M Tris-HCl buffer, which also served as the catalyst for the chemiluminescence reaction. Whereas the individual components, reagents and solvents employed by Imai have since been modified to achieve even greater sensitivity, the basic design for the peroxyoxalate-based chemiluminescence system has remained essentially unchanged.

The initial study by Kobayashi and Imai[23] achieved detection limits for dansylated amino acids as low as 10 fmol using $TCPO/H_2O_2$, levels of detection which indicated that chemiluminescent detection holds considerable promise for specific sensitive assays. Kobayashi and Imai correctly noted several of the contributing factors that impinge upon the sensitivity of this method, including:

1. The need to eliminate pump noise from the reactor feeds.[34]
2. The sensitivity to the nature and purity of the solvent.[23, 34]
3. The degree of solvent and reagent mixing[23, 24] and reagent solubility.[23]
4. The flow rates and reaction volumes, including mixing chambers and the connecting hardware.[23, 24]
5. The reagent concentrations.[23, 24]
6. The detector cell geometry and volume.[23, 24]
7. The pH of the column eluent.[33]
8. The quenching of the chemiluminescence by the reagents and products (phenols).[25]

A number of these parameters were optimized for the detection of the dansylated amino acids and oligopeptides.[23, 26] Several subsequent studies from Imai and other laboratories have reported the use of peroxyoxalate chemiluminescence detection for fluorescamine-labelled catecholamines,[24] 4-fluoro-7-nitrobenzo-2-oxa-1,3-diazole (NBD-F)-labelled primary and secondary amines[28] and ammonium 7-fluorobenzo-2-oxa-1,3-diazole-4-sulfonate (SBD-F) labelled thiols.[29] Birks has reported chemiluminescence detection of polyaromatic hydrocarbons (PAHs) and nitro- and amino-substituted PAHs.[30a,b] Variation of the aryl oxalate reagent such as bis(2,4-dinitrophenyl)oxalate (DNPO) for dansylalanine,[25] along with a variety of other HPLC applications have appeared.[33–35]

Others have also examined the effects of the preceding list of factors on polyaromatic hydrocarbon analyses.[30, 36, 37] Sigvardson and Birks[30b] noted that a background emission was observed from TCPO and H_2O_2 in the absence of any fluorophore and independent of the nature of the solvent (acetone, acetonitrile or ethyl acetate). The background emission displayed maxima at 440 and 540 nm and could not be induced by direct excitation of putative impurities in these solvents.

Both Imai, Miyaguchi and Honda[34] and Sigvardson and Birks[31] showed that a general trend of increasing chemiluminescence efficiency with decreasing oxidation potential or singlet excitation energy exists for a variety of fluorescent derivatives. In both studies, the chemiluminescence yield was obtained from the chromatographic response of the fluorophore and was corrected for several of the factors affecting the chromatographic response. The correlations failed for selected fluorophores, indicating incomplete accounting for the deviating factors. Nevertheless, correlations within a narrowly defined series and the overall trend appear consistent with an electron-transfer mechanism of the CIEEL type for the chemical activation step. Additional studies are needed, however, in order to clearly establish the relationship between oxidation potential and the chemiluminescence efficiencies.

de Jong, Lammers, Spruit, Brinkman and Frei[38, 39] have systematically examined four commercial fluorescence detectors as chemiluminescence detectors. The effect of increasing the cell volume on the peak height and width was determined for DNPO and TCPO. An increase from 70 to 300 μl caused only a 10% increase in the band width and a sixfold increase in the peak height using TCPO. Doubling the volume to 600 μl had no further significant effect on the peak height, however. With DNPO, the increase in cell volume had very little effect on either the peak height or width. The origin of these two responses to the volume change was traced to the rates of the two oxalate reactions, the faster DNPO reaction being less sensitive to the volume of the cell.

The lack of significant band broadening with an increase in the cell volume for TCPO was attributed to a "chemical band-narrowing effect" caused by the relatively rapid chemiluminescence decay. The effect could be simulated by superimposing the first-order decay of the chemiluminescence on the flow pattern of the analyte mixture of two closely eluting fluorophores.

The fraction of light captured by the photomultiplier was approximated from the first-order decay of the chemiluminescence and the cell volume. The general expression obtained is shown in Eq. (5-34), a much simplified treatment of the TCPO chemiluminescence kinetics (*vide supra*):

$$\frac{S_m}{S} = 2^{(-t_d/t_{1/2})} - 2^{(-t_r/t_{1/2})} \tag{5-34}$$

where S_m is the "captured" chemiluminescence, S is the total light emitted, t_d is the delay time before the reactants enter the detector, t_r is the total time until the analyte leaves the detector and $t_{1/2}$ is the half-life of the chemiluminescent reaction. This expression, which does not take into account the rise portion of the chemiluminescence profile [r in Eq. (5-32)], predicts that approximately 25% of the emission should be available to the photomultiplier with a 5 μl cell in a miniaturized HPLC system using DNPO and H_2O_2. Yet, the measured captured emission was only 0.02%, a thousand times less than predicted. A part of this discrepancy is undoubtedly due to the incomplete kinetic expression used for the DNPO chemiluminescence. These studies also predicted an optimum pH of 4.3 for TCPO (Table 5-2), which is clearly at odds with the empirical observations of Honda, Sekino and Imai[25] (pH ~ 8.0) and Weinberger[36] (pH ~ 7.5). The greatest light yield and fraction of the light captured, however, is at pH 8.2, as indicated in Table 5-2,[39] consistent with the earlier studies.

Table 5-2 ▪ Influence of pH of the Mobile Phase on the Fraction of Captured Chemiluminescence Emission.[a]

pH[b]	$t_{1/2}$ (s)	S_m	S	$S_m/S \cdot 100$ (%)
8.2	4.9	27.4	274	10
6.9	5.8	26.0	307	8.5
4.3	32.9	18.7	1195	1.6
2.8	176.9	2.2	749	0.3
2.5	209.1	1.5	624	0.2

[a] Half-lives were measured by stopped flow, S_m was determined chromatographically and t_d was assumed to kt_0. S was calculated from Eq. (5-34).

[b] Detection system: 10 mM TCPO in ethyl acetate (1 ml/min); 1.5 M H_2O_2 in THF (2 ml/min); LC eluate (1 ml/min); flow cell, 50 μl.

Source: G. L. de Jong, N. Lammers, F. J. Spruit, R. W. Frei and U. A. Th. Brinkman, *J. Chromatogr.* **353**, 249–257 (1986).

Table 5-3 ■ Calculated Response for HPLC Detector in Percent CL Captured as a Function of Cell and Mixing Volumes.

Mixing Volume[a] (μl)	Cell Volume (μl)	$Y_{cap}/Y \times 100$ (%)[b]
20	5	0.7
	60	9.8
	300	42
40	5	0.8
	60	10.0
	300	41
100	5	2.1
	60	9.1
	300	37

[a]Reagent mixing chamber volume plus the volume of the lines connecting the mixer to the cell. Total flow rate is 1 ml/min. Tubing is 0.11 mm i.d. Mixing chamber is 20 μl.
[b]Calculated from Eq. (5-32) using conditions from the third entry in Table 5-1. M, the hypothetical maximum emission intensity, as well as r and f.

A similar treatment of the efficiency for light capture (Y_{cap}/Y) as a function of the mixing volume and the cell size is given for the pooled-intermediates model in Table 5-3. In principle, any chemical parameter that can be studied by a stopped-flow or other time-based method and any physical variable of the detector can be evaluated by this model. It is noteworthy that the only constants needed are the rise and fall rate constants in order to determine the efficiency of the light capture. To determine the maximum response of a chemical detector system, however, one must also determine

Table 5-4 ■ Effect of the Imidazole and Hydrogen Peroxide Concentrations on the Rate Constants and Light Capture for HPLC Detection in 75% Aqueous Acetonitrile[a]

[Imidazole] (mM)	r (s^{-1})	f ($10^{-2}\ s^{-1}$)	Y_{cap} (%)	Φ (10^4 Einstein/mol)	Φ_{cap} (10^4 Einstein/mol)
			5 mM H_2O_2		
10.0	8.3(1.1)	17.1(2.8)	31	7.2	2.2
5.0	1.9(0.1)	6.1(0.6)	18	12.5	2.2
2.5	0.99(0.03)	3.5(0.2)	11	13.5	1.5
1.25	0.47(0.01)	3.0(0.1)	8.3	9.8	0.8
0.625	0.29(0.01)	2.8(0.1)	6.5	4.6	0.3
			10 mM H_2O_2		
2.5	2.0(0.2)	3.4(0.6)	11	14	1.5
1.25	1.2(0.1)	2.2(0.1)	7	19	1.4
0.625	0.60(0.02)	2.1(0.1)	6.4	13	0.8

[a][TCPO] = 0.50 mM; [DPA] = 0.25 mM; pH = 7.0 at 25°C. See text for definition of r, f and Y_{cap}.
Source: Orlovií, M., Schowen, R. L., Givens, R. S., Alvarez, F., Matuszewski, B., Parekh, N. *J. Org. Chem.* **54**, (1989) in press.

the total quanta captured, which is a function of the quantum efficiency for chemiluminescence and the light capture efficiency as given in Eq. (5-33).

As noted in Table 5-3, taking a narrow slice with the 5 μl cell of the intensity/time curve (see Fig. 5-6) leads to a highly variable (0.7 to 2.3%) capture of the CL emission. Standard CL detector cells of ~ 60 μl give a small, but relatively constant fraction of the emission signal. Larger cells (300 μl), as noted by de Jong *et al.*[39] improve the collection efficiency for a particular reagent combination and flow system design and appear to show little variation in resolution with changes in the mixing volume. For the DNPO system, the sensitivity to the reaction parameters and detector design is much greater than that outlined for TCPO.

Variation of the imidazole and H_2O_2 concentrations will greatly alter the capture efficiency as indicated in Table 5-4. The response for a specifically configured detector, however, will depend not only on this parameter but also on the absolute quantum efficiency as a function of the imidazole concentration. From this analysis, it is apparent that the maximum response is at the highest imidazole concentration and the lower of the two H_2O_2 concentrations explored in this study (the first and second entries in Table 5-4).

Conclusions

Peroxyoxalate chemiluminescence continues to hold promise as a method for sensitive detection of certain classes of analytes. We have presented a semiquantitative model for defining and optimizing the parameters important for the application to HPLC detector technology and have applied it to a selected number of possible configurations and chemiluminescence reaction situations. Future studies in these laboratories will be directed toward a clearer understanding of the reaction parameters and quantitative determination of the detector parameters for optimum sensitivity. Chemiluminescence detection methodology should continue to be an intriguing and useful approach to attaining high sensitivity with low noise in those selected systems where it will be applicable.

Acknowledgements

Generous support for the research carried out at the University of Kansas by the Center for Bioanalytical Research, Oread Laboratories, Inc. and the Kansas Technology Exchange Commission is gratefully acknowledged.

References

1. Chandross, E. A. *Tetrahedron Lett.*, 761–766 (1963).
2. Rauhut, M. M., Bollyky, L. J., Roberts, B. G., Loy, M., Whitman, R. H., Iannotta, A. V., Semsel, A. M., and Clarke, R. A. *J. Am. Chem. Soc.* **89**, 6515–6522 (1967).
3. Tseng, S.-S., Mohan, A. G., Haines, L. G., Vizcarra, L. S., and Rauhut, M. M. *J. Org. Chem.* **44**, 4113–4116 (1979).

4. Rauhut, M. M. *Acc. Chem. Res.* **2**, 80 (1969).
5. Kopecky, K. R., Van den Sande, J. H., and Mumford, C. *Can. J. Chem.* **46**, 25 (1968).
6. McCapra, F. *Chem. Commun.*, 155 (1968).
7. McCapra, F. *Pure Appl. Chem.* **24**, 611 (1970).
8. Schuster, G. B. *Acc. Chem. Res.* **12**, 366–373 (1979).
9. Schmidt, S. P., and Schuster, G. B. *J. Am. Chem. Soc.* **102**, 306–314 (1980).
10. Adam, W., and Cueto, O. *J. Am. Chem. Soc.* **101**, 6511-6515 (1979).
11. McCapra, F., Perring, K., Hart, R. J., and Hann, R. A. *Tetrahedron Lett.*, 5087–5090 (1981).
12. Schuster, G. B., and Schmidt, S. P. *Adv. Phys. Org. Chem.* **18**, 187–238 (1982).
13. Lechtken, P., and Turro, N. J. *Mol. Photochem.* **6**, 95–99 (1974).
14. Williams, D. C., Huff, G. F., and Seitz, W. R. *Anal. Chem.* **49**, 432 (1976). See also Scott, G., Seitz, W. R., and Ambrose, G. *Anal. Chim. Acta* **115**, 221–228 (1980).
15. Maulding, D. R., Clarke, R. A., Roberts, B. G., and Rauhut, M. M. *J. Org. Chem.* **33**, 250–254 (1968).
16. Cordes, H. F., Richter, H. P., and Heller, C. A. *J. Am. Chem. Soc.* **91**, 7209 (1969).
17. DeCorpo, J. J., Baronavski, A., McDowell, M. V., and Saalfeld, F. E. *J. Am. Chem. Soc.* **94**, 2879–2880 (1972).
18. Chang, M.-M., Saji, T., and Bard, A. J. *J. Am. Chem. Soc.* **99**, 5399–5403 (1977).
19. Catherall, C. L. R., Palmer, T. F., and Cundall, R. B. *J. Chem. Soc., Faraday Trans. II*, **80**, 823–836, 837–849 (1984).
20. Alvarez, F. J., Parekh, N. J., Matuszewski, B., Givens, R. S., Higuchi, T., and Schowen, R. L. *J. Am. Chem. Soc.* **108**, 6435–6437 (1986).
21. Hanaoka, N., Givens, R. S., Schowen, R. L., Kuwana, T. *Anal. Chem.* **60**, 2193–2197 (1988).
22. Orlovií, M., Schowen, R. L., Givens, R. S., Alvarez, F., Matuszewski, B., Parekh, N. *J. Org. Chem.* **54**, (1989) in press.
23. Kobayashi, S., and Imai, K. *Anal. Chem.* **52**, 424 (1980).
24. Kobayashi, S.-I., Seking, J., Honda, K., and Imai, K. *Anal. Biochem.* **112**, 99–104 (1981).
25. Honda, K., Sekino, J., and Imai, K. *Anal. Chem.* **55**, 940–943 (1983).
26. Miyaguchi, K., Honda, K., and Imai, K. *J. Chromatogr.* **303**, 173 (1984).
27. Miyaguchi, K., Honda, K., Toyo'Oka, T., and Imai, K. *J. Chromatogr.* **352**, 255–260 (1986).
28. Watanabe, Y., and Imai, K. *J. Chromatogr.* **309**, 279–286 (1984); *Anal. Chem.* **55**, 1786 (1983).
29. Toyo'Oka, T., and Imai, K. *J. Chromatogr.* **282**, 495–500 (1983).
30. (a) Grayeski, M. L., and Weber, A. J. *Anal. Lett.* **17**, 1539 (1984).
    (b) Sigvardson, K. W., and Birks, J. W. *Anal. Chem.* **56**, 432–435 (1983).
31. Sigvardson, K. W., Kennish, J. M., and Birks, J. W. *Anal. Chem.* **56**, 1096–1102 (1983).
32. Sigvardson, K. M., and Birks, J. W. *J. Chromatogr.* **316**, 507–518 (1984).
33. Honda, K., Miyaguchi, K., and Imai, K. *Anal. Chim. Acta* **177**, 111–120 (1985).
34. Imai, K., Miyaguchi, K., and Honda, K. In K. vanDyke, ed., *Bioluminescence and Chemiluminescence: Instruments and Applications*, Vol. II. CRC Press, Boca Raton, FL, 1985, Chap. 5, pp. 65–76.
35. Imai, K. In M. A. DeLuca and W. D. McElroy, eds., *Methods in Enzymology: Bioluminescence and Chemiluminescence*, Vol. 133, Part B. Academic, New York, 1986, Chap. 35, pp. 435–449.
36. Weinberger, R. *J. Chromatogr.* **314**, 155–165 (1984).
37. Grayeski, M. L., and Seitz, W. R. *Anal. Biochem.* **136**, 277–281 (1984).
38. de Jong, G. L., Lammers, N., Spruit, F. J., Brinkman, U. A. Th., and Frei, R. W. *Chromatographia* **18**, 129–133 (1984).
39. de Jong, G. L., Lammers, N., Spruit, F. J., Frei, R. W., and Brinkman, U. A. Th. *J. Chromatogr.* **353**, 249–257 (1986).

6

Photochemical Reaction Detection in HPLC

James R. Poulsen and John W. Birks

Department of Chemistry and Cooperative Institute for Research in Environmental Sciences (CIRES), Campus Box 449, University of Colorado, Boulder, CO 80309

Introduction

Detection Problems in HPLC

As with other reaction-detection methods for HPLC, the goal of photochemically based reaction schemes is that of enhancing the detectability of a class of compounds or of a particular target analyte. These improvements in detectability are realized as increased selectivity and/or sensitivity toward the analyte relative to other more conventional detection methods. Within the practical limitations of current chromatographic capability, spectroscopic detectors frequently exhibit insufficient selectivity to detect analytes present at low concentrations in complex matrices. Detectors based on changes in the bulk properties of the eluent, such as the refractive index (RI) detector, are even more susceptible to interference problems. Furthermore, compounds that do not possess strong chromophores are detected insensitively. Since the resolving power of the HPLC system is limited, and in any case one would like to minimize the analysis time, it is often efficacious to resort to a selective detection method in order to relax the demand for complete resolution of the analyte peak. Selective detection adds another dimension to the separation of a compound from the matrix in which it is found. While several reviews on reaction detection have emphasized the advantages of post-column reactions in HPLC,[1-10] much less attention has been focused on the application of photochemical reactions to HPLC detection.[1, 11-13]

Post-column reaction detection for HPLC is inherently suited to the analytes that pose the greatest problems for current methodology. HPLC is often used as an alternative to gas chromatography (GC) when the analytes are reactive, thermally labile or involatile. In these cases, GC methods involve time-consuming and tedious pre-column derivatization steps. These additional steps often reduce the overall precision and accuracy of analyses. By developing detection methods based on the reactivity of the analytes, the problems presented to other methods such as GC are transformed into desirable properties. The mild conditions under which HPLC separations are carried out allow many of these reactive compounds to be separated without pretreatment.

In this chapter, we hope to bring the reader up to date with developments in this approach to HPLC detection through an extensive review combined with the discussion of several selected systems in more depth. Where possible, the performance of photochemical reaction detection methods will be compared to detectors that are commercially available. In addition to their photochemistry, the design considerations for these detectors will be reviewed briefly. Hopefully, our treatment of this topic will encourage the investigation and application of other photochemical reactions to HPLC detection. In this regard, the organic photochemistry literature has only begun to be utilized.

As always, one should exercise caution when evaluating the sensitivity of these systems in terms of the reported detection limits.[14, 15] As is common in the chromatographic literature, many different definitions of the "detection limit" and methods to calculate it have been used in the original references.[16, 17] No effort has been made to standardize these reported detection limits other than the conversion of those reported in terms of sample concentration to amount injected on-column.[16] Otherwise, they are presented as reported in the original references and are useful primarily in gaining a general idea about a detection method's sensitivity rather than in choosing between two systems of similar sensitivity. Other analytical figures of merit, for example the determination limit, also have become ambiguous through nonuniform usage.[15,17] Adherence to the IUPAC terminology has been recommended convincingly.[14–16] At the very least, it is important to carefully define these terms, their methods of computation and all the revelant observables (e.g., is noise reported as full peak to peak or root mean square). Detailed experimental conditions also should be specified. However, conversion of detection limit values to "chromatographic reference conditions" to correct for peak-width differences, particularly when the standard deviation of the peaks is computed without consideration for asymmetry effects, seems a bit excessive.[16] By their nature, detection limits are subject to considerable uncertainty.[15]

Reporting the amount of analyte required to produce a specified signal-to-noise ratio is the simplest and most common way to convey the sensitivity of a detection scheme. In fact, a signal-to-noise ratio (full peak-to-peak noise) of two is taken as the "minimum detectability" (MD) by the ASTM. As discussed by Foley and Dorsey, the MD refers only to the detection step and this terminology should not be used when other steps in the analytical procedure reduce the analyte concentration at the detector.[16] Although the noise measurement is directly related to the standard deviation of the blank,[15] the colloquial usage of the term "detection limit" when referring to an arbitrarily selected signal-to-noise ratio bothers many analysts, particularly those outside the chromatographic field. Unfortunately, the statistically based calculation methods developed for other techniques often fail to account for the unique problems encountered as one approaches the lower limit of sensitivity in chromatography. Furthermore, the data must fit the statistical model that the calculation method is based upon.[14] Among the common errors are the application of models that assume a linear relationship between the amount of analyte injected and the detector response to nonlinear data and the extrapolation of data to detection limit values far from the calibration region. In this way, divergent detection limit values, which have little practical significance and often exaggerate the sensitivity, are generated by calculation methods that accept the same defintion and statistical confidence levels.[14, 16]

While many reasons exist for the abundance of reporting methods, it would be beneficial to all to adopt a standardized, and preferably simple method for chromatographic detection limit reporting. Perhaps the best option is the adoption of a specified signal-to-noise ratio for chromatographic detection limits, analogous to the minimum detectability of chromatographic detectors. As always, these values should be reported under conditions practical for the chromatographic separation of the analytes. If any extrapolation is involved, the range of the data used and an indication of the linearity of the system should be specified.

Photochemical Reactions for Analytical Applications

Photochemical reactions have long been used to enable the determination of photoactive analytes in complex matrices. In some cases irradiation alone is sufficient to produce the detectable compound, although it is often necessary to add reagents in order to stabilize or aid in the generation of the photoproduct. For analytical purposes photochemical reactions are extremely useful, as photons selectively provide energy to the compounds which absorb them and leave no residue in the absence of a reactive analyte. Most compounds will return to the ground state without undergoing a reaction. While this implies that many compounds are not amenable to photochemical reaction detection, it is also the basis for enhancements in the detection selectivity of those that are.

Many of the methods developed for "batch" analysis (static systems) can be adapted to flow systems such as HPLC or flow injection analysis (FIA). For example, many of the assays developed for phenothiazine tranquilizers in biological fluids take advantage of the enhanced fluorescence of their oxidized products.[18–23] Photooxidation is one of the simplest and most effective means of generating these products.[23] Both chemical[24] and photochemical[25–27] oxidations of phenothiazines to fluorescent sulfoxide derivatives have been applied to HPLC and significantly improve the detectability of these analytes. Other compounds for which batch photochemical methods have been adapted to dynamic flow systems include diethylstilbestrol, tamoxifen, clomiphene, benzodiazepines (e.g., clobazam), *N*-nitroso compounds, reserpine, cannabinoids and the ergot alkaloids. In fact, a wealth of batch-mode reaction detection techniques remains to be used for detection in HPLC.

Reaction Detection and HPLC—Compatibility Considerations

A common limitation of reaction detection schemes, especially when applied to HPLC, is the generation of a "background" response by the reagent(s). Derivatization reagents that are reactive enough to complete the derivatization reaction on a time scale suited to on-line addition in HPLC rarely are stable enough in solution and sufficiently selective toward the analyte to create a practical post-column reaction detector. In a post-column reaction mode, any mobile-phase component or contaminant that reacts with the reagent will create an interfering response to the detector. Even if a constant level of the interferent is present in the system, minor fluctuations in the relative flow rates of the reagent and eluent streams will cause the background response to change. This will result in baseline noise for short term fluctuations and drift in the case of flow changes, which have a longer periodicity. In other words, for a reagent to be suited to an on-line post-column reaction system its specificity for the analyte must be extremely high relative to the other compounds present in the eluent. Obviously, the reagent itself must not produce a detector response unless it can be transformed to an unresponsive compound by an additional reagent or the reaction product can be separated from the reagent in an on-line reaction/extraction system.[28-34]

Clearly, few derivatization reagents meet these requirements. Furthermore, post-column addition of reagent chemicals usually requires expensive pulse-free pumps and can result in complex systems that are difficult to operate. While solid-phase and immobilized reagent introduction systems for HPLC do reduce both the cost and complexity,[35-53] one would prefer to avoid the use of chemical reagents altogether. Enzyme catalyzed reactions are a possible exception. Immobilized enzymes have proven to be especially well suited to solid-phase reactors in dynamic flow systems, because the reagent is not consumed during the reaction.[48-50,54-57] By developing reaction detectors that do not require any chemical reagents, the need for recharging the reagent reservoirs, synthesis and purification of special chemicals, expensive pumping systems and the background problems caused by decomposition and reagent response are eliminated. As a "reagent," photons achieve this goal, since they can be added through simple photoreactors by light sources which are relatively inexpensive. Even so, reactions involving reagents such as *o*-phthalaldehyde and fluorescamine (fluram) for the detection of primary and secondary amines, and many other schemes that require post-column addition of chemical reagents, work amazingly well.[1, 10, 58] Although it is preferable to use a simple off-the-shelf detector whenever it meets the analyst's needs, the operation of even the more complicated reaction detectors is well within the capabilities of the average chemist.[59]

Reagent suitability for HPLC detection also is limited in terms of reaction kinetics, stability, solubility and the number of reagents that can be introduced. Hence, the advantages of photons as a "reagent" for reaction detection in HPLC are even more significant than in the static methods. It is imperative to minimize both the peak dispersion and complexity of a post-column reaction detector. The most significant advantages of photochemical reaction detection schemes for HPLC are:

1. Photochemical reactions are selective toward analytes that are capable of forming a detectable product through a photoreaction *and* absorb the output of the photoreactor source (at least in the case of unsensitized schemes).
2. There are no interfering residues or decomposition products of the reagent.
3. Light can be introduced without additional pumps or mixing devices.
4. Light is introduced without an addition solvent, so there is no analyte dilution beyond what is caused by dispersion processes occurring within the photoreactor. For the same reason, mixing noise is eliminated.
5. Photoreactions (at least their initial steps) have a tendency to be rapid. This allows the use of shorter reactors which contribute less to the peak variance.
6. Light sources do not exhibit the stability problems encountered with chemical derivatization reagents in solution.

7. Eluent selection is not limited by reagent solubility. However, the mobile phase should not absorb light.
8. Even in the case where reagents that participate in the photoreaction must be added, they often may be spiked into the mobile phase. Since the analyte and reagent both are in their ground states prior to entering the photoreactor, there is little probability for reaction to occur during the separation. In contrast, the reactivity of most chemical derivatization reagents precludes this approach.

These unique properties of photons often eliminate the problems associated with other reaction detection reagents. In the ideal case, a response to the detector will result only through the absorption of light and subsequent photoreaction of any analyte or sensitizer and not from a constituent of the matrix which happens to coelute with the analyte. For most of the photochemical reaction detection schemes reported to date there are some compounds which natively respond to the detector that is used. If this proves to be a problem, these compounds can be identified by running a chromatogram of the sample with the photoreactor lamp turned off. Such screening procedures are usually much simpler for photochemical reaction detectors relative to systems which use post-column reagents, because no alterations of the system are necessary. Another potential mechanism for interference is the reaction of a coeluting compound with an intermediate in the photochemical reaction or through its direct quenching of the excited state. The method of standard additions may provide a way to circumvent this type of interference.

Post-column vs. Pre-column Derivatization

Pre-column derivatization is another commonly used approach to increasing the detectability of analytes in HPLC. For analytes that are difficult to separate, the derivatization may enhance the chromatography as well as the detection. On the other hand, derivatizing compounds that occur in homologous series may make them even more similar and thus harder to resolve. Another significant limitation of pre-column derivatization reactions is artifact formation. If multiple products are formed from an analyte during the reaction, or detectable species are generated from matrix constituents, the resulting chromatograms are much more difficult to interpret. Although steps usually are taken to remove the excess derivatization reagent, a peak caused by residual reagent often obscures a portion of the chromatogram.

The reproducibility of reaction yields and extraction efficiencies also is of the utmost importance if quantitative results are to be obtained. Conditions under which these steps are executed must be well controlled, particularly the reaction time and temperature. The time necessary to drive the reaction to completion and execute any subsequent extraction steps compares unfavorably with on-line post-column systems. In addition, automation of pre-column steps is more difficult than for comparable post-column systems.

Complete reaction and the formation of a unique product are not necessary in the post-column mode provided that the identity of the product(s) and the rate of their formation is reproducible.[6, 10] Since the separation takes place before the reaction, a quantitative relationship can exist between the detector response and the analyte concentration as long as the extent of conversion and the branching ratios between products remain constant. Of course, the rapid formation of a single, stable and sensitively detected product, in a high yield, is still the most desirable situation. However, sharp optima in terms of sensitivity (signal-to-noise ratio) often are observed at reaction times which correspond to partial conversion of the analyte to the detectable product. One generally can attribute this to the subsequent decomposition of the product, dispersion of the analyte peak in the post-column reactor and/or an increase in the photochemical background at the longer reaction times necessary for more complete conversion of the analyte. Even when the reaction products are stable, operating at partial conversions can preserve integrity of the chromatographic separation by reducing the band broadening in the reactor as well as saving analyst time. Under first-order conditions, a 97% conversion of the analyte will require a reactor over five times larger than a 50% conversion. Any processes that destroy the product or reduce its concentration will compete more effectively with the reaction that produces it at longer reaction times. Since the flows of the HPLC eluent are well controlled, the reaction time (t_{rxn}) also is held constant. This represents a major advantage over pre-column reactions.

Attempts at using photochemical reactions in a pre-column mode have not been entirely successful.[60-62] In the case of tamoxifen however, pre-column photochemical derivatization did result in a method capable of monitoring clinical levels of the parent drug and its most important metabolites.[61] Photochemical reactions often generate multiple products and short-lived intermediates. Complex chromatograms will result if this occurs in a pre-column reaction. Once again, the post-column mode will overcome these obstacles.

Post-column photochemical reaction detection in HPLC is not entirely free of limitations and compromises. As with all detectors for HPLC, particularly reaction detectors, the eluent optimal for the chromatographic separation may not be the best for the detection step. Factors such as pH, the presence of buffers, ion-paring reagents and oxygen as well as the solvent composition may have important effects on the reaction. Other factors important in the optimization of photochemical reactions include the wavelength and intensity of the light source and the residence time and temperature in the photoreactor. In most cases, compatibility problems are less significant for post-column photochemical reactions than for reagent-based reaction-detection methods.

Classification of Photochemical Reaction Schemes

For the purpose of discussion, the photochemical reactions that have been applied to HPLC detection will be classified into four basic categories. In many cases the reactions will exhibit features that bring out the oversimplification inherent in such a classification scheme. In other examples the identity and mechanism of formation of the photoproduct is, at best, speculative. However, even such an arbitrary grouping facilitates a review of these schemes. At this point the basic classification scheme will be presented with a review of each general reaction type's application to HPLC detection presented in greater detail later in the text.

In the first reaction type, the analyte is photolyzed into smaller fragments which are more detectable than the parent compound. Usually, additional reagents are not required for these reactions. However, several of the photolytic reactions generate products that are best detected through reagent-based colorimetric reactions. In effect, these schemes couple two reaction detectors in series and as such can be extremely selective. The most widely known photolytic detection scheme is the photoconductivity detector (PCD).[63]

Modification or rearrangement of the analyte's molecular structure to create or enhance its response to a "conventional" detector is an alternative approach. Usually, improved detectability in one of the spectroscopic or electrochemical detectors is the goal. The generation or enhancement of a fluorophore is the most common example of this approach to date. It is possible to subdivide this category based on the molecularity of the initial step of the photoreaction. Both intramolecular and intermolecular reactions have been used to improve detectability in HPLC. In the latter case, the analyte reacts with a mobile-phase constituent. For example, one of the solvent components, oxygen, or a reagent added to the mobile phase may participate in the reaction. If reagents are necessary, they may be "spiked" into the eluent or introduced post-column.

The third class of reaction schemes are *indirect* detection methods. Here, a detectable compound that is not structurally related to the analyte is photochemically generated. The analytes function as photocatalysts in the production of the detectable surrogate and photochemically amplify the signal. Such amplification occurs at least in the sense that several of the "surrogate" molecules, those which provide the basis for detection, are produced by each analyte molecule. These schemes may require the introduction of additional reagents which function as traps, donors or sensitizers. However, the reagents for the photochemical reaction often can be added to the mobile-phase prior to the separation. Spiking a reagent into the mobile-phase is preferable to adding it post-column in any case where the chromatographic separation is not adversely affected. Nonionic reagents usually have little effect on reverse-phase separations when added at millimolar or lower concentrations. Pre-column derivatization reagents designed to respond to photochemically amplified schemes offer an additional route to the sensitive and selective detection of problematic analytes.

The final category is perhaps the most novel of all the applications of photochemistry to HPLC reaction detection. These detection schemes involve the use of light-absorbing reagents to detect analytes which do not absorb light themselves. Compounds that have no strong chromophore are the analytes for which the limitations of conventional HPLC detectors are most severe; hence, these systems provide unique detection capabilities. They also serve as excellent examples of the remarkable potential that well known photochemical reactions offer with regard to reaction detection in HPLC. At present, only a few detection schemes have been based on sensitized photochemistry, primarily because of the difficulty in finding a suitable sensitizer. In fact, even the schemes that are useful ultimately are limited by background noise caused by photochemical reactions of the sensitizer in the absence of an analyte. Once again, these sensitizers often are added to the chromatographic eluent.

Another approach to improving HPLC detection which utilizes photoexcitation deserves mention. Krull *et al.* have reported the direct electrochemical detection of short-lived excited states of several UV absorbing compounds.[12, 64–66] Here, an excited state of the analyte has enhanced detectability relative to the ground state. The best analytes for the *photoelectrochemical detector* (PED) are carbonyl compounds which are known to efficiently intersystem cross to the triplet state.[67] This is in agreement with the observation that the PED response tracks with the phosphorescence quantum yield.[64] It also is possible that some of the response results from transient intermediates formed from the analytes. Many of these carbonyl compounds also undergo rapid hydrogen abstraction reactions in the presence of alcohols.[68]

The PED is a commercially available, thin-layer amperometric flow cell modified so that it can be irradiated with a high-intensity UV lamp.[64] Although the PED primarily has been used in the oxidative mode, rigorous deoxygenation still is necessary to prevent oxygen quenching of the excited states and/or its reaction with detectable intermediates. The mobile-phase reservoir is heated and continuously purged with helium. A zinc column between the reservoir and the injector removes any residual oxygen. Samples also are deoxygenated prior to injection.

At this point, the reported detection limits are in the low nanogram range for aryl ketones and aldehydes and as such are not improved relative to UV absorption detection of the same compounds. Detection of alkyl aldehydes and ketones is 2 to 3 orders of magnitude less sensitive. It is possible that the PEDs limited sensitivity results

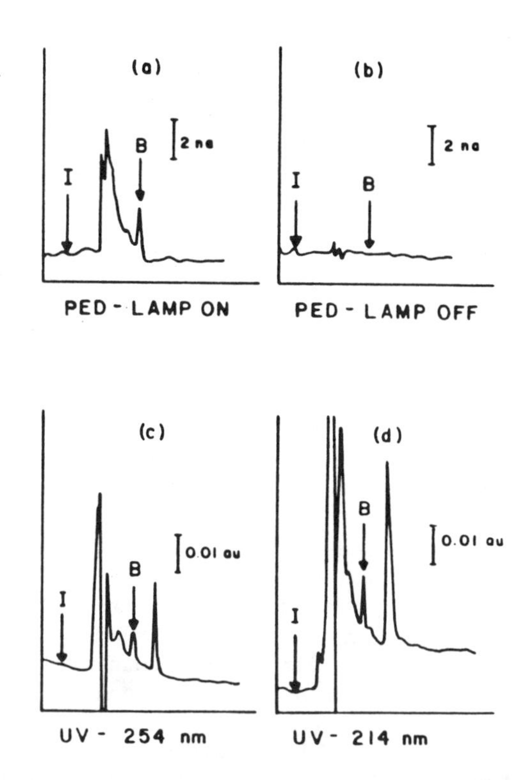

Figure 6-1 ■ Comparison of PED and UV detection for whole almond hydrolysis solutions. I is the time of injection and B is the benzaldehyde peak. The ability to identify compounds through the change in their response with the lamp turned off is emphasized. From W. R. LaCourse and I. S. Krull, *Anal. Chem.* **59**, 49 (1987).

from the electrode quenching the excited states of the analytes. However, the PED is quite selective when the electrode is poised at a low potential. For example, the determination of benzaldehyde in several foodstuffs at sub-parts-per-million to parts-per-thousand levels with PED detection has been demonstrated.[65] In this study, the PED produced results which were comparable to and in quantitative agreement with UV absorption at 220 and 254 nm. Figure 6-1 is a comparison of PED and UV detection of benzaldehyde in a whole almond extract.[65]

While the PED does not depend on a "permanent" photochemical transformation of the analyte, it may lead to broader application of photochemical principles to chromatographic detection. Photochemical reactions are initiated by changes in the electronic structure, and consequently the reactivity, of compounds between their ground and excited states. Similarly, the PED exploits the enhanced oxidizability or reducibility of the excited states of an analyte molecule. Since no additional steps are required to produce a unique product, the PED is not limited to analytes which undergo characteristic photoreactions. However, when an electroactive photoproduct or intermediate is formed from an analyte, it will add to the response. Such a mechanism may be important for analytes which undergo fast excited-state hydrogen abstraction or electron-transfer photoreactions.[69] By varying the electrode potential, the selectivity of the system can be tuned to a specific application. The authors also suggest the use of a dual-electrode detector to compare the irradiated and unirradiated response in real time. Such a system would provide additional qualitative information.[64] In the future, perhaps the unique properties of semiconductor electrodes will improve the detectability of excited states by preventing electrode quenching.

Photochemical Reactors for HPLC

Design Limitations

Photochemical reactors for on-line, post-column detection of photoactive compounds have evolved to the point that it is possible to carry out these reactions with very little loss in chromatographic efficiency. For relatively fast reactions, tubing can be crocheted into a cylindrical configuration that conveniently fits around the lamp and minimizes band broadening.[70] Flow segmentation, either with air or an immiscible solvent, allows the use of longer reaction times.[27, 71–73] Choice of the light source, a suitable reactor and, when necessary, filters to select the excitation wavelength varies with the application.

In designing any post-column apparatus for HPLC one must pay particular attention to its effect on the chromatographic integrity of the system. Poorly designed reactors, those with excessive dead volumes or poor flow characteristics, will cause the efficiency of the separation to be lost. These design requirements are even more severe for post-column apparatus attached to microbore HPLC systems.[34, 46, 74–77] Photochemical reactors have the additional constraint that they must be transparent to the radiation emitted by the excitation source and have a geometry that is suited to it.

Three basic reactor designs are available for post-column reaction detection in HPLC. These are the open-tubular reactor (OTR), including its improved cousin the deformed open-tubular reactor (DOTR), the packed-bed reactor (PBR) and the segmented-flow reactor (SFR). SFRs employ either an inert gas (GSFR) or a solvent that is immiscible with the eluent (SSFR) to partition the column effluent into discrete segments before entering the reactor. Each reactor type has advantages and disadvantages which complement each other to cover the range of HPLC applications. It is necessary to have a basic understanding of the peak dispersion processes which occur in these reactors in order to select the proper reactor and optimize its performance for a given application. The following section is a basic review of band-broadening considerations and other practical aspects which limit the range of applicability of each type of reactor in HPLC. For more extensive and mathematically detailed treatments of dispersion in post-column reactors and other chromatographic components, the reader is directed to the literature.[1, 71, 78–95] In addition to treating band broadening in chromatographic systems, these articles serve as a guide to the relevant engineering literature.

Band Broadening in Post-Column Reactors

Peak Variance Calculations. The variance of a Gaussian chromatographic peak [Eq. (6-1)], expressed in volume (σ_v^2) or time (σ_t^2) units,

$$\sigma^2 = \left(\frac{W_{0.50}}{2.354}\right)^2 = \left(\frac{W_{0.134}}{4}\right)^2 \tag{6-1}$$

is the sum of the variances caused by each component of the chromatograph,

$$\sigma_{\text{total}}^2 = \sigma_{\text{inj}}^2 + \sigma_{\text{col}}^2 + \sigma_{\text{ct}}^2 + \sigma_{\text{reac}}^2 + \sigma_{\text{det}}^2 \tag{6-2}$$

$W_{0.50}$ is the width of the peak at half-height expressed in time or volume units, while $W_{0.134}$ is measured at 13.4% of the peak height. The additivity of variances caused by extra-column components [Eq. (6-2)] is important in selecting a suitable reactor. For example, the column (σ_{col}^2), injector (σ_{inj}^2), detection cell (σ_{det}^2) and connecting tubing (σ_{ct}^2) each contribute to the variance of a chromatographic peak. A well designed liquid chromatograph eliminates all unnecessary dead volume to minimize resolution loss through extracolumn effects. When one adds a post-column reactor to such a system, a significant addition (σ_{reac}^2) to the total variance is usually observed.

Any deviation from a Gaussian peak profile will cause the variance calculated directly from a peak-width measurement to be incorrect.[79, 96–99] Even though calculations from $W_{0.134}$ better account for asymmetry, the true variance of an asymmetric chromatographic peak still will be underestimated. While Eq. (6-1) is not valid for non-Gaussian peaks, it commonly is used for calculating the variance of chromatographic peaks which are seldom perfectly symmetric. To date, most reported variances for post-column reaction systems are the result of these simple calculations based on peak width. Although the reported variances are not additive, because of the asymmetry effects, they are a useful aid in visualizing the compromises and estimating the resolution loss involved in coupling a HPLC system with a reaction detector.

Dispersive processes, which occur in extracolumn components, often are exponential in nature and generate asymmetric peaks.[87, 88, 96–99] The variance of asymmetric peaks should be calculated as the statistical second moment.[96–98] Peak "tailing" on the column and overloading of the stationary phase by large injections are additional sources of peak asymmetry. When second-moment calculations are used, even non-Gaussian contributions to peak variance are additive [Eq. (6-2)]. Unfortunately, these calculations are not amenable to a simple graphical computation method.[96-98] Furthermore, the accuracy of some computerized calculations is limited by sensitivity to baseline drift and noise. An empirical equation developed from an exponentially modified Gaussian model of skewed peaks allows the true variance of experimentally observed chromatographic peaks to be estimated quite accurately.[99] This method only requires measurement of the peak width and asymmetry factor (B/A) at 10% of the peak height,

$$\sigma^2 = \frac{W_{0.10}^2}{1.764(B/A)^2 - 11.15(B/A) + 28} \tag{6-3}$$

Constraints on the volume of the post-column reactors are not as severe as those on detector flow cells. The limitations on the detector-cell volume are a result of the sampling frequency necessary to achieve an accurate representation of a chromatographic peak as well as the physical dispersion of the peak itself. Detector response averages the concentration within the probed volume of the cell rather than responding to the instantaneous concentration of analyte in the chromatographic peak. Excessive electronic time constants cause broadening for the same reason.[96, 100]

The maximum tolerable increase in peak dispersion is strongly dependent on the detection scheme. For extremely selective reaction detectors the loss in chromatographic resolution is much less significant than for their more universal counterparts. Selective detection, to some extent, reduces the level of resolution necessary to identify the analyte and determine its concentration within a given matrix. The additional variance caused by a reaction detector has a less significant effect on resolution when coupled to a low-efficiency column as opposed to a high-efficiency column. For this reason, it is less difficult to design reaction-detection apparatus for the former. In any case, the loss in resolution should be kept to a minimum by selecting a reactor which is properly matched to the kinetics of the reaction and the chromatographic system as a whole. Furthermore, dispersive effects cause the maximum concentration of the analyte peaks to decrease, lowering the sensitivity of the system. Careful optimization of the reaction conditions to reduce the necessary residence time is often the simplest and most effective way to reduce peak dispersion. Shih and Carr have derived an equation to predict the optimum length and flow rate for post-column reactors under first-order reaction conditions that also accounts for the effect of changing flow rate on the efficiency of the other components of the HPLC system.[101]

Open-Tubular Reactors (OTR). OTRs are the simplest of the post-column reactors for HPLC. They also suffer from the most severe peak dispersion of any of the reactor designs. The equation used by Deelder and co-workers to calculate dispersion in OTRs,[82] can be rearranged to predict the change in volume variance ($\Delta\sigma_v^2$), per unit of residence time (t_{res}):

$$\frac{\Delta\sigma_v^2}{t_{res}} = \frac{\kappa d_t^2}{96 D_m} F^2 \tag{6-4}$$

D_m is the diffusion coefficient of the analyte in the mobile phase, F is the volumetric flow rate and d_t is the inner diameter of the tube. It should be noted that this equation is an approximation that is valid for the nonturbulent flow conditions typically observed in unpacked tubes at the mobile-phase velocities used in HPLC. Deviations from the laminar-flow profile that forms in a straight open tube are accounted for with the correction factor κ. Laminar flow in post-column reactors degrades the performance of the system, because it enhances axial dispersion. κ is the ratio of the variance caused by the coiled reactor to the variance observed for a straight tube of the same length and inner diameter.[79] The value of κ can be determined experimentally or estimated from the Dean and Schmidt numbers.[79, 82] These dimensionless parameters describe the flow conditions in open tubes. κ is less than 1 for HPLC reactors.[82] H_m, the height equivalent of a theoretical plate (σ_v^2/L), can be calculated for a coiled OTR as

$$H_m = \frac{\kappa d_t^2 u}{96 D_m} \tag{6-5}$$

where u is the linear velocity of the liquid passing through the OTR.[78] It is easy to see the detrimental effect of higher flow rates on the peak variance of an OTR. The difference between the flow rate at the tubing wall and that at the center of the tube increases in magnitude with the linear flow velocity. In addition to the plate height, the length of tubing necessary to produce sufficient residence time also will increase with the flow rate through an OTR. For this reason, it is best to present the reactor dispersion as the change in volume variance per unit residence time.

Flow in OTRs used for HPLC is not perfectly laminar ($\kappa = 1$), because the reactors, especially photoreactors, usually are coiled. A secondary flow is induced in a fluid flowing through a coiled tube which flattens the parabolic flow profile and thus reduces dispersion relative to a straight tube of the same inner diameter ($\kappa < 1$). Secondary flow increases the effective diffusion coefficient in the radial direction. The effect of this radial mixing on peak variance has been predicted theoretically and established by experimental results from a variety of systems.[80, 82–86, 88, 89] Radial mixing through secondary flow is enhanced by minimizing the diameter of the coils as well as the tubing inner diameter. Tight coiling of the OTR around the lamp also increases the photon flux through the reactor, because more of the lamp's surface is covered. One of the simplest approaches to reducing band broadening in OTRs involves deformation of the reactor tubing to increase the tortuousity of the flow path and maximize the secondary flow.[70, 84, 85, 89, 102–106]

Secondary flow decreases the value of κ in Eq. (6-4), and in the extreme case causes the transition from laminar flow to flow which appears to be turbulent to occur. At flow rates practical for HPLC, turbulent flow is not observed in straight capillary tubing. Turbulent flow in DOTRs is even more surprising in light of the observation that coiled open tubes actually stabilize laminar flow, at least in the sense that the transition to fully turbulent flow occurs at higher Reynold's numbers.[84, 85] However, the flow rates through post-column reactors in normal-bore HPLC systems are significantly greater than the flow through capillaries used as open-tubular HPLC columns. For this reason, the reduction of dispersion by the secondary flow is more significant in post-column reactors. Furthermore, the optimal DOTR designs are much more tightly coiled than the capillary open-tubular columns for which secondary flow effects have been found to be negligible.[86] The deformation of the DOTRs occurs in three dimensions, simultaneously twisting and bending the flow path. Coils of twisted metal

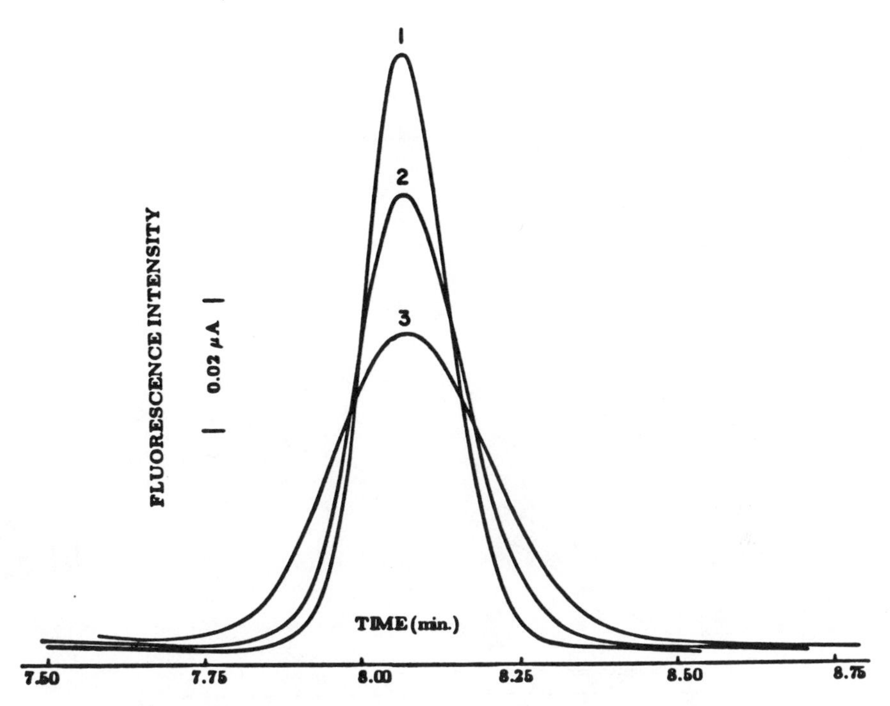

Figure 6-2 ■ Demonstration of reduced band broadening in crocheted post-column reactors. The time axis corresponds to direct coupling to the fluorometer (peak 1). To factor in the reactor volumes, add 3.03 and 2.84 min to the times for peaks 2 and 3, respectively. Phenanthrene was eluted from a Zorbax (Dupont) C_{18} column with methanol : water 95 : 5, delivered at a flow rate of 0.76 ml/min. Fluorescence of phenanthrene was monitored with a Kratos FS-970 fluorometer ($\lambda_{ex} = 280$, $\lambda_{em} > 389$ nm). (1) Direct coupling to the fluorometer. (2) Crocheted reactor with a volume of 2.36 ml. (3) Coiled reactor with a volume of 2.17 ml. From J. R. Poulsen, K. S. Birks, M. S. Gandelman and J. W. Birks, *Chromatographia* **22**, 231 (1986).

tubing have been shown to have better flow characteristics than coils of normal tubing.[85] Once again, optimally designed PTFE DOTRs achieve much sharper directional changes than are possible with metal tubing. In Fig. 6-2 actual chromatographic peaks that have passed through (1) no post-column reactor, (2) a 2.36 ml crocheted DOTR and (3) a 2.17 ml OTR coiled in 8 cm loops are compared.[70] The tubing in reactors 2 and 3 had the same specified internal diameter (0.30 mm). Dispersion caused by the post-column reactor is significantly reduced by crocheting the reactor tubing ($\sigma^2_{crocheted}/\sigma^2_{coil} = 0.25$).

Other reports have focused on the flow rate dependence of band broadening in DOTRs and find that dispersion becomes flow-rate-independent at relatively low linear velocities. Flow-rate-independent dispersion is expected for the plug-like flow profile formed under turbulent conditions.[78, 105, 106] Optimal DOTR designs produce the minimum variance per unit reaction volume for a tubular reactor of a given diameter.[70, 78, 105, 106] No mathematical treatment currently is available for predicting

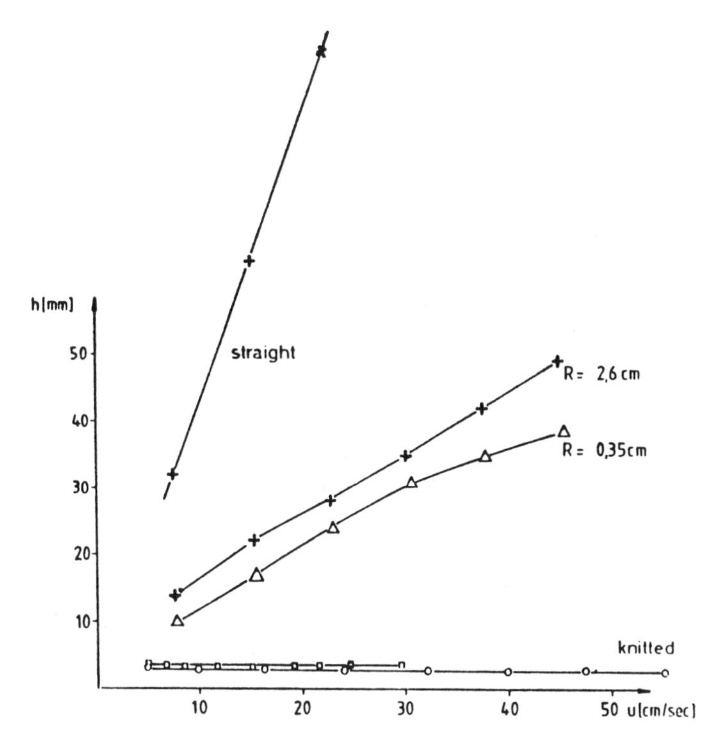

Figure 6-3 ■ Peak broadening in straight, coiled and knitted open tubes. This comparison demonstrates the dramatic difference in the dependence of the height equivalent of a theoretical plate on the flow rate between these reactor designs. Tube diameter 0.36 mm, length 1000 cm (□ = 4080 cm). Methanol eluent, benzene sample. Rheodyne 7125 injector, 1 μl sample loop; UV detector (homemade), cell volume 0.5 μl, time constant 500 or 40 ms for fast peaks. From H. Engelhardt and B. Lillig, *J.H.R. C. and C.C.* **8**, 531 (1985).

the amount of dispersion caused by DOTRs, their pressure drop or their optimal geometric configuration. Therefore, experimental values must be used to compare DOTRs to other reactors. Flow rate dependence data for DOTRs indicate that it is possible to operate under turbulent conditions throughout the range of flows used in normal-bore HPLC. In Fig. 6-3 the dependence of the plate height on the linear velocity through straight, coiled and "knitted" OTRs is compared.[105] The most important factors in designing a DOTR are narrow-bore tubing, tight coiling and deformation of the tubing in three dimensions.[103] In essence, the amount of curvature and number of directional changes within the reactor are maximized.

Another significant factor in the design of a post-column reactor is the pressure drop. For a straight OTR this can be calculated from the Poiseuille equation:

$$\frac{\Delta P}{t_{\text{res}}} = \frac{512\eta F^2}{\pi^2 d_t^6} \qquad (6\text{-}6)$$

Some authors have pointed out the advantages of extremely narrow coiled OTRs in terms of dispersion.[88] In practice however, high back pressures and problems with clogging limit this approach. Tubing of this inner diameter ($\lesssim 40\ \mu m$) also poses a problem in terms of the physical dimensions necessary to produce a reactor of sufficient volume for normal-bore HPLC applications. The secondary flow, which decreases the peak dispersion in a DOTR, also reduces its permeability. An increase in pressure drop over that calculated from Eq. (6-6) is predicted and observed in tightly coiled tubular reactors.[82, 84, 85, 89] The crocheted DOTR in Fig. 6-2 caused an increase of about 40% in the pressure drop relative to the coiled OTR.[70] Other geometrically deformed OTR designs generate similar increases in the pressure drop.[78, 105, 106]

Equation (6-4) also implies that the broadening in a tubular reactor will vary with the eluent composition. The diffusion coefficient of an analyte (D_m) is inversely proportional to the solvent viscosity (η). For the common reverse-phase solvent combinations this generally means that broadening in an OTR will increase with the percentage of water in the eluent. Similarly, increased mobile-phase viscosity also will lead to higher back pressure in the reactor [Eq. (6-6)].

A special type of OTR has been developed recently to add post-column reagents without additional pumping systems. Hollow fiber reactors (HFR) consist of semipermeable capillaries (sulfonated hollow fiber membranes) immersed in a concentrated solution of the appropriate post-column reagent.[107–109] Since the flux through the membrane is determined by the concentration gradient across it, the reagent concentration in the eluent will increase much faster than the analyte concentration will decrease. HFRs already have been used for adjusting the pH to alter the spectral properties of phenols,[107] the addition of fluorescamine and ninhydrin to detect amines,[107] the addition of *o*-phthalaldehyde for amino acid detection[108] and the detection of β-lactamase inhibitors through post-column alkaline degradation.[109] Broadening caused by HFRs will be equal to an OTR of the same inner diameter and length. While elimination of the reagent addition pump is the principal advantage of an HFR, analyte dilution by reagent addition solvents and dispersion in mixing devices are prevented as well. These novel reactors are not likely to be used for photochemistry, but they are certain to interest researchers who are developing other post-column reaction detectors.

Segmented-Flow Reactors (SFR). Another approach to reducing band broadening in photoreactors is flow segmentation.[27–34, 71, 72, 90–94] The operating principle in a SFR is the forced formation of a plug-like profile in the column effluent as it flows through the reactor. Within each segment a "bolus" flow is induced (Fig. 6-4).[91] The flow streams characteristic of the bolus flow determine the mixing dynamics within the segments. Since the discrete segments are kept from contacting one another, the laminar flow typical of OTRs is disrupted. SFRs do not require the use of extremely narrow-bore or crocheted tubing. Hence, SFRs allow very long reaction times to be used without excessive band broadening or back pressure. Either an inert gas (GSFR) or an immiscible solvent (SSFR) may be used to effect the segmentation of the eluent. Three-phase systems have been reported, especially for reaction/extraction detection schemes,[28–30] but offer no significant advantages to justify their additional complexity.[30]

Selection of the segmenting agent depends on the application. When high temperatures are combined with large back pressures, gas segments may break up and render

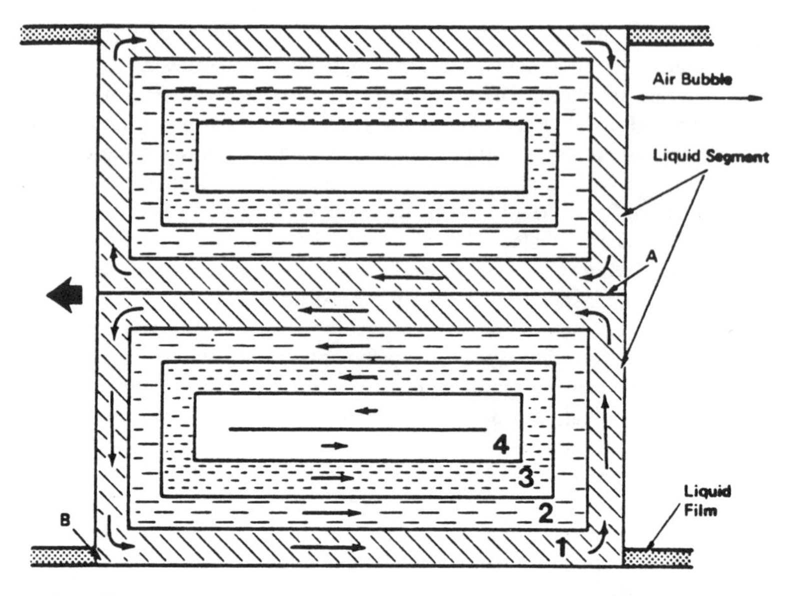

Figure 6-4 ■ Diagrammatic representation of the flow patterns within a moving liquid segment (bolus flow) in a segmented flow reactor (SFR). From L. R. Snyder and H. J. Adler, *Anal. Chem.* **48**, 1022 (1976).

the segmentation ineffective. Solvent segments are not compressible, so they are better suited to these conditions. SFRs usually are constructed from wider-bore tubing than nonsegmented OTRs, so the back pressure of the reactor itself will be low. The incompressibility of solvent-segmented streams also tends to suppress noise resulting from flow and temperature fluctuations.[31] On the other hand, partial extraction of the analyte or photoproduct into the solvent segments can be a significant problem. In fact, it is possible to use solvent segmentation/extraction to add selectivity for certain analytes, particularly those which form ion pairs.[28–34, 75, 76] Obviously, sensitivity will suffer if the analyte or its reaction products partition into the waste stream. When extraction presents a problem an inert gas is the preferred segmentation agent.

Extraction into the segmenting solvent drastically changes the environment of the analyte. Incompatibility with reverse-phase solvents, a problem common to mass spectrometers and some of the GC detectors which have been adapted to HPLC, can be avoided by extracting the analyte into the nonpolar segmenting agent.[76] Perhaps extraction in SSFRs will prove to be useful for optimizing solvent-dependent photochemical reaction-detection schemes.

Dispersion in SFRs is more difficult to treat theoretically than in either OTRs or PBRs. Of the two major sources of band broadening, the reaction-time-independent variance caused by the phase separator has made the largest contribution to peak dispersion in most reported systems.[29, 31–33, 79, 82] For this reason, considerable effort has been expended to improve these devices. Typical phase separator variances have been reduced from 7000–9000 μl^2[29–32, 79, 82, 94] to 150–600 μl^2.[27, 72, 73] By reducing this variance contribution, the residence time where SFRS become advantageous relative to other reactor designs also is reduced. Since nonsegmented systems do not have phase separators, they always will be more effective than SFRs at residence times where their total variance is less than the phase separator variance.

The phase separators used for HPLC fall into three categories. Separation of the phases may result from density differences between the phases, differing affinities for materials in the desegmentation device, the passage of one of the phases through a selective membrane or some combination thereof.[27, 32, 72, 73, 75, 110-113] In many cases, desegmentation systems originally designed for FIA have been adapted to HPLC. *Electronic desegmentation*, that is, gating the detector to discriminate against the segmentation medium, eliminates the dispersion caused by the phase separator.[71, 79, 82, 114] Noise, resulting from the difficulty in keeping the segmentation device and detection electronics synchronized, has thus far prevented wide use of this approach in HPLC. Lack of precision in controlling the segmentation frequency or either of the flow rates has a deleterious effect on electronic phase separators. In addition, wetting of the reactor and detector cell causes the borders between gas and eluent segments in GSFRs to be diffuse.[114] One must consider that the volume of the detector cell will place an upper limit on the segmentation frequency in this mode of operation.[71] The variance of an electronically desegmented system actually can be larger than that of a physically desegmented system, a result of the requirement that the volume of the eluent segment be greater than the volume of the detector cell.

Another concern is the efficiency of a phase separator. Inefficient phase separators lose a large portion of the analyte stream to waste during the desegmentation process. When the phases are physically separated, some of the eluent stream inevitably will be sacrificed to prevent noise caused by the passage of small amounts of the segmenting agent through the detector cell. Once again, electronic desegmentation techniques would eliminate this problem.[71, 79, 82, 114] In the future, it is possible that computer-assisted electronic desegmentation systems will be improved to the point where their performance is competitive with the best phase separators.

The second variance contribution is residence-time-dependent and is caused by a partial breakdown of the plug-like flow profile of an ideal SFR. Leakage between segments occurs when the reactor tubing is wetted by the eluent (Fig. 6-5).[90] The solvent film left on the wetted tubing provides the medium for mass transfer of the analyte from one segment to the next. Higher flow rates increase the film thickness.[115]

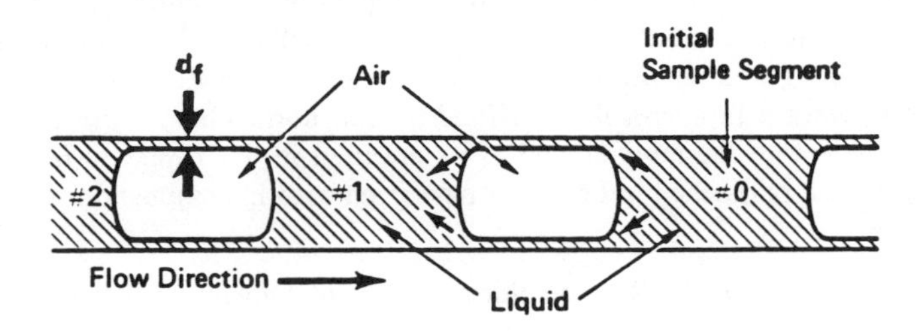

Figure 6-5 ■ Representation of segmented flow through an open tube. From L. R. Snyder and H. J. Adler, *Anal. Chem.* **48**, 1017 (1976).

The amount of dispersion this causes depends, in part, on the mixing efficiency within the segments. Poor mixing within the segments, a result of the bolus flow streams, prevents complete equilibration between the segment and film. This slows the transfer of the analyte between the film and the segment and between the flow streams within the segment itself, producing nonuniform analyte distributions. These mixing dynamics complicate quantitative treatments of dispersion in SFRs.[91] Coiling an SFR enhances mixing within the segments and reduces this contribution to band broadening. Reductions in dispersion resulting from secondary flow in an SFR are much less dramatic than in OTRs. The effectiveness of narrow-bore tubing and tight coiling are limited by other practical aspects of the system.[33] Severely asymmetric peaks commonly are observed when the residence time in an SFR is short. This is explained through an asymmetric residence-time distribution function.[71, 79] As the residence time increases, the peak is spread over a greater number of segments and becomes more symmetric.

A semiempirical equation that was developed by Snyder[71] to describe dispersion in nonideal GSFRs is

$$\sigma_t^2 = \left[\frac{538 d_t^{2/3}\left(F_1 + 0.92 d_t^3 n\right)^{5/3} \eta^{7/3}}{\gamma^{2/3} F_1 D_{W,25}} + \frac{1}{n}\right]\left[\frac{2.35\left(F_1 + 0.92 d_t^3 n\right)^{5/3} \eta^{2/3} t_{\text{res}}}{\gamma^{2/3} F_1 d_t^{4/3}}\right] \tag{6-7}$$

In this equation γ is the surface tension, F_1 is the liquid flow rate, n is the segmentation frequency and $D_{W,25}$ is the diffusion coefficient of the analyte in water at 25°C. While it is quite complex, Eq. (6-7) provides a basis for identifying the important variables in optimizing a GSFR for HPLC.[71] The segmentation frequency often is the most significant variable in terms of optimizing the performance of an SFR. Depending on the tubing diameter and flow rate, the minimum in the predicted variance occurs at frequencies ranging from 7 to $\approx 100\ \text{s}^{-1}$.[71] Unfortunately, these high segmentation frequencies are difficult to achieve in real HPLC systems.[78]

No comparable mathematical expression for broadening within SSFRs currently is available. The dispersion dynamics in SSFRs are even more complex than those in the GSFRs. Since peak dispersion within the reactor tubing typically is secondary in importance relative to that which is caused by the phase separator, empirical evaluation of the reactor design usually will suffice. High segmentation frequencies, which keep the size of the eluent segments to a minimum, judicious choice of the reactor material and the use of the proper segmentation agent are the best ways to control the reaction-time-dependent dispersion in SFRs for HPLC.

SFRs also have been adapted to microbore HPLC systems.[34, 75, 76] Even more than in normal-bore HPLC, the design of the phase separator is of critical importance. To date, these miniaturized reactors have not been applied to photochemical reactions. Since the advantages of reagent-free photochemical reaction detection are probably even more significant in microcolumn HPLC than in normal-bore HPLC, it is only a matter of time before the miniaturized SFRs are used in this capacity. Both gas and solvent segmented photochemical reactors have been used in conjunction with normal-bore HPLC.[27, 72]

Packed-Bed Reactors (PBR). The third method of reducing band broadening in post-column reactors is to pack the reactor with small particles. The packing material may be inert or may actually participate in the post-column reaction. Retention of the analyte or reaction products is not desirable. Mass transfer in PBRs is the result of convective mixing and molecular diffusion, as in the HPLC column itself. Therefore, the optimal design of a PBR will parallel that of an HPLC column. PBRs have not been used as photochemical reaction cells, though they currently are being investigated as a source of immobilized photochemical reagents.[116] One of the primary reasons for this is the difficulty in efficiently irradiating PBRs.

Assuming that the analyte and its reaction products are unretained by the PBR packing material, the volume variance per unit residence time caused by such a reactor can be calculated as[79]

$$\frac{\Delta\sigma_v^2}{t_{\text{res}}} = \left(\frac{2\Gamma D_m t_{\text{res}}^2}{L^2} + \frac{A t_{\text{res}}^{0.67} d_p^{1.33}}{L^{0.67} D_m^{0.33}} \right) F^2 \qquad (6\text{-}8)$$

Γ and A are empirical constants that best fit experimentally observed PBR variances when they both are assigned a value of 0.80.[79] Realistically it is unlikely that a wide range of analytes will be completely unretained by any single packing material. Reaction band broadening will occur if the products formed in a PBR have different affinities for the packing material than the analyte. A catalytic-bed reactor provided a model system for the investigation of this effect.[51-53]

In order to minimize the variance caused by a PBR, the particle size and column diameter should be as small as possible. In terms of dispersion, the packing quality and uniformity of the particle size frequently are the facts that limit PBR performance. The narrow size distribution and uniform shape of HPLC packing materials usually is not achieved by the inert packings used in PBRs. Since it is difficult to pack long columns ($\gtrsim$ 30 cm) with small particles, either the column diameter must be increased, the flow rate reduced or several packed beds must be coupled in series for sufficient residence times to be generated. Higher dispersion results from the first approach, while the second increases the analysis time and the third approach raises the back pressure. The packing materials for PBRs often have particle sizes that are significantly larger ($\gtrsim$ 20 μm) than those in the analytical column to allow dry packing of the reactor.[79] While more dispersion occurs when these materials are used, one does not need the expensive slurry packing apparatus necessary to produce efficient columns with 3 to 10 μm particles. Low reactor back pressure is an advantage of larger particles.

The flow-rate dependence of dispersion in PBRs is analogous to that in HPLC columns. However, the minimum in the van Deemter plot of plate height vs. mobile-phase velocity will be higher than the practical optimum flow rate for a PBR. Flow rates greater than the minimum in the van Deemter plot for an HPLC column generally are employed to reduce the time required to carry out separations. In other words, one would like the maximum number of theoretical plates per minute from the HPLC column, but in the PBR the residence time per theoretical plate should be maximized. In both, the volume equivalent of a theoretical plate should be minimized by proper design. Obviously, the flow rate will be a compromise between that optimal for the separation and that for the PBR. In practice, the reactor should be designed to accommodate the flow rates that are best for the separation.

The pressure drop in a PBR can be calculated from the Darcy equation,

$$\frac{\Delta P}{t_{\text{res}}} = \frac{\eta L^2}{k_0 t_{\text{res}}^2 d_p^2} = \frac{\eta u^2}{k_0 d_p^2} \qquad (6\text{-}9)$$

The permeability constant (k_0) typically is about 2×10^{-3} for PBR packing materials.[79] Once again, the viscosity of the mobile phase and its flow rate play major roles in determining the maximum reactor volume that is practical in terms of pressure drop.

Another reactor packed with particles is the so-called *single bead string reactor* (SBSR). Designed primarily for FIA, this reactor reduces dispersion without large back pressure increases. The particles in the SBSR are nearly as large as the diameter of the tube itself.[118, 119] Except when very low back pressures are required, as is the case when reagents are added with peristaltic pumps, SBSRs do not appear to offer advantages over normal PBRs for HPLC applications.

Materials for Photoreactor Construction. Stainless-steel capillary OTRs, which have been applied to thermally activated post-column reactions, clearly are unsuitable for photochemistry. On the other hand, Pyrex and quartz tubing—the obvious choices for the construction of photoreactors, because of their transparency—are limited by availability, fabrication difficulties, fragility and expense. These reactors also exhibit peak broadening caused by hydrogen bonding to silanol groups on the reactor surface. Treatment of the reactor with a silanizing reagent reduces this problem. While coils of both Pyrex and quartz tubing have been used as photoreactors, the maximum accessible reaction times are quite short if the flow is not segmented. Pyrex has the additional property of filtering out radiation having wavelengths shorter than $\approx$ 320 nm, a severe restriction for organic photochemistry. Another problem encountered when using these materials is the difficulty in maintaining tight connections between the reactor and HPLC system when the back pressure is high. Although stainless-steel tubing is the most capable in terms of connectability and withstanding back pressure, flexible HPLC tubing (e.g., PTFE, Teflon, FEP) also performs quite well in this regard.

PTFE tubing, originally proposed as a material for photoreactor construction by Scholten *et al.*,[26] has proved to be an excellent substitute for quartz and Pyrex. Even though the direct transmission of UV light by PTFE is not very efficient, diffuse radiation transfer through the pores of the tubing enhances the transmission enough for it to be applied to photoreactor construction. Furthermore, a multiple internal reflectance mechanism, the so-called light pipe effect, increases the effective photon flux inside such a reactor.[26] PTFE tubing also is durable, resistant to corrosive reagents, relatively inexpensive and can be obtained with a narrow and uniform bore.

Transmission of UV light by PTFE tubing was demonstrated through a comparison of the conversion of clobazam, a benzodiazepine anxiolytic agent and several phenothiazines to fluorescent products. In Pyrex reactors no reaction took place.[26] This result was anticipated, because the reactions are initiated best by light with wavelengths of 230 and 270 nm for clobazam and phenothiazines, respectively. Pyrex does not transmit these wavelengths. However, reductions in the optimal reaction time also were observed in PTFE OTRs relative to quartz capillary OTRs.[26] The multiple internal reflectance mechanism was invoked to explain this observation. In the same study, improvements in peak shape through reductions in tailing were observed in the PTFE photoreactors when compared to the quartz and Pyrex reactors.

PTFEs flexibility is another important advantage. This material can be deformed by crocheting, knitting, knotting or tight coiling to induce secondary flow and reduce the dispersion of chromatographic peaks.[70, 78, 102–106] "Stitched" stainless-steel capillary DOTRs have been developed for nonphotochemical reactions.[78, 105] Fabrication of a quartz or Pyrex DOTR would be extremely difficult. Narrow tubing inner diameters, sharp bending of the flow path and deformations in three dimensions are features common to all the optimal DOTRs. Flexible tubing is the simplest and, for photochemistry, the best material for DOTR construction.

Along with its desirable properties, PTFE exhibits a permeability to oxygen and other gases such as hydrogen and methane.[120, 121] In many applications this presents no problem, but in the case of anaerobic reactions PTFE photoreactors must not be in contact with oxygen.[122, 123] It is not difficult to build a reactor housing that can be purged of oxygen.

PTFE tubing can be used in segmented reactors as well. The key consideration in material selection for SFRs is the wettability of the tubing by the mobile-phase. PTFE tubing is wetted efficiently by organic solvents and may cause problems in SSFRs and reaction/extraction systems where the analyte partitions into the nonpolar phase.[32, 115] Quartz and Pyrex capillaries also are suitable for SFRs, because the efficiency does not depend as strongly on the diameter of the tubing. The larger bores of SFRs prevent the back pressure related plumbing problems and allow larger reaction volumes to be irradiated by the photoreactor lamp. Both quartz and Pyrex are wetted by polar solvents. One should select a reactor material that is not wetted by the phase which contains the analyte.

For PBRs to be applied to photochemical reaction detection, a transparent reactor must be designed. Though more difficult in terms of packing and plumbing, quartz and Pyrex tubes can be used for this purpose. Back pressure becomes a limiting factor at much lower levels in comparison to PBRs packed in stainless-steel columns, because of the problem in maintaining tight connections. Gübitz, Aischinger and Birks have developed transparent PBRs for use with immobilized photochemical reagents.[116]

Comparisons of Reactor Performance. Comparisons between the three reactor types have reached conclusions which vary considerably as to the best reactor design for a given residence time. Advocacy of one particular type of reactor plays no small role in these comparisons. Among others, Lillig and Engelhardt[78] have noted: "Sometimes very impractical optimal conditions, such as segmentation frequencies of 46 sec^{-1} or capillaries with inner diameters of 40 μm have been discussed in these reports." Most of the published comparisons between the three reactor types have concluded that unsegmented OTRs are limited to fast reactions ($t_{res} \lesssim 0.25$ to 1 min), with PBRs the best for intermediate residence times ($t_{res} \approx 0.5$ to 5 min) and SFRs for slower reactions ($t_{res} \gtrsim 5$ min).[73, 79–82] Recent advances in the design of the tubular reactors, namely DOTRs, and improved phase separators have changed these conclusions significantly. DOTRs extend the range of reaction times that can be used without segmentation,[70, 105, 106] while the improved phase separators enhance the performance of SFRs at shorter reaction times.[27, 72, 73]

Just as an advocate of a particular reactor design may tend to overemphasize its merits, technical problems regarding its use and preparation may be ignored. For these reasons, recommendations regarding reactor selection at different residence times should be viewed only as broad guidelines. There are significant overlaps between the reactors in terms of applicability, particularly for residence times in the 1 to 5 min range. Personal preferences of the analyst and the availability of equipment and materials also may play a significant role in selecting a photoreactor. Nevertheless, general guidelines for reactor selection such as those of Lillig and Engelhardt, summarized in the following text, are useful.[78]

Their comparison emphasized conditions that are commonly used in normal-bore HPLC, including total flow rates through the reactor of 1.0 ml/min and segmentation frequencies of 2 to 5 s^{-1} for SFRs. Peak broadening data consisted of theoretical predictions for the dispersion occurring in PBRs and the time-dependent variance in SFRs, while the contributions of the DOTRs [knitted open tube (KOT)] and the phase separator of the SFRs were experimentally determined. The best reported values for each were used. They conclude that for residence times of 1 min or less, DOTRs with a 0.25 mm i.d. are superior to the other reactor designs. The high back pressure in these narrow DOTRs limits the maximum residence time. Reaction times of more than ~ 5 min are accommodated best by a segmented system. In the 1 to 5 min region reactors of all three types can be applied successfully.

To describe the performance of the three reactor types in the moderate residence time domain, they were compared in terms of volume variance normalized for reactor volume and back pressure. In this way, the design parameters for reactors which cause equal variance could be compared. Experimental data for DOTRs and predicted variances of PBRs show that 0.25 mm i.d. DOTRs are comparable to 2.8 to 4.0 mm i.d. PBRs packed with 15 μm particles. To match the performance of these reactors with an SFR, segmentation frequencies of 5 s^{-1} would have to be used in combination with electronic desegmentation.[78] An additional consideration is the generation of asymmetric peaks in SFRs at short residence times.[71, 79] The back pressure of both the PBR and SFR will be lower than the 0.25 mm DOTR. DOTRs of 0.35 mm i.d. are roughly equivalent in performance to 2.8 to 4.0 mm i.d. columns, packed with 20 μm material. SFRs with the best reported phase separators can produce similar results with segmentation frequencies of 2 s^{-1}. Within this set of reactors the PBR (2.8 mm i.d., 20 μm packing) will produce the highest back pressure.[78] As the residence time increases, the SFR will cause the least dispersion and back pressure.

Segmentation apparatus does add to the complexity of a reaction detector. For this reason, PBRs, DOTRs and even OTRs often are used at residence times where a SFR would reduce the dispersion. Design and construction of DOTRs has progressed rapidly from the "black art" which characterized early attempts. Simple and reproducible methods for producing optimal DOTRs have been published.[70, 78, 103–106] While there still is a small investment of time in constructing these reactors, it is justified by their simple operation and excellent performance. Access to HPLC column packing equipment is required for the production of low dispersion PBRs. Packing materials which can be dry packed simply will not perform as well as DOTRs and SFRs. Furthermore, many of the PBRs used for post-column reaction detection have fallen short of the theoretical efficiencies due to particle sizes and distributions that are larger than the optimum, poor packing quality and analyte/reaction-product retention by the packing material. The packing material's pH range and resistance to corrosive reagents also may be of concern with some reaction schemes.[74]

Given the difficulty in irradiating PBRs, in addition to the performance advantages of narrow DOTRs at short reaction times, it is unlikely that PBRs packed with inert materials will serve as photochemical reactors in normal-bore HPLC. PBRs are of interest primarily in systems where the packing material participates in the reaction. However, the potential for using them in conjunction with laser excitation for microcolumn HPLC and as a source of immobilized photoactive reagents remains. In other applications, DOTRs and SFRs probably will continue to be the reactors of choice for post-column photochemistry.

Light Sources

A large majority of the photochemical reactions used in detecting organic compounds are initiated by light in the 200 to 400 nm wavelength region. Any time high-intensity, short wavelength UV light is produced, precautions should be taken to prevent exposure of the eyes and skin to the source. In order to maximize the efficiency and selectivity of the reaction, a lamp which generates radiation near the maximum absorption (λ_{max}) of the analyte or sensitizer should be selected. Furthermore, to reduce the possibility of interference and photochemically generated background, the ideal lamp output is free of radiation that is not absorbed by the analytes or sensitizer, since it will serve no useful purpose. Filters can be used to remove undesirable radiation from the source output before it reaches the photoreactor. The removal of such unwanted radiation by suitable filters is extremely important when chemical traps that are photochemically active themselves are used. For example, in detection schemes based on singlet oxygen trapping, the photoreactor source and acceptor molecule must be selected to minimize photoexcitation of the acceptor and its subsequent self-photo-oxidation.[124]

Lamp intensity is the second major consideration in selecting a photoreactor source. By increasing the photon flux, the reaction time necessary for sufficient sensitivity may be reduced. Rapid reactions make it much easier to design photoreactors that are compatible with HPLC. On the other hand, high lamp intensities may initiate photodecomposition of the analyte or photoproduct. Photochemical reactions that are initiated by high-intensity sources tend to exhibit sharper optima which occur at shorter reaction times than those observed for the same reactions carried out with low-intensity lamps. When photodegradation reactions are competitive with product formation, low-intensity lamps may yield the best overall sensitivity. In Fig. 6-6, a comparison between an 8 W fluorescent lamp with a PTFE DOTR (IV) and a 200 W high-pressure xenon-mercury arc lamp with a PTFE OTR (II) shows the large difference in their optimal reaction times. This comparison demonstrates that for the photoreduction fluorescence (PRF) detection scheme of Gandelman *et al.* the low-intensity lamp is superior.[125] PRF detection of nonabsorbing analytes is discussed later in this chapter. (The arc lamp and its housing are identical to that which appears in Fig. 6-7 later in this section.) It is possible to irradiate the reactor with more than one lamp, but this approach to increasing the photon flux is limited by the reactor geometry and the cost of the lamps.

Secondary considerations in choosing a lamp are the cost of the source and its ease of operation. For example, arc lamps generate large amounts of heat and, if exposed to oxygen, produce ozone along with high-intensity light. They are both more expensive and a challenge in terms of reactor housing design.

At this point we will briefly survey a few of the lamps available for use in photoreactors and mention some of their merits and disadvantages. By necessity, the comparisons between lamps are somewhat vague. To be more exact, one would need to specify particular lamps and their operating conditions. This type of information usually is available from the vendors.

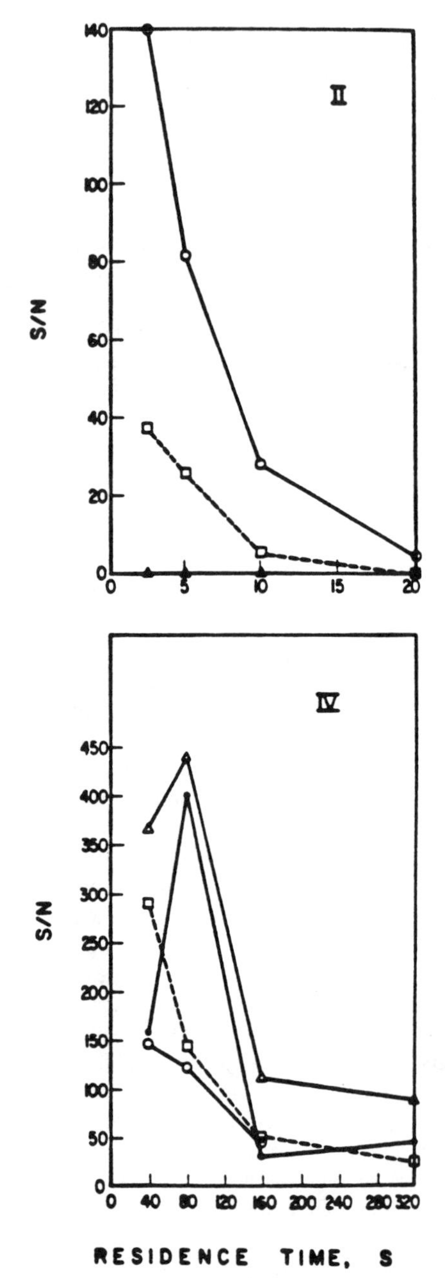

Figure 6-6 ■ Variation of the signal-to-noise ratio with residence time in the photochemical reactor for various concentrations of anthraquinone-2,6-disulfonate. ○ $= 1.2 \times 10^{-3}$ M; □ $= 6.0 \times 10^{-4}$; ▽ $= 2.0 \times 10^{-4}$; ● $= 4.0 \times 10^{-5}$. Reactor II was a 20 cm coil of 0.33 mm i.d. PTFE capillary irradiated with a 200 W high-pressure xenon-mercury arc lamp. Reactor IV was a 12.5 m long crocheted PTFE capillary, also 0.33 mm i.d., irradiated by an 8 W fluorescent poster lamp. From M. S. Gandelman, J. W. Birks, U. A. Th. Brinkman and R. W. Frei, *J. Chromatogr.* **282**, 193 (1983).

Arc Lamps. Many of the photochemical reaction-detection schemes for HPLC have utilized high- or medium-pressure arc lamps. The most common of these lamps contain mercury, xenon or a mixture of the two. Lamps with power consumption ratings in the 100 to 500 W range typically are used in photochemical reaction-detection applications, although xenon lamps of up to 5000 W are available. High-pressure lamps emit broad-band radiation which covers the entire UV-visible spectrum. Virtually all molecules that contain significant chromophores will be excited by such sources. Atomic emission lines of the element(s) in the lamp are superimposed upon the continuum spectrum. Atomic emission lines significantly enhance the intensity of the lamp at these wavelengths, in some cases by over an order of magnitude. These intense lines are particularly useful in applications that require a monochromatic excitation source. Filtering the output of such a lamp to improve selectivity is possible, but has proved to be unnecessary in many of the reported HPLC applications. Stability of the high-pressure arc lamps is limited by "arc wander," as the arc is confined to the central region of the lamp.

High-pressure mercury lamps have an emission throughout the UV-visible spectrum with 11 strong lines between 270 and 600 nm. Due to self-absorption, the 253.7 nm line is missing and there is very little output around this wavelength. The intensity falls off dramatically below 230 nm.

The continuum emission of xenon lamps is less intense throughout most of the UV spectral region than that of the high-pressure mercury lamps. However, a more gradual decrease in the continuum emission at short wavelengths (200 to 250 nm) and the presence of emission lines above 800 nm provide more intensity in these regions of the spectrum. Xenon lamp emission includes 254 nm radiation, which is absent from high- or medium-pressure lamps that contain mercury.

When xenon and mercury are mixed in a high-pressure arc lamp the characteristic emission lines of both elements are observed, as well as the self-absorption by mercury at 254 nm. While the continuum intensity is slightly lower than the mercury lamp over most of the UV spectrum, it incorporates the main advantages of the xenon lamp, namely, the extended UV range and the intense emission in the near infrared.

The emission of large quantities of infrared radiation can be a significant drawback of the high-pressure arc lamps. High temperatures in the photoreactor increase the probability of polymerization and other side reactions that decrease the yield of the desired photoproduct(s). Control of the lamp temperature also is important, since its output will vary with the temperature. A final consideration is the design of the reactor housing. Arc lamps produce ozone through the photolysis of oxygen, so the housing must be vented to a fume hood. Furthermore, the lamp has the potential to explode. Safety considerations dictate the use of a housing that will withstand such a blast. All of these features were incorporated into the housing shown in Fig. 6-7.[126]

As the pressure in a mercury discharge lamp is decreased, the line emission becomes predominant. Distribution of the energy between the various lines also has a strong dependence on the lamp pressure. To a lesser extent, the energy distribution is affected by the operating conditions such as lamp current, temperature and the wattage of the lamp. Under normal conditions the majority of the radiant energy from a medium-pressure mercury lamp (80 to 85%) is in the form of lines with wavelengths greater than 300 nm. Once again, self-absorption eliminates the 254 nm line. Low-pressure mercury

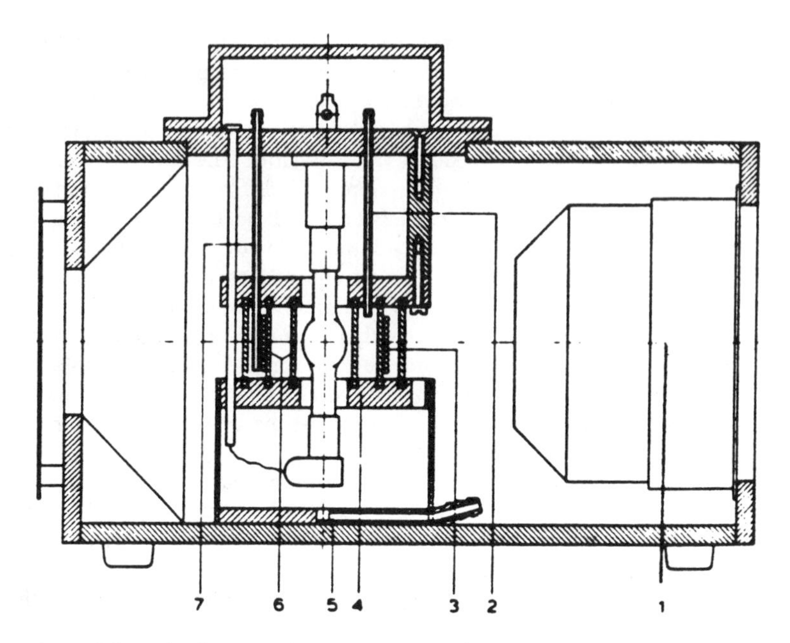

Figure 6-7 ■ Schematic diagram of a water-cooled photochemical reactor housing. (1) Fan. (2) Outlet for circulating liquid (cooling). (3) Capillary. (4) Filter compartment. (5) Additional cooling via pressurized air. (6) Quartz cylinders. (7) Liquid inlet. The light source is a 200 W Hg-Xe lamp operated at a current of 10 A. The PTFE reaction coil is cooled by a continuous flow of water through the jackets formed by two quartz and one Pyrex cylinder. From M. F. Lefevere, R. W. Frei, A. H. M. T. Scholten and U. A. Th. Brinkman, *Chromatographia* **15**, 459 (1982).

discharge lamps emit greater than 80% of their radiation in the form of the 254 nm mercury line.[127] Generally, the stability of low-pressure mercury lamps is better than the higher-pressure arc lamps, because the low-pressure discharge is diffuse and not as susceptible to wander.[128]

Deuterium and Hydrogen Lamps. Low-pressure d.c. discharge lamps containing hydrogen or deuterium commonly are used as sources for variable wavelength UV absorption detectors. The emission from such a lamp is a very stable continuum throughout the 165 to 400 nm region with a maximum at ≈ 230 nm. The intensity decreases sharply as the wavelength increases. Hydrogen lamps are 3 to 5 times less intense than the deuterium lamps and have shorter operational lifetimes.[129] Neither hydrogen nor deuterium discharge lamps have been used widely as sources for post-column photoreactors, because even the deuterium lamp is much less intense than a mercury or xenon–mercury arc lamp. Without adequate filtering, the short wavelength radiation from these lamps would have a high probability of initiating photochemical background reactions as it is absorbed by the common HPLC eluents.

Incandescent Lamps. The emission from an incandescent source, such as the tungsten filament lamp, is primarily in the visible and infrared region, so it is not that useful for most organic photochemistry. Typically, the broad continuum is fairly intense and very stable, so tungsten lamps are the preferred visible excitation source. A wide range of lamp configurations and wattages is available.

Any unreferenced spectral information presented in the preceding section was adapted from tables and spectra in Murov's *Handbook of Photochemistry*.[127]

Pencil Lamps. *Pencil lamps* are low-pressure metal vapor lamps that are designed to fit in small spaces. Mercury is the metal most commonly used in such lamps. Like other low-pressure mercury discharge lamps, these sources emit the mercury lines with a large majority (≈ 90%) of the energy in the form of 254 nm radiation. Phosphors are available to alter the emission maximum to other wavelengths. However, the phosphors also change the nature of the source from line emission to broad-band radiation and cause the lamp intensity to decrease. While these lamps do not have as great a total output as the high- or medium-pressure arc lamps, they still are quite intense (e.g., 3.9 $\mu W/cm^2$ at 100 cm for the 254 nm line of the Pen-Ray™ model #11 SC-1 from UV Products).[130] Perhaps the greatest advantages are their moderate price and ability to fit into small areas without generating large amounts of heat. 254 nm radiation is absorbed by a wide variety of organic compounds; in fact, many of the single-wavelength UV absorption detectors use these lamps as their source.

Similar lamps containing other metals are available as well. For example, zinc (214 nm) and cadmium (229 nm) pencil lamps are sold by UV Products, Inc. and BHK, Inc.[130, 131] These lamps are more expensive, operate at a higher arc temperature and produce less intense radiation than the mercury vapor lamps. To control heat transfer from the arc to the surroundings, the zinc and cadmium pencil lamps are encased in a vacuum jacket so they are larger in diameter than the mercury pencil lamps. The emission intensities for the strongest lines are 0.12 $\mu W/cm^2$ at 214 nm and 0.10 $\mu W/cm^2$ at 308 nm for the zinc lamp, and 0.34 $\mu W/cm^2$ at 229 nm and 0.52 $\mu W/cm^2$ at 326 nm for the cadmium lamp, respectively. These specified minimum intensities for the first 500 hr of use are measured at a distance of 100 cm from the source.[130] A magnesium lamp with a maximum emission at 206 nm is available from LKB (Bromma, Sweden).

A pencil lamp housing designed by Gandelman and Birks for use with PTFE OTRs and DOTRs is shown in Fig. 6-8.[122] This housing is purged of air in order to carry out anaerobic reactions in the oxygen-permeable reactors. The flow of nitrogen cools the lamp and housing as well.

Fluorescent Lamps. Fluorescent lamps are available with a variety of spectral characteristics. Poster lamps (black light, e.g., Sylvania #F8T5/BLB) generate a broad-band spectrum centered at 366 nm. These lamps are readily available, economical and generate very little heat. Fluorescent lamps are larger in diameter and longer than the pencil lamps and have the power supply built into the lamp holder. Designing a suitable housing, in particular one which utilizes a fluid for temperature control, is more difficult relative to a pencil lamp source. For slower reactions, the extra area of the lamp can be an advantage in accommodating the larger reactors.

Fluorescent lamps that emit at longer wavelengths are easy to obtain. The small fraction of organic compounds which absorb the output of these white lamps limits their utility, however.

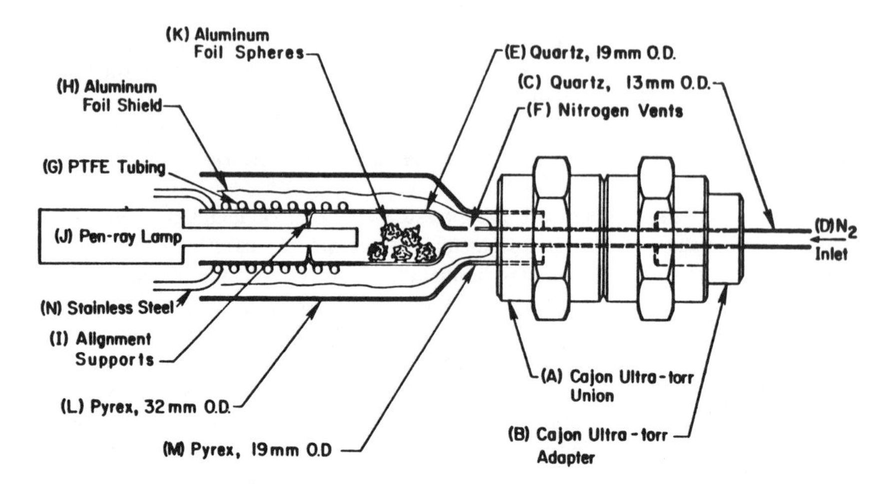

Figure 6-8 ■ Schematic diagram of the Pen-Ray photochemical reactor for photoreduction/fluorescence detection. From M. S. Gandelman and J. W. Birks, *Anal. Chim. Acta* **155**, 159 (1983).

Germicidal Lamps. Germicidal and bactericidal lamps also are useful sources of UV radiation. These lamps are available from several sources in a variety of intensities and configurations. Some are similar in size and shape to fluorescent and poster lamps. Housings and reactors can be designed to accommodate all of these lamps. Through their flexibility in excitation wavelength, photoreactors designed in this way could be applied to a wide range of photoreactions by simply changing the reactor source.

Lasers. Although lasers have been used to increase the sensitivity and versatility of conventional HPLC detectors such as refractive index (RI)[132] and fluorescence,[133, 134] they have not been widely applied to photochemical reaction detection. Lasers are particularly useful for microcolumn HPLC where minimizing the detector cell volume is critical. Other detectors based on polarimetric methods[135] and indirect measurements of absorption, such as photoacoustic spectroscopy and the thermal lens effect, also use lasers as excitation sources.[136] In photochemical reaction detection, the primary application of the laser has been in conjunction with the photoionization detector (PID).[137]

Although lasers are prohibitively expensive for many potential applications, their intense radiation and ability to focus on small reaction volumes are significant advantages relative to other light sources. This is especially true for potential applications of photochemical reaction detection to microcolumn HPLC. Wavelength tunability, at least in the case of dye lasers, also provides interesting possibilities. The monochromatic nature of laser radiation is important to many of the laser-based detection schemes.

Potential applications of lasers to photochemical reaction detection include extremely fast reactions which produce fluorescent products. Such schemes could use the excitation source of a laser-induced fluorescence detector for the photoreactor as well. In a system designed to detect quinones through the fluorescence of their photoreduc-

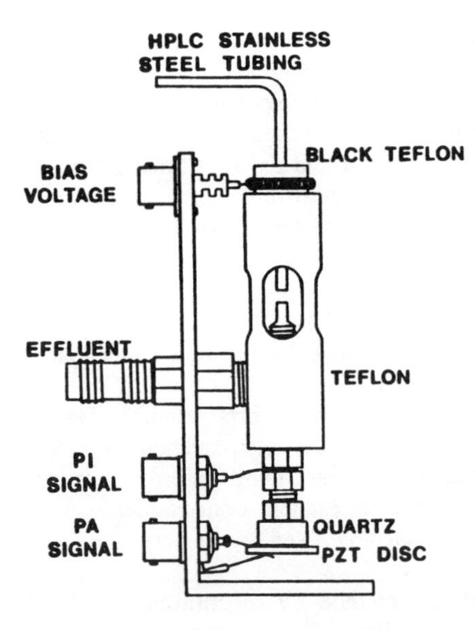

Figure 6-9 ■ Schematic diagram of the windowless HPLC flow cell for PID, photoacoustic and UV absorption detection. From E. Voigtman and J. D. Winefordner, *J. Liq. Chromatogr.* **5**, 2113 (1982).

tion products, the reaction was fast enough to generate the product within the detector cell of a conventional fluorescence detector.[123] While the sensitivity was lower relative to the same detection scheme with a photoreactor in place, the increased photon flux of a laser could make such a dual purpose laser detection system feasible. Photoreduction of quinones for detection in HPLC is discussed in greater detail later in this chapter. A multipurpose laser detection cell for HPLC is shown in Fig. 6-9.[137] By combining PID, fluorescence and photoacoustic detection in one cell, the capabilities of the laser are utilized more fully.

The disadvantages of using lasers in terms of cost and complexity are obvious, especially since the cheapest and simplest of the lasers—the helium/neon laser—produces 633 nm radiation which is not absorbed efficiently by most organic compounds. In the near future, due to their cost, lasers probably will be reserved for special applications where their spatial and intensity advantages are absolutely necessary. Furthermore, their range of application in reaction detection is likely to be limited to schemes that are capable of detecting a wide variety of analytes or to special applications where the system is selective toward an analyte of extreme interest. However, given the unique capabilities of lasers, especially when combined with the interest in miniaturizing HPLC systems, one can anticipate increasing applications as excitation sources for photochemistry. This certainly will be true if the price of lasers capable of producing radiation in the UV region decreases significantly.

Photolytic Reactions

The photolytic reactions that have been applied to HPLC detection typically have the homolytic fission of a bond as the light-induced first step. Two free radicals are produced in such a reaction. Subsequent reactions of the radicals determine the identity of the final products. Residence times in HPLC photoreactors are long enough that the free radicals formed in the initial step are not present at a significant level by the time the analyte peak reaches the detector cell. Often several different products will form during these reactions. Product yields, both absolute and in relation to each other, may be strongly dependent on the mobile-phase composition and reaction time. Extensive studies of the photoreactor products are rare; often only the desired product is identified and investigated in the interest of optimizing the sensitivity. In some cases, even the detectable compound is only tentatively identified.

Photolysis/Griess Reaction for *N*-nitroso Compounds. Iwaoka and Tannenbaum first applied photochemical reaction detection to HPLC for the detection of *N*-nitroso compounds in foodstuffs.[138] This system evolved from the batch experiments of Daiber and Preussmann[139] and the subsequent adaptation of this batch method to a flow system.[140] The analytes were photolyzed with a fluorescent lamp to produce nitrite which subsequently was determined through the Griess reaction. Long residence times in the quartz photoreactor caused severe band broadening. Even so, the detection limits were in the 10 to 100 ng range.

Shuker and Tannenbaum have improved this method by substituting a 400 W metal-halide lamp for the low-intensity fluorescent photoreactor source and a photoreactor constructed from coiled narrow-bore PTFE tubing for the quartz OTR used in the previous work.[141, 142] The lamp emission overlaps with absorption bands characteristic of the analytes, 300 to 380 nm for nitrosamines and 380 to 430 nm for nitrosamides, respectively. While these compounds have stronger absorptions at shorter wavelengths (220 to 250 nm), selectivity against the ubiquitous nitrate anion is improved by exciting these longer wavelength transitions.[140] A reaction time of 2 min is sufficient to achieve detection limits in the low to middle nanogram range for both nitrosamines and nitrosamides with the improved photoreactor. The post-column reagents necessary to detect nitrite with the Griess reaction consist of sulfanilamide and an aryl amine (*N*-1-naphthylethylenediamine). These reagents are introduced downstream from the photoreactor in an acidic solution. Sulfanilamide reacts with the photolytically generated nitrite to form a diazonium salt. An azo dye that absorbs at 541 nm is formed when the arylamine reacts with this diazonium salt.

Since the *photohydrolysis detector* (PHD) is actually two reaction detectors in series, it is quite complex (Fig. 6-10).[141] Photolysis of the *N*-nitroso compounds occurs in the first reactor with the mobile phase as the reaction medium. In the second reactor, a stainless-steel coil immersed in a water bath and heated to 65°C, both the diazotization and chromogenic reaction take place. Equal flows of the chromatographic effluent and the mixed reagent stream are combined in a tee piece prior to this thermal reactor. Solvent pH in the reactor is a compromise between the optima for the two reactions. A cooling unit after the second reactor prevents baseline stability problems in the UV detector.[141]

N-nitroso compounds are extremely carcinogenic. Their formation in the gut from nitrite additives in foods as well as naturally occurring nitrogenous precursors is suspected. One of the goals for this detection scheme is the detection of these analytes at trace levels in biological fluids such as gastric juice (Fig. 6-11).[141]

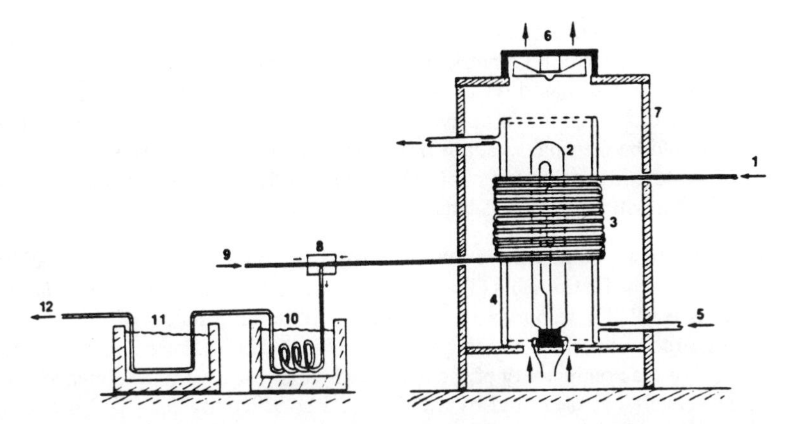

Figure 6-10 ■ Diagrammatic representation of the photohydrolysis detector. (1) HPLC column outlet. (2) 400 W discharge lamp. (3) PTFE reactor (16 m long, 0.25 mm i.d.). (4) Pyrex jacket. (5) Tap water inlet for cooling the reactor. (6) Fan for cooling the lamp. (7) Reflective box. (8) Mixing tee. (9) Griess reagent inlet. (10) Stainless-steel reaction coil (0.25 mm i.d., 1 m long). (11) Stainless-steel cooling coil (0.25 mm i.d., 0.5 m long). (12) To UV detector. From D. E. G. Shuker and S. R. Tannenbaum, *Anal. Chem.* **55**, 2152 (1983).

Detection of nitrite via the Griess reaction has been used in conjunction with other nitrite-generating reactions. Singer, Singer and Schmidt adapted the Griess reaction to the detection of nitrite produced by the post-column cleavage of nitrosamides and other *N*-nitroso compounds by hot, dilute hydrochloric acid.[143] The total residence time in the post-column reaction system was ~ 3 min. Nitrosamines do not respond, but other *N*-nitroso compounds were reported to have 0.5 nM detection limits. As in

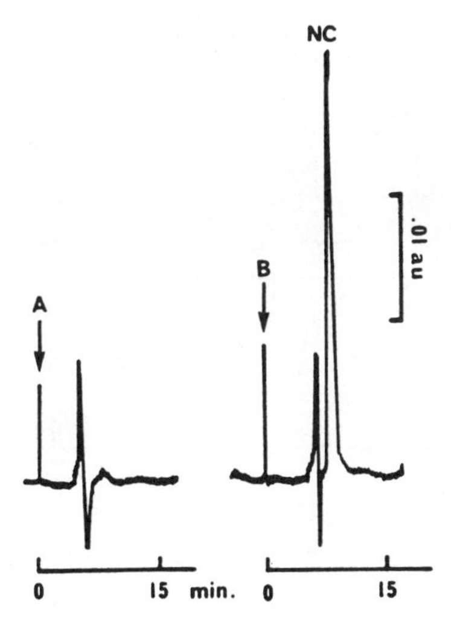

Figure 6-11 ■ Detection of *N*-nitrosocimetidine (NC), the nitrosation product of cimetidine (Tagamet), in gastric juice. Trace A: Normal human gastric juice. Trace B: The same gastric juice spiked with authentic *N*-nitrosocimetidine. Conditions: Mobile phase, $NH_4H_2PO_4$(30 mM, pH = 6) : CH_3CN 20 : 0.9 (v/v) at 0.5 ml/min, colorimetric reagent at 0.5 ml/min, detector at 541 nm. From D. E. G. Shuker and S. R. Tannenbaum, *Anal. Chem.* **55**, 2152 (1983).

refs. 138 and 141, a Technicon AutoAnalyzer was adapted to carry out this reaction-detection scheme. Spectrophotometric and spot test methods for the detection of gunpowder residues, also based on the Griess reaction, have been developed.[144] Here, nitrite is generated through the basic hydrolysis of nitrate esters ($R—O—NO_2$) in the residues. It should be possible to adapt this procedure to HPLC as well.

Another photolytic scheme for *N*-nitroso compounds, where the nitrite is detected electrochemically, will be discussed later in this section.[145]

Photoconductivity Detector. A detector based on photolytic reactions has been available from Tracor Instruments (Tracor, Inc., Austin, TX) since 1978. The *Photoconductivity detector* (PCD) responds to compounds containing halogens, nitrogen and sulfur through their photolysis to ionic products (Fig. 6-12).[63] These ions are detected by the change in the conductivity of the irradiated eluent relative to a reference stream (unirradiated eluent), as measured by a differential conductivity cell. No post-column reagents are required. However, care must be taken to remove ionic contaminants from the mobile phase. On-line deionization is effected with a pre-column pumping system which cycles the eluent through a mixed bed anion/cation exchanger. A limitation of the PCD is its incompatibility with buffered eluents. Buffers cause high backgrounds and rapid depletion of the ion exchange beds. Halogenated and light-absorbing solvents should be avoided as well. Conductivity increases caused by dissolved gases such as CO_2 are prevented by purging the mobile phase with helium or nitrogen.[63]

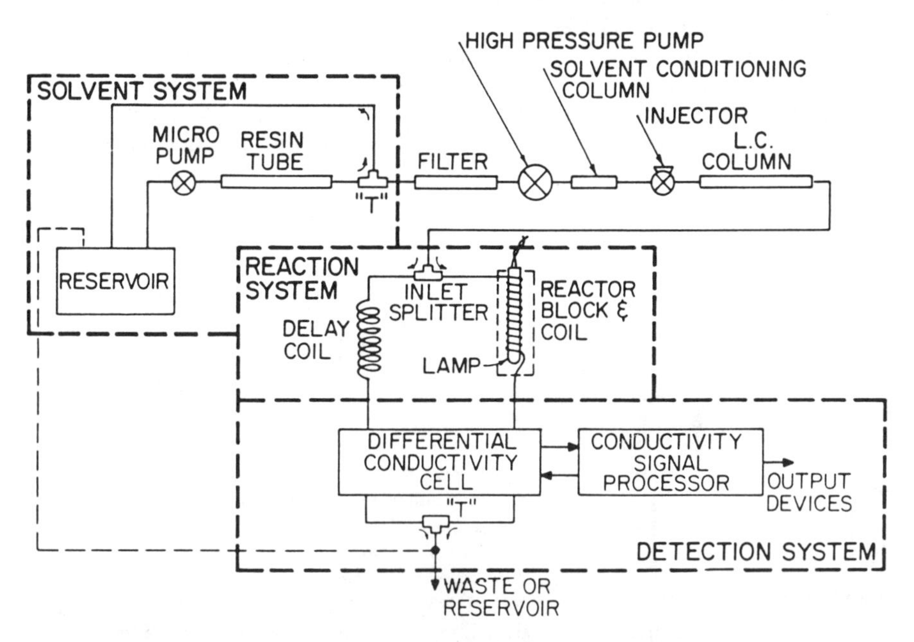

Figure 6-12 ■ Block diagram shows the relationship of the photoconductivity detector to the other components of the HPLC system. From D. J. Popovich, J. B. Dixon and B. J. Ehrlich, *J. Chromatogr. Sci.* **17**, 643 (1979).

The commercially available photoreactor for the PCD consists of a quartz OTR which is coiled around a low-pressure, metal-vapor discharge lamp. Either a mercury (254 nm) or zinc (214 nm) pencil lamp can be used. With a reactor volume of about 120 μl, the peak broadening observed in the PCD is tolerable for most applications.

Ciccioli, Tappa and Guiducci have demonstrated that the quartz reactor of the PCD can be replaced with a Teflon FEP coil without losses in sensitivity.[146] This minor alteration improves operational stability by allowing the addition of flow restrictors to control the split ratio and match the residence time in the photoreactor to the reference stream delay coil. Difficulty in maintaining tight connections with the quartz reactor prevents the use of flow controllers, because they increase the back pressure. Improvement in the symmetry of the chromatographic peaks is observed in PTFE reactors relative to quartz.[26, 146] Since PTFE tubing is suitable for PCD reactors, reaction times could be extended without excessive peak dispersion by using a suitable DOTR.[70, 106] However, one should note the results of Batley when using Teflon reactors in the PCD. He observed the release of F^- and H^+ from Teflon tubing irradiated with UV light.[147] Photodegradation of PTFE reactors also has been observed to be solvent-dependent[148]; hence, PTFE may not be universally applicable in the PCD.

PCD detection limits for halogenated compounds and nitrosamines are in the low to middle picogram range. Sub- to low nanogram detection limits are reported for many other sulfur and nitrogen containing compounds.[63, 149–152] Amines, particularly aromatic amines, are poor analytes. The lack of photolysis exhibited by aryl amines can be attributed to their high fluorescence quantum yields. Response to sulfur- and nitrogen-containing functional groups, other than amines, is probably caused by the formation and reaction with the solvent of SO_x and NO_x, respectively, to form strong acids.[63] Nitroaromatics also may be photoreduced to aryl amines[149]; however, these compounds are much less conductive than HNO_x. Ions produced by dissociation of the acids generated through the hydration of these gases cause a large increase in the conductivity of the irradiated eluent.

Application of the PCD to pharmaceutical agents which contain photolytically active groups demonstrates its utility with regard to clinical studies. In Fig. 6-13, PCD detection of chloramphenicol in serum is compared to UV absorption at 254 nm.[149] PCD response to this antibiotic is caused by a photolabile aromatic nitro group. Improvements in the detection selectivity are evident from the chromatograms.

Jasinski applied the PCD to determinations of nitrosamines in beer and malt samples with considerable success.[150] Relative to UV absorption at 254 nm, improvements of more than 2 orders of magnitude in the minimum detection limits were reported. Quantitative results with accepted gas chromatographic methods (thermal energy analyzer detection) were in agreement with those obtained from the HPLC/PCD system.

The PCD responds well to many pesticides.[146, 151, 152] Walters compared the PCD response of a wide variety of pesticides and fungicides to UV absorption at 220 nm. With this tandem detector arrangement, it was found that for most compounds PCD detection was at least as sensitive and often much improved relative to UV absorption.[152] In the same study, the use of acetonitrile as a mobile-phase polarity modifier was observed to cause an increasing background over time. Difficulties in predicting the responsiveness of compounds to the PCD, particularly halogenated aromatics, also were discussed. In this regard, dual detection facilitates the screening of compounds for

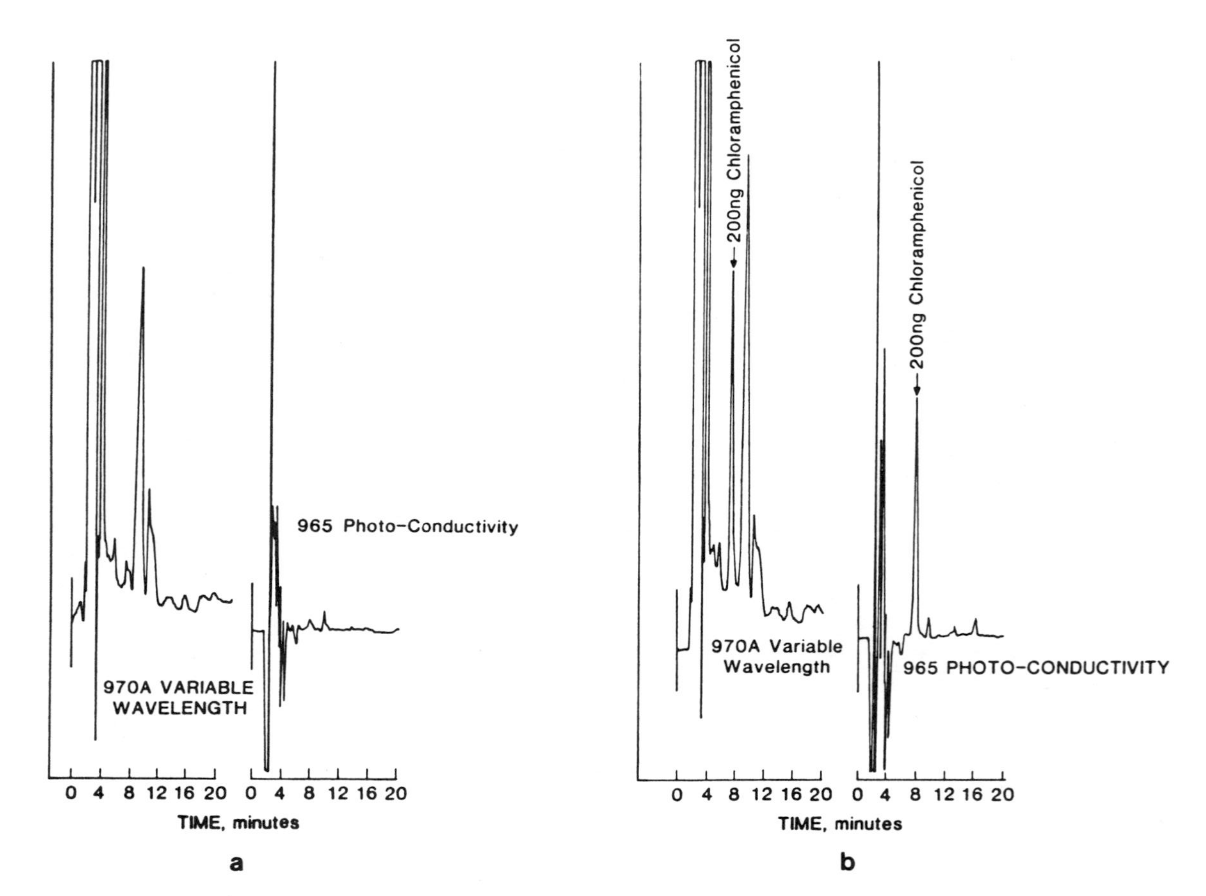

Figure 6-13 ■ Comparison between UV absorption at 254 nm and PCD with a zinc lamp. (a) Ethyl acetate extract of blank serum. (b) Ethyl acetate extract of serum spiked with chloramphenicol at the 500 ppb level. Conditions: Zorbax ODS column, 25 cm × 4.6 mm; mobile phase, 70 : 30 water : methanol at 1 ml/min. From W. A. McKinley, *J. Anal. Tox.* **5**, 209 (1981).

response to the PCD by identifying the analyte peak's retention time through its UV response. The PCD was capable of detecting a pesticide, captan, in both corn and strawberry extracts where UV detection lacked sufficient sensitivity.

These selected applications are intended to demonstrate the ability to sensitively and selectively detect a variety of compounds of analytical interest. Matrix interferences and low concentrations—problems common in clinical, environmental and food contaminant applications—are solved, at least in part, by the PCD. The commercial availability and response of the PCD to analytes important to several research disciplines indicate that it should be used more frequently in the future.

Photolysis / Fluorescence Detection of Chlorophenols. The PCD responds to halogenated organics through their photolysis to the corresponding haloacid. In some cases, other compounds with improved detectability are produced along with the haloacid. For example, a chlorinated phenol (Ar—Cl) would generate hydrochloric acid and phenol through the reaction sequence

$$\text{Ar—Cl} + h\nu \rightarrow \text{Ar}\cdot + \text{Cl}\cdot \quad (6\text{-}10)$$

$$\text{Cl}\cdot + \text{H—R} \rightarrow \text{HCl} + \text{R}\cdot \quad (6\text{-}11)$$

$$\text{Ar}\cdot + \text{H—R} \rightarrow \text{ArH} + \text{R}\cdot + \text{other products} \quad (6\text{-}12)$$

The other products include hydroxyphenols, dihydroxybiphenyls, polymers and other photodegradation products.[148, 153, 154] HPLC eluents, such as methanol, can serve as the hydrogen atom donor (H—R) in many free radical reactions. Other aromatic halides will undergo similar photolytic halogen elimination reactions.

Photolytic reactions of halogen-containing aromatic compounds can be used to enhance fluorescence detection. This method utilizes the larger organic fragments rather than the ionic products of the halogen radicals. Werkhoven-Goewie *et al.* converted chlorinated phenols to unsubstituted phenol with a 200 W Xe-Hg lamp in a PTFE OTR.[148] After the removal of the chlorine substituents from the aromatic ring [Eqs. (6-10) and (6-12)], the quantum yield of fluorescence increases dramatically. The lack of fluorescence from chlorinated phenols is a result of the intramolecular heavy-atom effect. Chlorine and other halogens increase a molecule's probability for intersystem crossing to the triplet state with a corresponding decrease in its fluorescence yield (as discussed in Chapter 1). Low nanogram detection limits were reported for monochlorinated phenols at a reaction time of 36 s, with no post-column reagent addition or flow segmentation. However, it was advantageous to remove oxygen from the eluent to minimize polymerization of the photoproducts. The linear range was 2 to 3 orders of magnitude.[148] Polychlorinated phenols were detected with better sensitivity by UV absorption. Longer reaction times were mentioned as a partial remedy for lack of sensitivity of the photochemical reaction methods toward these compounds. The reported detection limits, especially in combination with the enhanced selectivity, represent improvements relative to UV absorption at 220 nm of about a factor of 3 to 10 for the monochlorinated phenols.[148, 155] Electrochemical detection of these compounds was investigated; however, interferents in the preconcentrated urine samples caused the electrodes to foul.[148] Post-column reduction of the nonfluorescent Ce(IV) ion to fluorescent Ce(III) by phenolic compounds is another means of detecting these environmentally important analytes. The sensitivity and selectivity of the cerate fluorescence method are similar to the photolytic/fluorescence method.[156, 157]

Photooxidation Products of 9,10-Dibromoanthracene. Photochemical elimination of halogen atoms from aromatic compounds also has been observed in the case of 9,10-dibromoanthracene. In an unsegmented PTFE DOTR, irradiated with an 8 W fluorescent black lamp, significant yields of 9,10-anthraquinone and other dehalogenated products were observed.[158, 159] This work, which focused on determining whether quinones were formed from polycyclic aromatic hydrocarbons (PAH) in an aerobic photoreaction, utilized the photoproduct heart-cutting method described at the end of this chapter.

Photooxidative Cleavage of Methotrexate. The detectability of the antineoplastic agent methotrexate (2,4-diamino-N^{10}-methylpteroylglutamic acid) and its principal metabolities are improved after post-column photolysis to highly fluorescent 2,4-diaminopteridine derivatives.[160, 161] The 0.4 to 1.0 ng detection limits represent an ≈ 20-fold improvement in sensitivity over UV detection.[161] In order to carry out this reaction, H_2O_2 is spiked into the mobile phase (12 mM) and the effluent from the column is irradiated in a 90 cm PTFE coil by a germicidal lamp (254 nm). Residence times of 1 to 4 s are sufficient to effect nearly quantitative photolysis of the analytes. An interesting feature of the reaction is the role of the chemical oxidant. Presumably the H_2O_2 is photolyzed to hydroxyl radicals which subsequently cleave the reactive carbon–nitrogen bond. The optimization of the reaction detector and its interaction with the chromatographic separation are described. The strong solvent and pH dependences do not prevent the application of this detection scheme to practical separations. It is possible to determine clinical levels of methotrexate and two of its metabolites in plasma and urine samples without preconcentration steps.[160, 161] Less than 10 min are required to carry out the chromatographic separation (Fig. 6-14).[160]

Photolysis/Electrochemical Detection of Nitrosamines. Snider and Johnson designed a flow system to screen samples for the presence of nitrosamines.[145] As in the photolysis/Griess reaction systems which were described previously, the nitrosamines were photolyzed to form nitrite. The photolysis reaction took place under alkaline conditions in a quartz OTR that was irradiated by a 500 W xenon arc lamp. Nitrite was detected in an amperometric flow cell operated in the reductive mode.

Two ion exchange columns were used in this scheme; however, these columns do not separate the nitrosamine analytes from one another. The first column removes nitrite from the sample prior to the photoreactor, and the second traps and concentrates the nitrite produced from *N*-nitrosamines in the photoreaction while allowing the other photoproducts to elute. A rather complicated solvent switching system is required for this screening method. The photoreaction is carried out in dilute sodium hydroxide. After photolysis, the nitrite is eluted from the anion exchanger with 0.01 M $HClO_4$ which is mixed with 9.0 M HCl prior to entering the electrochemical detector cell. Between runs the scrubber column must be flushed with $HClO_4$. A residence time of 30 s in the photoreactor with an injection volume of 800 μl yields a detection limit of about 1 ng for *N*-nitrosodipropylamine. As configured, the photoelectroanalyzer is something of a hybrid between FIA and HPLC.[145] Sample throughput is quite low, on

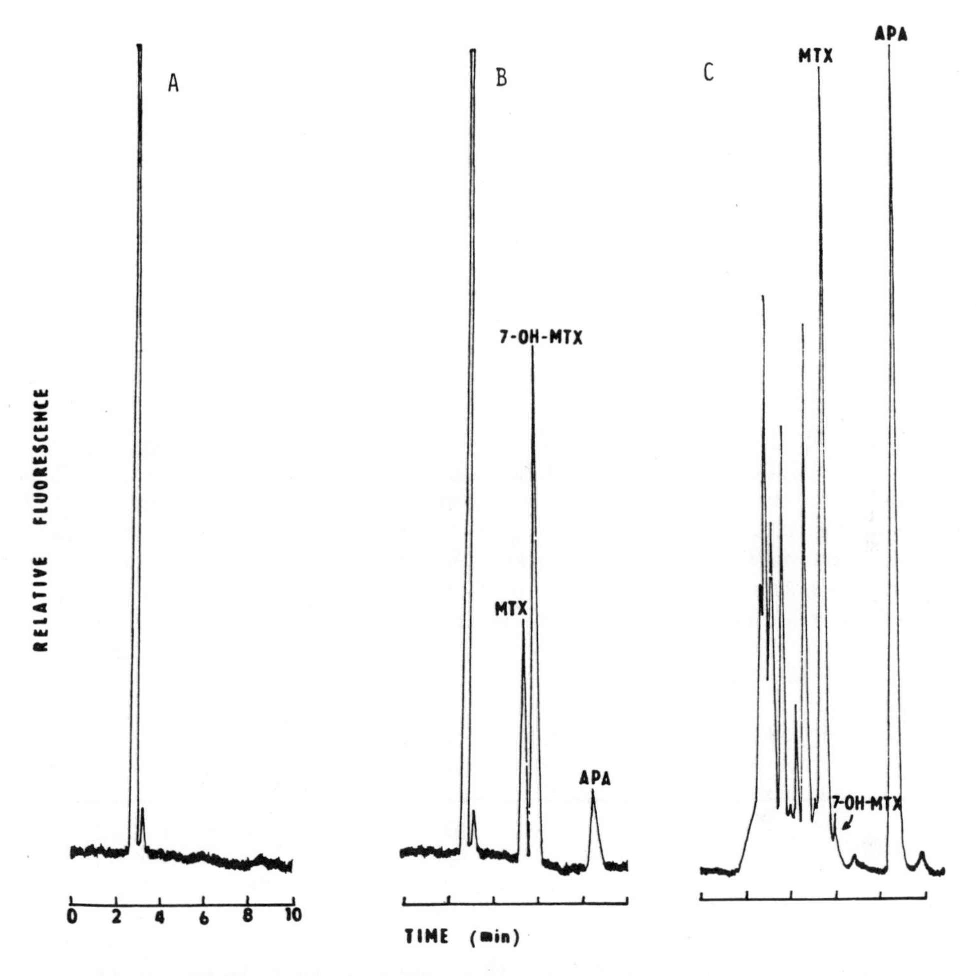

Figure 6-14 ■ Chromatograms of plasma and urine samples of a patient with an ascitic tumor treated with methotrexate (MTX). (A) Chromatogram of plasma before MTX administration. (B) Plasma 24 h after administration of 50 mg of MTX into the ascitic tumor (concentrations of MTX and its metabolites found in the sample: MTX, 192 ng/ml; 7-hydroxymethotrexate (7-OH-MTX), 417 ng/ml; 2,4-diamino-N^{10}-methylpteroic acid (APA), 95 ng/ml). (C) Chromatogram of urine from the same patient 4 hr after MTX administration (concentrations found: MTX, 11 μg/ml; 7-OH-MTX, 0.3 μg/ml; APA, 22 μg/ml). From J. Salamoun and J. František, *J. Chromatogr.* **378**, 173 (1987).

the order of 30 min/sample without including the isolation procedure. The requirement for aqueous mobile phases and the extreme pHs involved limit the adaptability of this system to reverse-phase HPLC.

Generalized Photolysis/Electrochemical Detection. Application of the photolysis/electrochemical detection method to HPLC has been reported by Krull *et al.*[12, 162–168] With post-column reaction apparatus similar to that of Snider and

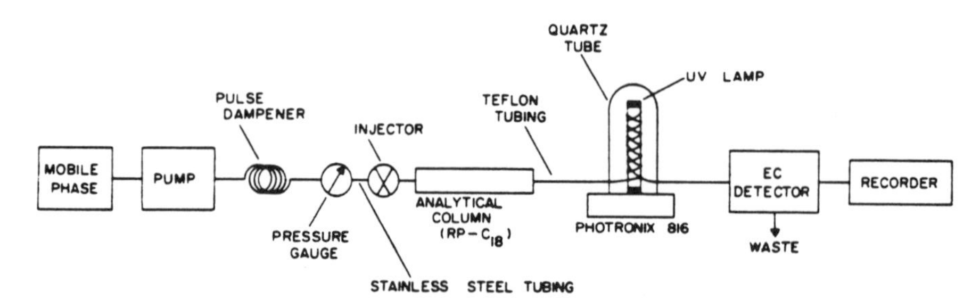

Figure 6-15 ■ Schematic diagram of HPLC-*hν*-EC instrumentation. From I. S. Krull, X.-D. Ding, C. Selavka, K. Bratin and G. Forcier, *J. Forensic Sci.* **29**, 449 (1984).

Johnson,[145] they have screened a wide range of analytes for electrochemically oxidizable photolysis products. In the *HPLC-photolysis-electrochemical* detector (HPLC-*hν*-EC), the eluent is irradiated by a low-pressure mercury lamp (Photronix #816 batch irradiator) in a Teflon OTR located after the analytical column.[162] Several of the common reverse-phase HPLC solvents are compatible with this detection scheme. After photolysis the electroactive products are detected in a thin-layer amperometric flow cell, generally operated in the oxidative mode. Either a single or dual-electrode HPLC detector may be used. Early studies with coiled Teflon OTRs were hampered by excessive band broadening and led to the development of new DOTR fabrication methods.[106] In addition to the usual reasons for preferring oxidative detection over the reductive mode, the permeability of the Teflon reactors to oxygen limits the range of reductive potentials. The reactor housings used for the HPLC-*hν*-EC detector to date have no provision for deoxygenation (Fig. 6-15).[162]

HPLC-*hν*-EC responds to a wide variety of therapeutic agents and drugs of abuse,[163] nitro compounds,[162] organothiophosphate pesticides,[164] beta lactams such as the penicillin family of antibiotics[165] and organoiodine compounds.[69, 166, 167] Especially with the dual-electrode flow cell, this detection method provides data rich in qualitative information. By comparing the relative response at two electrodes held at different potentials, screening the sample with the photoreactor turned off and varying the electrode potentials, an analyte can be identified with a reasonable amount of certainty. In Fig. 6-16, dual-electrode chromatograms of a serum extract, determined to contain ≈ 45 ng (50 μl injection) of phenobarbitol, are shown with lamp on and off.[163] Electrode potentials were optimized through hydrodynamic voltammograms performed in a plug-injection mode.

While the HPLC-*hν*-EC detector response is caused by photolytic products of the analyte which have oxidative electroactivity, in most cases the mechanism of their formation and even their identities are unknown. In order to better understand HPLC-*hν*-EC detection, a cell which can perform cyclic voltammograms on the photolyzed solutions has been developed.[69, 167] After batch photolysis in the HPLC photoreactor, the electrochemical properties of the photoproducts can be determined rapidly. This system can be used to screen compounds for response to HPLC-*hν*-EC detection and to optimize the reaction conditions. An electrochemical cell with direct electrode irradiation for performing cyclic voltammetry of excited states and

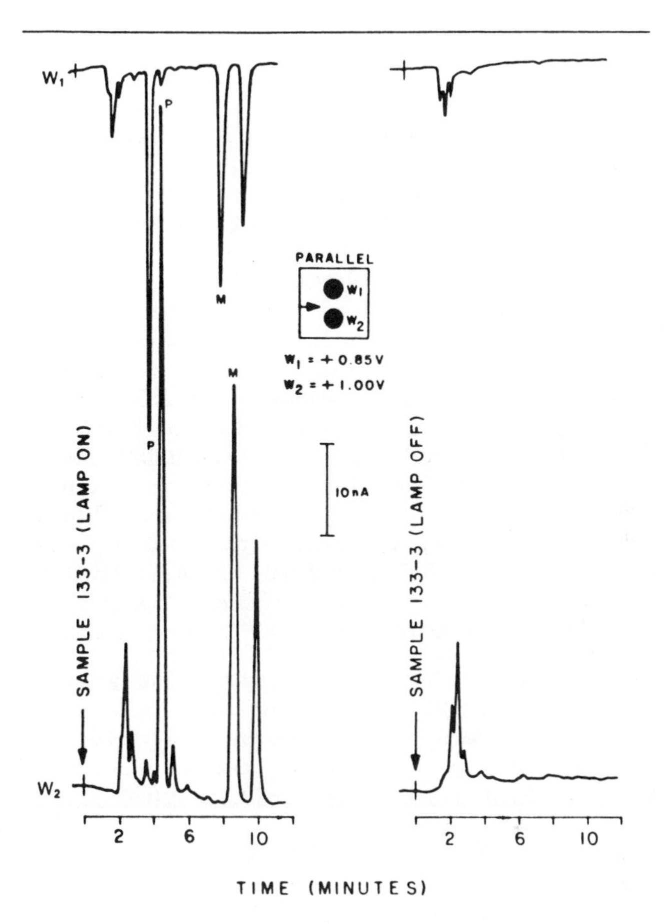

Figure 6-16 ■ Chromatogram of an extract of human serum displaying HPLC-*hν*-EC determination of phenobarbital. Conditions: Waters Radial Pak™ C_{18} column, 5 μm, 5 mm × 100 mm; mobile phase, MeOH : 0.2 M NaCl (40 : 60) at 1.6 ml/min. Peaks: P, phenobarbital; M, mephobarbital (internal standard). From C. M. Selavka, I. S. Krull and I. S. Lurie, *J. Chromatogr. Sci.* **23**, 499 (1985).

transient intermediates also is described.[69] Even at the present level of understanding, the wide linear range (often 3 orders of magnitude or more), high sensitivity (mid picogram to low nanogram detection limits) and selectivity of the HPLC-*hν*-EC system are impressive. The selectivity has been demonstrated through the detection of a variety of interesting analytes in serum, urine and blast-residue extracts.

A promising route to expanding the range of applicability of the HPLC-*hν*-EC detector is the development of derivatization reagents designed to enhance the photolytic response. Colgan *et al.* report that a silver picrate reagent will produce picryl ethers from epoxides and some organohalogen compounds.[168] These pre-column reactions were performed with both homogeneous and solid-phase reagents. The reagent's three aryl nitro groups allow sensitive detection of the derivatives by reductive EC, UV absorption at 220 nm and HPLC-$h\nu$-EC. Sensitivity of the two electrochemical methods is comparable, while the detection limits by UV absorption range from 4 to nearly 20 times greater. A review of derivatizations for HPLC/EC includes several other derivatizing reagents which contain nitro aromatic functionalities.[169] Though originally developed for UV absorption or reductive EC detection, they should enhance the response to HPLC-$h\nu$-EC and the PCD as well. Dinitrophenyl hydrazine derivatizations of aldehydes and ketones are excellent examples of the adaptation of classical qualitative organic reagents to new ends.

A strength of the publications involving HPLC-$h\nu$-EC is the inclusion of interlaboratory comparisons and method validation of the quantitative results obtained with this detector. All too often, publications presenting new detection methods will limit applications to *designer chromatograms*, that is, the separation and detection of prepared mixtures of stock chemicals with known retention times and responsiveness. Particularly for detectors that are touted as being extremely selective, it is important to demonstrate their capabilities with "real" samples, namely, the analytes for which they are designed in the matrices that they typically are found. Rigorous quantitative comparisons, such as those published by Krull and colleagues, are rare. This work, and the excellent agreement between methods, adds needed credibility to claims that post-column reaction-detection methods are suitable for routine analysis.

Photolysis of Organophosphorous Compounds. A two-step reaction scheme for detecting organophosphorous compounds has been developed by Priebe and Howell.[170, 171] In the first reaction, the analytes are photolyzed in the presence of ammonium peroxydisulfate to produce orthophosphate. This *photomineralization* reaction takes place in a fused-silica OTR irradiated by a 450 W xenon arc lamp.[171]

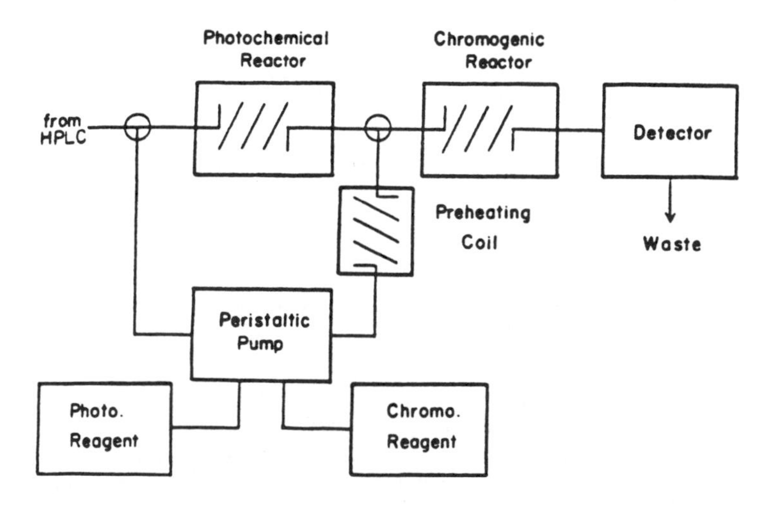

Figure 6-17 ■ Schematic diagram of the post-column reaction-detection system for organophosphorous compounds. From S. R. Priebe and J. A. Howell, *J. Chromatogr.* **324**, 53 (1985).

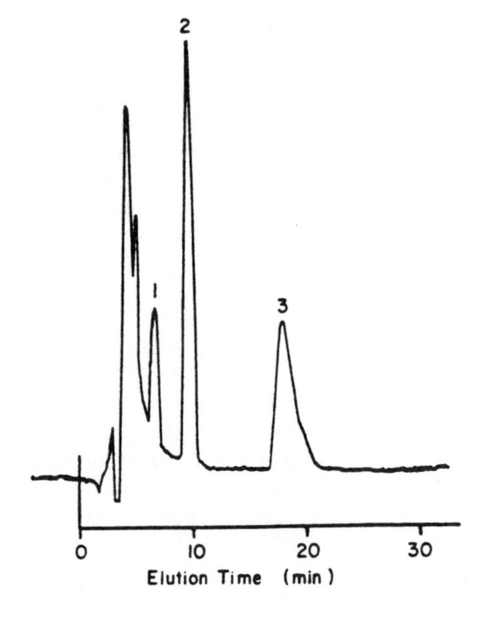

Figure 6-18 ■ Separation and detection of organophosphorous pesticides in a tomato sample. Peaks: (1) dylox, 4.7 ppm; (2) dimethoate, 4.0 ppm; (3) dichlorvos, 4.7 ppm. From S. R. Priebe and J. A. Howell, *J. Chromatogr.* **324**, 53 (1985).

Ammonium peroxydisulfate is added post-column to enhance the rate of photolysis as well as to protect against some organic interferences (Fig. 6-17).[171]

After the photolytic reaction, a mixed reagent solution consisting of disodium molybdate, potassium antimonyl tartrate and ascorbic acid dissolved in dilute sulfuric acid is introduced. Orthophosphate reacts with these reagents to form reduced heteropolymolybdate, a compound that absorbs long wavelength light ($\lambda_{max} = 885$ nm). All of the post-column reagents are added by one multichannel peristaltic pump.

Once again, this reaction was adapted from well documented static methods.[170] After optimizing the reaction detector in a plug-injection mode, it was coupled to HPLC.[171] Band broadening was quite serious and primarily was caused by the 90 s residence in the photoreactor rather than the 7 s long chromogenic reaction. Substitution of a suitable DOTR for the fused silica photoreactor used in this work would ameliorate the dispersion problem. Organic mobile-phase modifiers severely interfere with the photolysis reaction. In order to circumvent this problem, separations with totally aqueous eluents were developed.[171] Applications involving both urine and produce extracts demonstrate the excellent selectivity of this detector (Fig. 6-18).[171] Detection limits were in the low to middle nanogram range.

Photolysis Coupled to the *o*-Phthalaldehyde Reaction. Many pesticides and herbicides have moieties that can be photolyzed to primary amines.[172, 173] Given the utility of the *o*-phthalaldehyde/mercaptoethanol (OPA-MERC) reaction in detecting primary amines, a coupled photolysis/OPA-MERC reaction detector offers an attractive way to enhance both the selectivity and sensitivity of the detection of these compounds. The reaction of the OPA-MERC reagents with a primary amine produces a highly fluorescent isoindole derivative. Furthermore, the reaction is extremely fast, very selective toward primary amines and the nonfluorescent reagents are stable in solution.[1] For these reasons, the OPA-MERC reaction is ideally suited to post-column derivatization in HPLC. Previously, the OPA-MERC reaction had been used to detect some of the same compounds after post-column alkaline hydrolysis to primary amines.[174, 175] Primary amines were not produced from phenylurea herbicides by chemical hydrolysis, but phenylurea herbicides do respond to the photolysis/OPA-MERC scheme.[172, 175] However, some of the fluorescence may be caused by fluorophores formed photochemically rather than photolysis to amines and reaction with the OPA-MERC reagents.[173] Detection limits in the low nanogram to high picogram range have been reported.[173]

Photochemically Active Complexing Agent for Metal Ions. Shih and Carr report the development of a photochemically active precolumn derivatization reagent, *n*-butyl-2-naphthylmethyldithiocarbamate (BNMDTC), for the detection of metal ions.[176, 177] These complexes are stable and well suited for the preconcentration of trace metals. Divalent ions form 2:1 complexes, while trivalent ions form 3:1 complexes. Since the complexes are separable by reverse-phase HPLC, both the separation and detectability are improved. Irradiation of the BNMDTC complexes in a 0.69 ml Teflon OTR with a 175 W medium-pressure mercury lamp causes their decomposition. The overall reaction is

$$M(BNMDTC)_n + h\nu \rightarrow n\text{BNM-amine} + nCS_2 + M^{n+} \quad (6\text{-}13)$$

n-butylnaphthylmethylamine (BNM-amine) is one of the products that forms during photolysis.[177] As other aromatic amines, BNM amine is highly fluorescent. A possible, and potentially more selective and sensitive alternative to fluorescence detection of the BNM amine is peroxyoxalate chemiluminescence.[178] Photolysis/fluorescence detection provides a linear range in excess of 2 orders of magnitude with detection limits in the area of 10^{-8} M for 200 μl injections of the metal ion complexes.[177] This represents over an order of magnitude improvement relative to detecting the nonfluorescent complexes by UV absorption at 221 nm.[176] Selectivity also is enhanced by photolysis/fluorescence detection.

A special advantage, which is found in both the UV absorption and photolysis/fluorescence modes of detecting the BNMDTC complexes, is that the spectral properties of the complexes are not strongly dependent on the metal ion. Thus, it is possible to detect a wide variety of BNMDTC metal ion complexes under the same conditions. A sample chromatogram demonstrates the ability to separate and detect these complexes (Fig. 6-19).[177] Another unique feature of this system is the use of a pre-column reagent as a photochemically activated "release tag." Not only does the pre-column reaction facilitate the preconcentration and separation of metal ions, but it has optimal properties for reaction detection as well. Namely, the reaction is photochemically activated, requires no additional reagents and is reasonably fast. This work demonstrates that pre-column derivatization can be compatible with post-column reaction detection. Hopefully, more examples of this approach will be developed.

Photoionization Detection in HPLC. Photoionization detectors (PID) for gas chromatography (GC) have provided a sensitive and selective alternative to the analyst. Since GC is not limited by detector availability to the extent that HPLC is, the fact that such a system could find a commercial niche in GC inspired its adaptation to HPLC. For GC-PID the vacuum UV resonance lines of a low-pressure, rare-gas

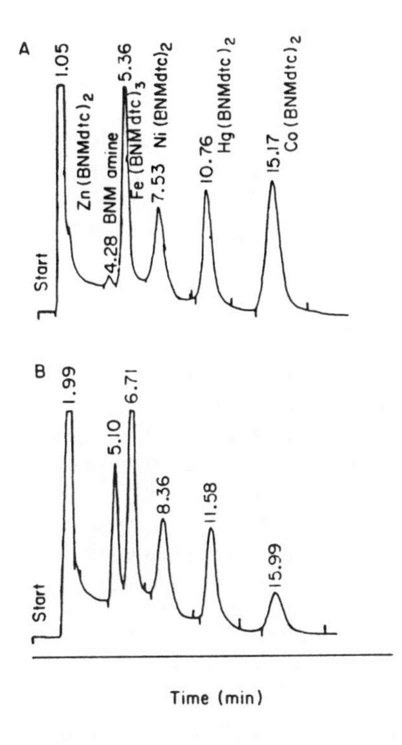

Figure 6-19 ■ Chromatograms of BNMDTC complexes with and without photochemical reactor. Conditions: μBondapak C_{18} column (10 μm packing), 90 × 4.6 mm; mobile phase, 95 : 5 methanol : water with 1 mM Tris (pH 8.25) at 1 ml/min. Samples: 1×10^{-6} M of each component, 200 μl injection. Detector A, IBM 254 nm fixed wavelength detector, 0.01 AUFS; detector B, Shoeffel FS-970, 220 nm, 1×10^{-6} AFS. From Y.-T. Shih and P. W. Carr, *Anal. Chim. Acta* **159**, 211 (1984).

discharge lamp provide the ionizing radiation. Molecules with nonbonding electrons and π systems are the primary analytes. As is typically the case in HPLC detection, the physical properties of the eluent complicate the adaptation of the PID from GC to HPLC.

Photoionization is a more practical means of detection in the gas phase than in condensed media. Ion mobilities are much lower in solution and the probability for recombination of the ion/electron pairs or capture of the electrons by other molecules is increased. In a practical sense, this means that it is much more difficult for the electrodes to collect the photocurrents efficiently. Furthermore, the energy required to photoionize molecules in the liquid phase is significantly lower relative to the gas phase due to solvation effects. The decrease in the ionization potential precludes the use of low-pressure rare-gas discharge lamps because they will ionize the HPLC eluents as well as the analytes. These aspects of photoionization and the subsequent collection of the product ions have led to the development of two distinct types of photoionization detectors for HPLC.

The first approach to coupling HPLC with PID involves the vaporization of the eluent as it exits the column.[179, 180] In this way, the standard PID available from H · NU Systems (Newton Highlands, MA) can be used with the normal photoionization sources.[180] Driscoll *et al.* report detection limits in the 3 to 700 ng range for a variety of organic compounds by this method.[180] The HPLC flow is split and vaporized in a GC oven before entry into the PID. A makeup flow of helium is added to the vaporized eluent. This system is suitable for both normal and reverse-phase chromatography. Clearly, this approach is limited to nonionic compounds which are relatively volatile. Such compounds usually can be separated and detected more efficiently by GC.

Liquid phase photoionization detection cells also have been reported.[137, 181–185] Excitation source selection is extremely important, as the photoionization of the analyte must be carried out in the presence of the eluent. Locke and co-workers built a PID which uses the continuum emission from a xenon discharge lamp operated at 200 torr. The lamp has a short wavelength cutoff of ≈ 160 nm.[181] This system is limited to normal phase chromatography by the photoionization of water. However, it is quite sensitive (low picogram detection limits for some compounds) and is reported to have a linear range of 6 orders of magnitude.[181]

Two-photon ionization by lasers also can be used for liquid phase PID.[137, 182–185] Nitrogen (337.1 nm) and XeCl excimer lasers (308 nm) produce radiation which is not absorbed by the common HPLC eluents. Laser PIDs are amenable to both reversed and normal phase chromatography, because water is not ionized. The high power densities of these pulsed lasers are required for the two-photon mechanism to produce significant photoionization yields. In the windowless multiple detector cell, designed by Voigtman *et al.* (see also Fig. 6-9), it is possible to monitor fluorescence, photoionization and photoacoustic response simultaneously.[137, 182, 183] PID response in this system is comparable to UV absorption at 254 nm for most of the analytes.[137] Another two-photon PID cell with better sensitivity has been described.[184, 185] In this cell, the stainless-steel inlet and outlet capillaries serve as the ion collection electrodes.[185] Since scattered light does not present a problem in the PID mode, quartz windows are used to seal the cell and prevent solvent evaporation. Detection limits in the 5 to 30 pg range were reported for several polycyclic aromatic hydrocarbons (PAHs) and halogenated PAHs. The liquid phase PID system ranges from 0.65 to 17 times as sensitive as UV absorption at 254 or 337 nm for the analytes which were investigated.[185] Two-photon PID schemes are more selective than those based on the absorption of vacuum UV radiation. In order to be detected, the molecule must absorb strongly in the 300 to 330 nm region. For this reason, the two-photon laser PID applications primarily have involved PAHs. Since the analytes need not be volatilized, the two-photon PID is applicable to the types of analytes which are most important to HPLC.

New Directions. Another interesting possibility is the application of semiconductor electrodes to photolytic detection schemes. The unique properties these electrodes offer are the possibility of photoexciting either the electrode, the analyte or both simultaneously. Because there are no charge carriers within the band gap of a semiconductor, semiconductor electrodes can be selected so that the excited states of the analyte will not be quenched as efficiently as they are by metal electrodes. In the case where the electrode must be excited for the detection of the analyte, the effective volume of the flow cell can be reduced by decreasing the area of the electrode irradiated by the light source. Such an electrode would ease the problem of peak dispersion in electrochemical and reaction-detection flow cells for microcolumn HPLC. When both the electrode and the analyte are excited, a large amount of energy is available to drive the reaction.

In a practical sense, response to a broad class of compounds often is more useful than the unique photoconversion of a specific analyte. This is particularly evident in the photoconductivity detector case. Its response to several analytes whose determinations are difficult without the improved selectivity and sensitivity of the PCD illustrates its potential.[63, 146, 147, 149–152] To date, the PCD is the only commercially available photochemical reaction detector for HPLC. The HPLC-*hν*-ECD system appears to have great potential as well. Better characterization of the response to compounds which contain photolyzable functionalities will increase the value of these sensitive reaction detection schemes. The photolytic reactions specific to certain substituent groups such as *N*-nitroso compounds and carbamate pesticides are of interest as well. If the photochemical reaction produces products that respond to reagent-based reaction-detection methods, extremely high selectivity can be achieved.

Photochemical Transformation of the Analyte to a More Detectable Compound

As mentioned previously, this class of reactions can be subdivided based on the molecularity of the initial step. In cases where the mechanism is in doubt this subclassification is somewhat tenuous.

Intramolecular Reactions

Of the intramolecular photochemical reactions which have been applied to HPLC detection, a vast majority are photocyclizations and other rearrangements that result in cyclic products. One of the most studied groups of organic reactions, especially from a theoretical standpoint, photocyclization reactions also may be applied to the detection of some analytes. These reactions usually are specific to a narrow class of compounds or a particular analyte of interest. For example, in the reaction

hν

H

H

O_2

(6-14)

stilbene is photochemically converted to dihydrophenanthrene.[186] Dihydrophenanthrenes are oxidized to highly fluorescent phenanthrene derivatives by a variety of oxidants including molecular oxygen.[186–190] Zweig has reviewed the literature involving photochemical reactions which generate fluorescent products.[191] A few of these photofluorescence reactions already have been adapted to HPLC detection, and some of the others may be suited to this purpose as well. Increased conjugation and rigidity of the product's ring system, relative to the analyte, are common features of the photofluorescence reactions. Rigid ring systems have lower rates of nonradiative deactivation.

Detection of Stilbene Derivatives. While stilbene itself is of little interest, the antineoplastic agent tamoxifen, the ovulatory stimulant clomiphene and the synthetic estrogenic agent diethylstilbestrol (DES) [3,4-bis-(*p*-hydroxyphenyl)-3-hexene] are derivatives of stilbene. Enhanced detection of all of these compounds and some of their metabolites, through photofluorescence reactions, has been demonstrated.[60–63, 192–198] Only *cis*-stilbenes are capable of undergoing this photocyclization reaction.[186, 189] However, a rapid photoisomerization between *cis*- and *trans*-stilbenes allows both isomers to be detected.[186, 189, 199]

The ability to detect trace residues of DES in biological matrices is essential to the enforcement of the ban on its use in promoting livestock weight gain. Goodyear and Jenkinson developed a batch photofluorescence method for DES prior to the discovery of its carcinogenicity.[192, 193] This method had low parts per billion (10^{-9}) sensitivity. Adaptation of this reaction to on-line, post-column photofluorescence detection in HPLC was investigated by Rhys Williams and co-workers.[194] With a HPLC photoreactor designed by Twitchett, Williams and Moffat[200] they achieved subnanogram detection limits for DES in spiked urine samples (Fig. 6-20).[194] Although metabolites of DES were not studied, it is likely that some would respond to this detection scheme. Hexestrol (HES), [3,4-bis-(*p*-hydroxyphenyl)-hexane] was shown not to undergo the photofluorescence reaction.[194] Since HES lacks the central double bond, one would anticipate this result.

The effectiveness of tamoxifen [(*Z*)-2-[*p*-(1,2-diphenyl-1-butenyl)-phenoxy]-*N*,*N*-dimethylethylamine] in the treatment of breast cancer made the development of methods that could monitor therapeutic levels of this compound and its metabolites a necessity. Triphenylethylenes, such as tamoxifen, will cyclize to phenanthrene derivatives through reaction (6-14). Mendenhall and colleagues found that batch irradiation of these compounds dramatically increased their detectability. Batch photofluorescence methods for stilbene derivatives are limited in the sense that the compound of interest must be separated by extraction from its structurally similar metabolites. Otherwise, these metabolities will undergo the cyclization reaction and fluoresce as well. An ion-pair extraction procedure allows tamoxifen to be determined by direct fluorometry in the irradiated extract.[60] A simpler extraction requires chromatographic separation of the irradiated extract; however, the important metabolites of tamoxifen as well as the parent compound can be determined.[60, 61] One of the tamoxifen metabolites, (*Z*)-2-[*p*-(1-*p*-hydroxyphenyl-2-phenyl-1-butenyl)-phenoxy]-*N*,*N*-dimethylethylamine, reacts to form the fluorescent phenanthrene derivative during the extraction. For this reason, the pre-column irradiation of the extract is preferred.[61] On-column injections of low picogram amounts of these compounds were detectable (100 ng/ml in plasma).

Pre-column irradiation of the samples does have the drawback of increasing the complexity of the sample. Photodegradation reactions can produce interfering compounds which render the separation more difficult. Post-separation irradiation procedures were developed for TLC[196] and HPLC.[197, 198] The post-column photoreactor for HPLC consisted of a quartz OTR sandwiched between two short wavelength UV

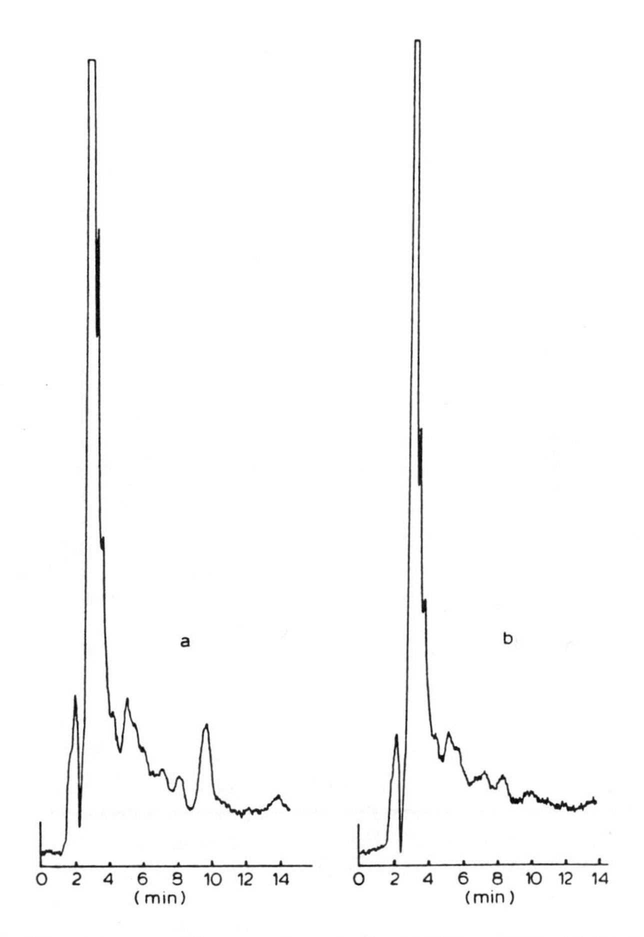

Figure 6-20 ▪ Chromatograms of human urine extracts: (a) 12.5 ng DES added before extraction ($\approx$ 2 ng on column). (b) Blank. DES peak elutes at $\approx$ 9.8 min. From A. T. Rhys Williams, S. A. Winfield and R. C. Belloli, *J. Chromatogr.* **235**, 461 (1982).

lamps. The irradiation time, less than one second, was much shorter than that used in the batch irradiation procedures.[197] Detection limits of 200 pg were reported for tamoxifen and its major metabolite *N*-desmethyltamoxifen. Other metabolites were detected with similar sensitivity in both normal and reverse-phase chromatographic systems. Post-column photochemistry was shown to reduce the analysis time, improve the control of the reaction time, reduce the potential for interference during the photoreaction and simplify the chromatography.[197]

Clomiphene (1-[*p*-(β-diethylaminoethoxy)phenyl]-1,2-diphenlchloroethylene) is another clinically important triphenylethylene derivative. Harman, Blackman and Phillipou developed a post-column photochemical reaction detector which was suited to clomiphene and its metabolites.[195] The 60 pg detection limits they report represent

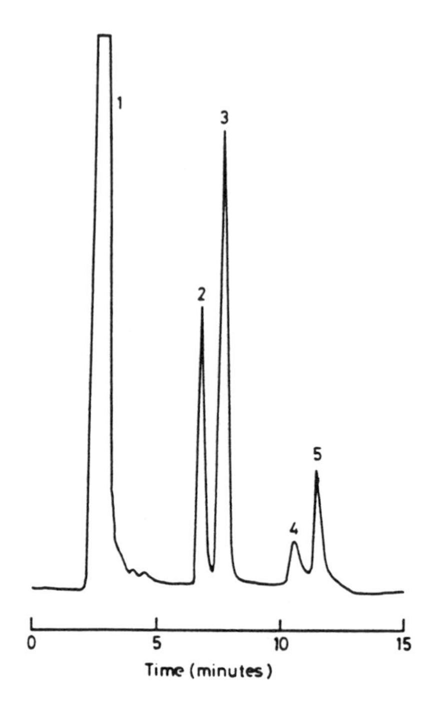

Figure 6-21 ■ HPLC chromatogram of a plasma sample from patient J. Peaks: 1, solvent peak; 2, *cis*-clomiphene; 3, *trans*-clomiphene, 4 and 5, clomiphene metabolites. From P. J. Harman, G. L. Blackman and G. Phillipou, *J. Chromatogr.* **225**, 131 (1981).

nearly a 100-fold improvement in sensitivity relative to UV absorption. Selectivity is vastly improved as well. A maximum fluorescence intensity was reached after a reaction time of ≈ 16 s in the PTFE OTR. The reactor was irradiated by a medium-pressure mercury lamp. The normal phase separation could resolve the cis and trans isomers of clomiphene as well as their metabolites (Fig. 6-21).[195] Only the cis isomer is pharmaceutically active and the cis and trans isomers have different rates of metabolism. Hence, a method must be able to distinguish between these isomers to be clinically useful.[195]

Frith and Phillipou attempted to perform this reaction pre-column and found that a complicated mixture of photoproducts results.[62] Furthermore, photoconversion between the cis and trans isomers and the inability to resolve the resulting mixture of phenanthrene photoproducts by HPLC, led them to the conclusion that the post-column photoreaction is vastly superior to the pre-column mode.[62]

Cannabinoid Detection. Drugs of abuse such as the cannabinoids (e.g., THC) and ergot alkaloids (e.g., LSD) have been detected through post-column photoreactions.[200] Cannabinol forms a product with enhanced fluorescence which improves its detectability compared to UV absorption. Relative to its weak native fluorescence, the emission is shifted to longer wavelengths. Irradiation of Δ^9THC produces a similar increase in intensity and red shift in its fluorescence emission spectrum.[201] Production of this fluorophore proceeds through a two stage, four-step mechanism which consists of a ring opening and hydrogen transfer, followed by a photoinduced dehydrogenation and

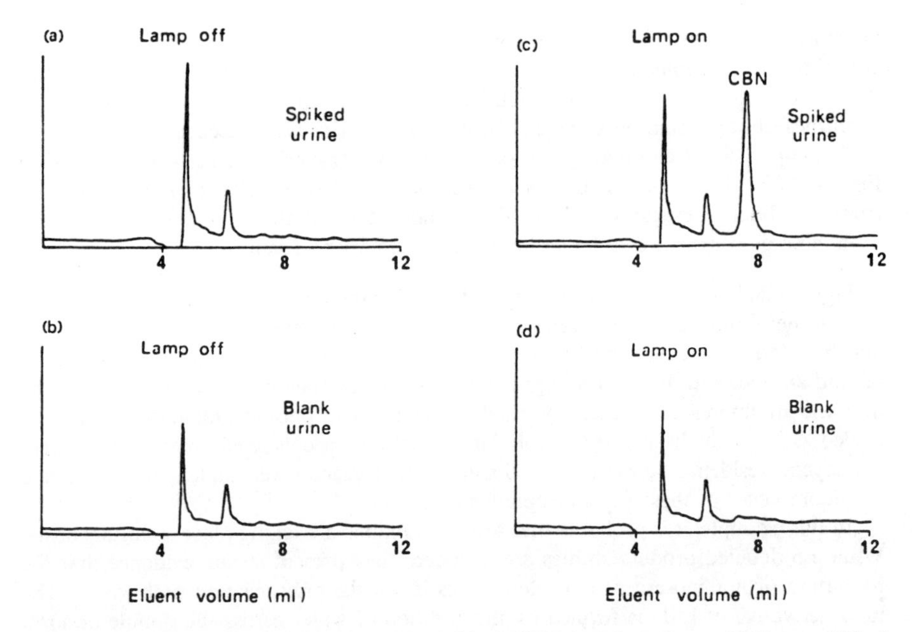

Figure 6-22 ■ Photochemical detection of cannabinol (CBN) in urine extracts. (a) Extract (500 μl) of urine containing 20 ng/ml of CBN. (b) Extract of blank urine. (c) and (d) Same as a and b, respectively, but chromatographed with UV irradiation. Conditions: Partisil PAC column; mobile phase, isooctane : dioxane 82.5 : 17.5; at 1 ml/min; detector, MPF-2A fluorometer, $\lambda_{ex} = 258$ nm, $\lambda_{em} = 362$ nm, slits 12 and 15 nm, sensitivity $\times$ 5. From P. J. Twitchett, P. L. Williams and A. C. Moffat, *J. Chromatogr.* **149**, 683 (1978).

ring closure[202]:

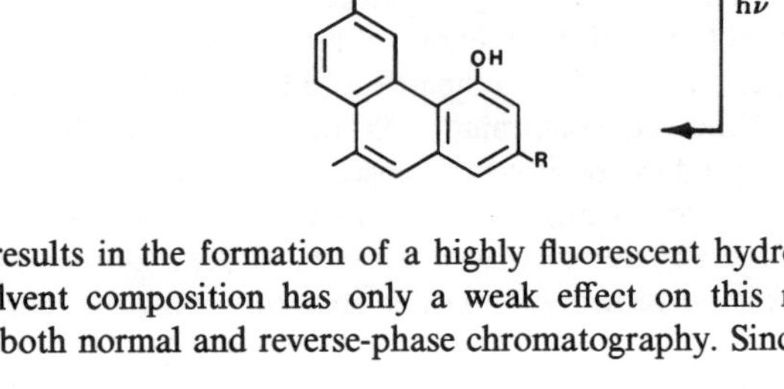

OH R $\xrightarrow{h\nu}$ OH HO R

$h\nu$ (6-15)

OH R

This ultimately results in the formation of a highly fluorescent hydroxyphenanthrene derivative.[202] Solvent composition has only a weak effect on this reaction, so it is compatible with both normal and reverse-phase chromatography. Since this reaction is extremely fast, 1 to 5 s under the conditions employed, very little loss of resolution occurs in the unsegmented OTR. A 100 W medium-pressure mercury lamp served as the photoreactor source for a 0.25 mm i.d. fused-silica capillary reactor. Subnanogram quantities of cannabinol in urine can be determined with this detection scheme.

A comparison of chromatograms of spiked and blank urine extracts is presented in Fig. 6-22.[200] The extraction procedure includes a hydrolysis step to free cannabinol from metabolic conjugates such as glucuronides and sulfates.[200] Conjugation of the cannabinoids is expected to be an important excretory mechanism.

Ergot Alkaloid Identification. Twitchett, Williams and Moffat also demonstrated the on-line photochemical conversion of LSD to a nonfluorescent product.[200] Scholten and Frei found that by running two chromatograms, one with the photoreactor lamp on and another with it shut off, ergot alkaloids such as LSD could be identified by the decrease in fluorescence intensity resulting from their photodegradation.[203] An air-cooled, 150 W high-pressure xenon lamp and an unsegmented quartz OTR were employed. Residence times of 20 to 50 s in the photoreactor were sufficient to eliminate the fluorescence of these psychotropic compounds.

In this example, it is probable that several reactions are taking place simultaneously. While no detailed product studies are included, they present strong evidence that the formation of nonfluorescent lumi derivatives is not the only reaction pathway.[203] The lumi derivative of LSD is formed by the addition of water across the double bond at the 9,10 position[204]:

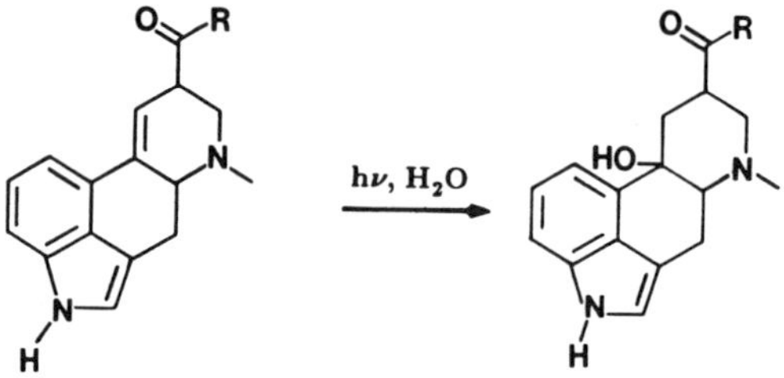

$\xrightarrow{h\nu,\ H_2O}$ (6-16)

Their argument is based on the fact that 9,10-dihydro analogs of LSD, as well as indole itself, exhibit similar decreases in fluorescence after irradiation.[203] A ring opening of the indole moiety probably is an important reaction pathway. The photodecomposition of other hallucinogenic drugs during the measurement of fluorescence spectra has been reported.[205]

This reaction scheme is designed specifically as a confirmatory test for ergot alkaloids and is not intended to improve the quantitative determination of these analytes. However, lamp on/off comparisons add an additional means of achieving selectivity and qualitatively confirming peak identity in photochemical reaction-detection systems that produce positive responses as well. In this way, it is possible to discriminate against compounds that have a native response to the detector and add the dimension of ratioing the response between the modes. By running a chromatogram with the photoreactor turned off, one is able to screen the sample for coeluting compounds which natively respond to the detector. If two detectors are available it is possible to split the column effluent and run both the lamp on and lamp off chromatograms simultaneously. In no way is the ability to perform quantitative determinations compromised by this procedure.

Benzodiazepines. The photochemical rearrangement of nonfluorescent benzodiazepine anxiolytic agents to fluorescent quinazolinone derivatives is an efficient and simple way to improve their detectability.[25-27, 72, 102, 206-209] Strojny and de Silva proposed the reaction of demoxepam, a 1,4-benzodiazepine derivative, based on their experience with a batch photooxidation/laser induced fluorescence detection scheme[207]:

(6-17)

Brinkman and co-workers reported GC-MS evidence for other photoproducts which were missing the chlorine substituent.[27] Photolytic dechlorination would help explain the high fluorescence efficiency of the photoproducts.

Once again, the post-column photochemical reaction schemes for benzodiazepines evolved from batch spectrofluorometric methods. The method of Koechlin and D'Arconte involved the selective extraction of chlordiazepoxide, followed by thermal hydrolysis to its lactam derivative in an acidic buffer. This lactam was photochemically converted to a fluorescent compound in an alkaline medium.[208] Their method can detect therapeutic levels of this drug up to 48 hr after oral administration. A modification of the procedure allows the lactam (demoxepam), a pharmacologically active metabolite of chlordiazepoxide, to be determined as well.[208] Variable levels of interferents were observed in plasma samples obtained from different subjects. Urine samples had even worse blank and background interference problems. Schwartz and Postma further modified this batch method to allow the determination of an additional metabolite, *N*-desmethylchlordiazepoxide.[209] The original assay for chlordiazepoxide did not discriminate between the parent compound and the *N*-demethylated metabolite. While Strojny and de Silva achieved a fivefold improvement in sensitivity ($\lesssim 10$ ng/ml detection limit for demoxepam) by using a laser induced fluorescence detector,[207] the interference problems encountered by all the batch photofluorescence measurements made it clear that an efficient method of separating these drugs from biological matrices prior to irradiation would considerably improve the reliability of their determination.

An air-segmented post-column photoreactor for HPLC solved these interference problems to a large extent.[27] The photoreaction, after optimization of the reaction time and solvent composition in both stopped and dynamic flow mode, produced a 100 pg

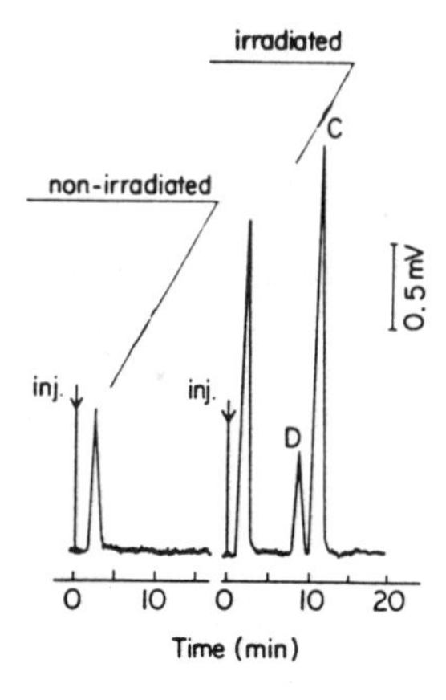

Figure 6-23 ■ Chromatogram of human serum sample spiked with 0.7 ppm of clobazam and desmethylclobazam, with and without irradiation. Conditions: Mobile phase, methanol : 0.01 M sodium acetate 1 : 1 at 0.9 ml/min. Detector conditions: Gain, 3; sensitivity range 10; slit widths, 10 nm; response, slow; recorder at 5 mV. Peaks: C, clobazam, D, desmethylclobazam. From A. H. M. T. Scholten, U. A. Th. Brinkman and R. W. Frei, *Anal. Chim. Acta* **114**, 137 (1980).

detection limit for demoxepam. This is an order of magnitude better than UV absorption detection. Since the PTFE OTR was air-segmented, very little band broadening occurred during the 110 s residence time. The reported linear range was 3 orders of magnitude.[27]

Clobazam, a 1,5-benzodiazepine anxiolytic agent, also is converted to a fluroescent compound upon irradiation.[25, 26, 72, 102, 206] Although a sensitive batch method that could determine this compound without interference from its main metabolite (desmethylclobazam) was developed,[206] HPLC methods were investigated as well.[25, 26, 102] Significant reductions in the optimal irradiation times were observed in the flow reactors compared to the batch methods. A residence time of 15 s in a PTFE OTR irradiated with a 200 W xenon-mercury arc lamp, gave 20 and 50 pg detection limits for clobazam and desmethylclobazam, respectively.[26] A 28 s residence time in an unsegmented quartz OTR irradiated by the same lamp was about three times less sensitive.[25] The ability of this system to separate clobazam and desmethylclobazam from matrix interferents was demonstrated with a spiked urine sample (Fig. 6-23).[25]

In addition to providing excellent detection capabilities for benzodiazepine tranquilizers, this work produced major improvements in the design of photochemical reactors for HPLC.[25–27, 72, 102] Clobazam served as one of the model compounds in the work of Uihlein and Schwab.[102] The ≈ 200 pg detection limits reported for the HPLC photofluorescence scheme were secondary in importance to the first reported use of a PTFE DOTR as a HPLC photoreactor. Subnanogram photofluorescence detection limits for the anthelmintic fenbendazole offered improvements over UV detection as well.[102] Benzodiazepines also respond to the HPLC-$h\nu$-EC detector, as described earlier, but with somewhat less sensitivity than the photofluorescence method.[163]

Intermolecular Reactions

Photoreduction of Quinones. Intermolecular reactions also may be used to produce a more detectable compound from the analyte. For example, quinones, particularly the K vitamins, have been detected by several workers through their reduction. Quinone photoreduction proceeds through the hydrogen abstraction reaction sequence[210, 211]:

$$Q + h\nu \rightarrow {}^1Q^* \quad (6\text{-}18)$$

$$ {}^1Q^* \rightarrow {}^3Q^* \quad (6\text{-}19)$$

$$ {}^3Q^* + CH_3OH \rightarrow \cdot CH_2OH + \cdot QH \quad (6\text{-}20)$$

$$\cdot CH_2OH + Q \rightarrow \cdot QH + CH_2O \quad (6\text{-}21)$$

$$2 \cdot QH \rightarrow Q + QH_2 \quad (6\text{-}22)$$

$$Q = \text{quinone analyte}$$
$$\cdot QH = \text{semiquinone radical}$$
$$QH_2 = \text{dihydroquinone(dihydroxy PAH)}$$

The lowest triplet state of the quinone abstracts a hydrogen atom from a compound which has a weak carbon-hydrogen bond—methanol is the *hydrogen-atom donor* (HAD) as shown—to produce the semiquinone radical and an α-hydroxy alkyl radical. HAD compounds typically contain nitrogen or oxygen with at least one carbon-hydrogen bond α to the heteroatom.[212, 213] A large majority of excited-state, quinone molecules undergo intersystem crossing to the triplet manifold.[214] High concentrations of the hydrogen-atom donor assure that (6-20) will be the predominant deactivation pathway for the triplet molecules.[211, 215] In the final step of this sequence, two semiquinone radicals rapidly disproportionate to a dihydroquinone and a molecule of the parent quinone.[215–218] This photoreduction pathway only will occur under anaerobic conditions. If oxygen is introduced, the dihydroquinone will be oxidized back to the parent quinone with the production of hydrogen peroxide[215, 219–221]:

$$QH_2 + O_2 \rightarrow Q + H_2O_2 \quad (6\text{-}23)$$

The reduced quinone, a dihydroxy(polycyclic)aromatic compound, is both highly fluorescent and electrochemically oxidizable. K vitamins are examples of naphthoquinone analytes which can be detected by photoreduction/fluorescence[123, 126, 222–224]:

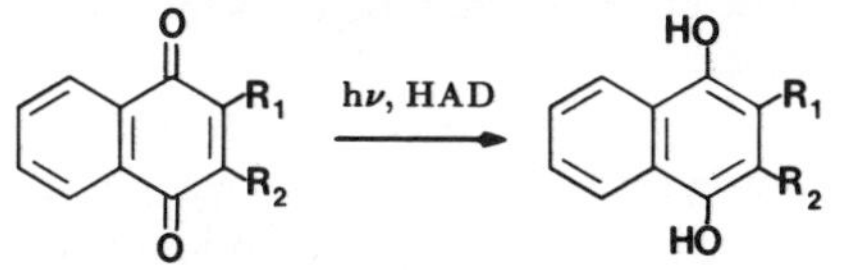

(6-24)

K vitamins also have been detected by electrochemical[225–227] and chemical[126, 228, 229] reductions, coupled with fluorescence at sensitivities which are comparable to the photochemical reductions. Direct reductive electrochemical[230–232] and dual-electrode serial reduction and reoxidation detection are possible as well.[233, 234]

The photochemistry of vitamins K_1 and K_2 is complicated by the cyclization and photodegradation of their hydrophobic side chains.[212, 222, 235–244] Cyclic products, such as the naphthochromenol

$$\text{(quinone)} \xrightarrow{h\nu,\ \text{HAD}} \text{(naphthochromenol)} \qquad (6\text{-}25)$$

can form through an intramolecular hydrogen abstraction that is followed by cyclization of the phytyl or prenyl side chain, respectively.[126, 238, 239, 243, 244] The naphthochromenol is the predominant reaction product under anaerobic conditions, while aerobic conditions favor the formation of orthoquinone methides, hydroperoxides, cyclic trioxanes, 2,3-epoxides and naphthoquinone derivatives resulting from oxidative cleavage of the side chain.[239–244] Cyclization renders some of the photoproducts of biogenic K vitamins less likely to reoxidize back to the parent quinone in the presence of oxygen as compared to the photoreduction products of other quinones (6-23).[222, 223, 237, 238] Even when oxygen and a high level of HAD substrates are present during the photoreaction, vitamin K_1 is a poor sensitizer for producing H_2O_2.[158, 159] The importance of reaction (6-25) will depend on the reaction conditions in the photoreactor. Photoreduction of K_1 to fluorescent products in aerobic solution has been reported; however, it is unlikely that the dihydroxynaphthalene derivative (6-24) is the primary photoproduct.[222, 223, 237] Menadione (2-methyl-1,4-naphthoquinone) remained nonfluorescent when it was irradiated under these aerobic conditions.[222] A TLC study of the photoproducts showed that the fluorescence formed during the irradiation of aerobic solutions of K_1 was caused by intermediates in the photoreaction.[222] Nevertheless, a batch-mode, photochemically generated fluorescence method dramatically improved both the detectability (5 ppb) and linear range for vitamin K_1 determinations without the requirement for oxygen removal.[223, 224] A drastic change in the relative yield, rate of formation and identity of the photoproducts is observed between the aerobic and anaerobic photoreactions of vitamin K_1.[239–243] Differences in the optimal excitation and emission wavelengths reported for fluorescence detection after reduction, may be accounted for by the relative yield of these two compounds [(6-24) and (6-25)] and other photoproducts.[126, 238] On the other hand, these variations may be caused by differences between the excitation sources and optics of the fluorometers themselves.

Table 6-1 compares the detection limits achieved by reducing K vitamins followed by fluorescence and electrochemical detection to direct UV absorption.[245, 246] While these systems vary greatly in complexity, our work, photoreduction fluorescence detection of quinones (PRFQ), demonstrates that a relatively simple reaction detector can be extremely sensitive.[123] The only modification of the HPLC system, besides the addition of a photoreactor, is the need to deoxygenate the mobile phase and photoreactor with nitrogen bubblers. No reagents are added to the eluent or introduced post-column. Furthermore, the reaction is fast enough that the 3 to 20 s residence time in the photoreactor adds very little to the analysis time, peak variance or back pressure. However, the mobile phase must include a hydrogen-atom donor (HAD) such as

Table 6-1 ▪ Detection of K Vitamins in HPLC

Reaction Scheme	Analytes	Reactor	Aerobic	Reagents	Detection Limits (pg)	$\lambda_{ex}/\lambda_{em}$	Ref.
Photoreduction/ fluorescence of quinones (PRFQ)	K_1 K_3	PTFE DOTR-17s	no	no	30 20	243.5/> 370	123
Photoreduction/ fluorescence	K_1	PTFE GSFR-146s	no	ascorbic acid[a] acetate buffer	150	320/420	126
Chemical reduction/ fluorescence	K_1	PTFE GSFR-89s	no	$NaBH_4$[b]	≈ 150	320/420	126
Chemical reduction/ fluorescence	K_1	PTFE DOTR-51s	no	$(CH_3)_4NB_3H_8$[b]	150	325/420	228, 229
Electrochemical reduction/ fluorescence	K_1 analogs K_1-ep[e]	ECD[c]	no	$NaClO_4$[d]	25–60 60	320/420	225, 226, 227
Electrochemical reduction/oxidation	K_1	ECD[c]	yes	acetate buffer[d]	50	—	233, 234
UV absorption	K_1	none	yes	no	550	248 nm	245

[a] Reagent(s) can be spiked into the mobile phase or added post-column.
[b] Reagent(s) added post-column.
[c] Model 5100-A coulometric electrochemical cell, Coulochem, E.S.A., Bedford, MA.
[d] Reagent(s) spiked into mobile phase.
[e] Phylloquinone-2,3-epoxide.

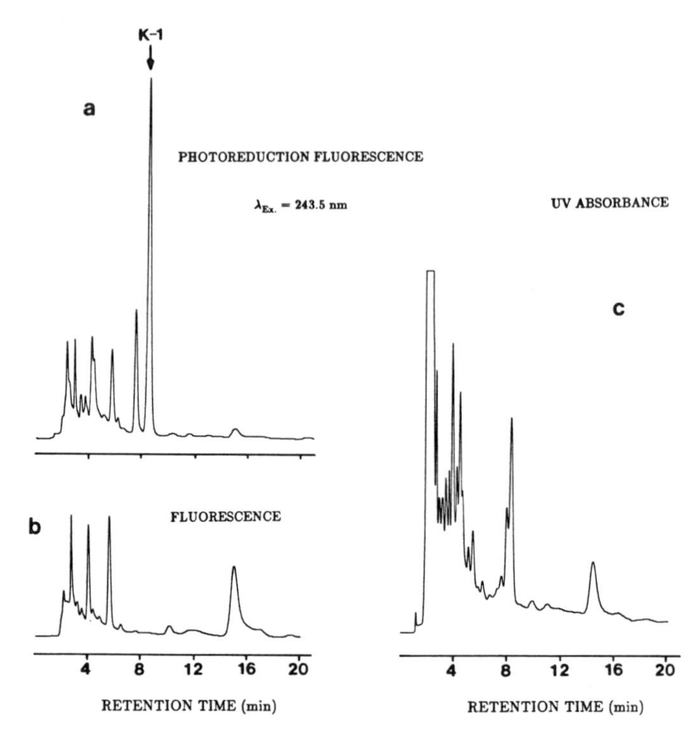

Figure 6-24 ■ Chromatogram of a plant extract (*capsella bursa pastorius*). HPLC conditions: Dupont Zorbax C_{18} column, 250 × 4.6 mm; mobile phase, methanol : 2-propanol 60 : 40 at 1.3 ml/min. In chromatograms (a) and (b) the PMT was set at 1000 V, $\lambda_{ex} = 243.5$ nm and $\lambda_{em} > 370$ nm. (a) PRFQ detection ($t_{rxn} = 16$ s). (b) Fluorescence detection (oxygenated mobile phase, no reactor). (c) UV absorption detection at 247 nm. Vitamin K_1 is not detected (the vitamin K_1 peak is buried under the two interferent peaks in the 8 to 8.5 min retention time region). From J. R. Poulsen and J. W. Birks, unpublished.

methanol or isopropyl alcohol. The photoreactor consists of a PTFE DOTR irradiated by the 254 nm radiation from a low-pressure, mercury-vapor lamp.

This system has been generalized to the detection of other quinones.[123] Many naphthoquinone and anthraquinone derivatives are sensitively detected. Samples that contain weakly retained quinones must be deoxygenated before injection in order to detect these compounds. Oxygen elutes as a broad peak early in the chromatogram and will destroy the desired photoproducts of quinones which coelute with it. Detection limits are in the 4 to 15 pg range for alkyl anthraquinones ($\lambda_{ex} = 257.5$ nm), while naphthoquinones are detected with slightly less sensitivity (15 to 60 pg, $\lambda_{ex} = 243.5$ nm).[123] The selectivity of the PRFQ detection scheme is demonstrated by the detection of vitamin K_1 in a plant extract (Fig. 6-24).[123, 159] No prefractionation is necessary for PRFQ detection of K_1, while UV absorption is completely swamped by interferents in the extract. The identity of the K_1 peak in the PRFQ chromatogram is confirmed by its

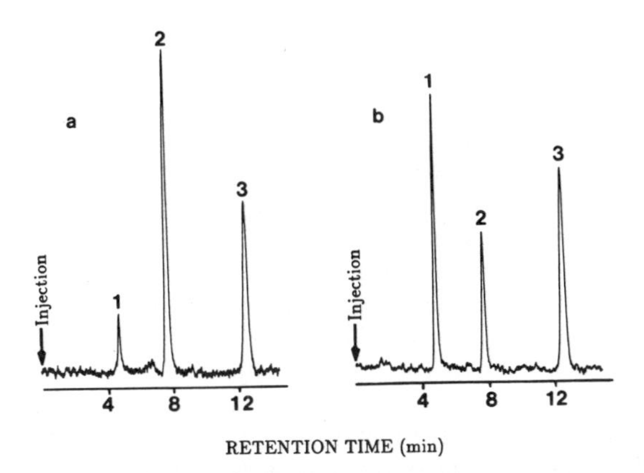

Figure 6-25 ■ Separation of a nitro-PAH mixture with PRFQ detection. Conditions: Dupont Zorbax C_{18} column, 250 × 4.6 mm; mobile phase, methanol : water 85 : 15 at 1.20 ml/min (t_{rxn} = 18 s); λ_{em} > 370 nm, PMT at 1180 V with a 2.5 s time constant. The sample was degassed before injection. (a) λ_{ex} = 257.5 nm chromatogram. (b) λ_{ex} = 243.5 nm chromatogram. Peaks: (1) 1-nitronaphthalene (11 pmol, t_R = 4.85 min), (2) 9-nitroanthracene (1.3 pmol, t_R = 7.71 min), (3) 1-nitropyrene (1.9 pmol, t_R = 12.54 min). From J. R. Poulsen, Ph.D. thesis, 1988.

disappearance from the fluorescence chromatogram run with oxygen present in the mobile phase. Other naphthoquinone derivatives behave in a similar way.

Nitro-PAHs by Photoreduction Fluorescence. Nitro-PAH compounds also are detected in the PRFQ. This response may be caused by either or both of two possible mechanisms. In the first, the nitro-PAH is converted to a quinone which is subsequently photoreduced.[247] Quinone formation from nitro-PAHs is much more efficient under aerobic conditions but has been observed in anaerobic systems. Photoreduction of the nitro-PAH to the corresponding amino-PAH is the second possibility.[248, 249] While the detection limits are not as spectacular as those achieved by reducing these compounds to amino-PAHs and detecting via peroxyoxalate chemiluminescence[250] or fluorescence,[251–255] they are comparable to UV absorption. Once again, selectivity is enhanced. A separation of a mixture of three nitro-PAHs is presented in Fig. 6-25.[159] Although the trend in response ratios between the two excitation wavelengths qualitatively agrees with the production of naphthoquinone from 1-nitronaphthalene and anthraquinone from 9-nitroanthracene, these ratios are smaller than would be expected were the quinones the sole photoproducts. It is likely that photoreduction to the amino-PAH is responsible for at least part of the increase in fluorescence.

Reserpine. The antihypertensive agent reserpine can be oxidized to more fluorescent compounds such as 3,4-dehydroreserpine and lumireserpine (3,4,5,6-tetradehydroreserpine) by light, oxidizing agents or acids.[256–258] Photoproduct studies demonstrated that 3,4-dehydroreserpine is an intermediate in the production of lumireserpine.[256] In the batch reactor very long reaction times (≈ 120 hr) were required for maximal formation of this product. 3,4-dehydroreserpine is formed much faster.

Lang and co-workers developed two HPLC reaction detection schemes for reserpine.[257, 258] In the first, reserpine was oxidized by a mixture of sulfuric acid and sodium nitrite (nitrous acid) in an air-segmented glass OTR.[257] A 200 pg detection limit was reported at a reaction time of about 6 min.

In the interest of reducing the reaction time, improving the efficiency of fluorophore development and reducing band broadening, a photoreactor was added to this system.[258] A 1 min reaction with nitrous acid as above, was followed by 1 min of irradiation in a PTFE photoreactor. Once again, the reaction system was air-segmented. This photochemically assisted reaction reduced the detection limit of reserpine to 80 pg. The PTFE GSFR was irradiated by the 254 nm line of a mercury-vapor lamp. Other wavelengths were eliminated by a cutoff filter. Unfortunately, no chromatograms were included in these publications.[257, 258] A 20-fold increase in the fluorescence of reserpine after batch irradiation also has been reported.[27]

The photooxidation mechanism of reserpine is thought to involve hydrogen transfer to molecular oxygen.[256] However, no definitive mechanistic studies under the conditions used in the HPLC photoreactor have been reported.

Phenothiazines. A final example of an intermolecular reaction which modifies the analyte's structure is the detection of phenothiazines through photooxidation. These pharmacologically important compounds contain a reduced sulfur that can be photooxidized to a sulfoxide. An increase in the fluorescence quantum yield results, because the ring system of the oxidized product is more rigid. It is likely that the reaction involves the production of singlet oxygen through oxygen quenching of the photoexcited phenothiazines. The singlet oxygen subsequently reacts with the sulfur and ultimately produces the sulfoxide. Such a reaction is termed a type 2b photooxidation, where the phenothiazine is both the sensitizer and the substrate.[259] Many of the phenothiazine drugs have other moieties that are readily oxidizable. This scheme demonstrates that a photochemical reaction can be effective for detection purposes even in the case where the mechanism is indirect and the products are not well documented. The end result probably is a complex mixture of photooxidation products.

In addition to the batch fluorometric methods mentioned previously,[18–23] phenothiazines have been determined with methods based on the difference in the UV absorption between the oxidized and unoxidized forms,[260, 261] cerometric titration to a colored product that is oxidized to the colorless sulfoxide by further addition of the titrant[262] and spectrophotometry after oxidation by molybdoarsenic acid,[263] among others. Since many of their metabolites are oxidized, selective extraction procedures usually are required to prevent interferences in the spectroscopic determinations.[264, 265] It often is important to determine the levels of the various metabolites as well as the parent drug, so an efficient separation technique coupled to a sensitive detector represents a large improvement over the batch methods.

Many phenothiazines and phenothiazine metabolites require derivatization prior to GC separation to enhance their volatility and prevent thermal decomposition.[265–267] For this reason, HPLC is the logical choice for their separation from biological matrices. Conventional HPLC detection is limited with regard to selectivity and sensitivity for these compounds. Although phenothiazines have strong UV absorption bands, the selectivity of UV detection is too low for routine clinical use. The weak native fluorescence of phenothiazine derivatives usually will not provide sufficient selectivity. Muusze and Huber developed a reaction detector based on the perman-

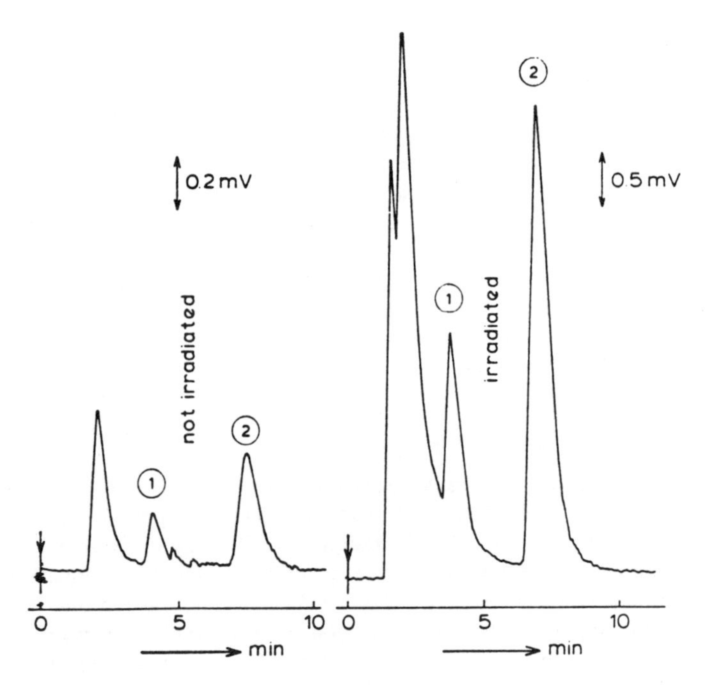

Figure 6-26 ■ Chromatogram of human serum spiked with (1) 2.3 ppm of mesoridazine and (2) 6 ppm of thioridazine. The wavelengths selected were optimal for thioridazine, viz., $\lambda_{ex} = 325$ nm and $\lambda_{em} = 435$ nm. $\lambda_{ex} = 340$ nm and $\lambda_{em} = 370$ nm for the nonirradiated and irradiated experiment, respectively. From A. H. M. T. Scholten, P. L. M. Welling, U. A. Th. Brinkman and R. W. Frei, *J. Chromatogr.* **199**, 239 (1980).

ganate/H_2O_2 oxidation of these compounds to fluorescent sulfoxides. With their home-built reaction system (PTFE OTR) and fluorescence cell, the sensitivity of UV absorption was equalled with considerably enhanced selectivity.[24]

Scholten *et al.* improved the detection limits for sulforidazine, mesoridazine and thioridazine by factors from 2 to 11.5 over their weak native fluorescence.[25, 26] Unsegmented PTFE and quartz OTRs in conjunction with a 200 W xenon-mercury arc lamp were used. No post-column reagents were required; however, in some cases ammonium peroxodisulfate was spiked into the mobile phase to enhance the rate of photooxidation. After being irradiated for 25 s in a PTFE OTR, the detection limits for the three phenothiazines were about 0.5 ng.[26] The gain in fluorescence intensity was demonstrated with a serum sample spiked with two of these drugs (Fig. 6-26).[26] Other phenothiazines, such as nedaltran, largactil, fenergan and levopromazine were detected at 40 to 100 pg levels after photooxidation in an air-segmented reactor.[27] The PTFE GSFR had a residence time of 35 s.

The metabolites of these pharmaceuticals often have an intact phenothiazine ring system or are oxidized at the ring sulfur. Thus an additional advantage of this reaction scheme is the ability to sensitively detect the parent compound under reaction conditions and at excitation and emission wavelengths that also are optimal for many of the metabolites.

Photochemically Amplified Detection Schemes

In contrast to the previous examples, photochemically amplified reaction-detection schemes result in detectable compounds that are not formed directly from the analyte. These surrogates for the analytes cause an indirect response in the detection cell. If side reactions do not cause decomposition of the analyte, then it may cycle through the reaction sequence many times and produce a large number of the surrogate molecules. Chemical amplification provides a pathway to improved selectivity as well as high sensitivity. The detector is biased against compounds which respond to it natively, that is, without undergoing the amplification reaction, as expressed by the selectivity factor (S_f). S_f is the ratio of the detector response factors of the surrogate (R_{sur}) and interferent (R_{int}) times the catalytic factor of the analyte ($N_{sur}/N_{analyte}$):

$$S_f = \frac{R_{sur}}{R_{int}} \times \frac{N_{sur}}{N_{analyte}} \tag{6-26}$$

Chemically amplified detection schemes for HPLC are not entiely new. For example, the catalytic activity of iodine and iodinated organic compounds on the reduction of Ce(IV) by As(III) causes the decolorization of Ce(IV). A post-column system based on this reaction can detect subnanogram quantities of iodinated organics, such as thyroid hormones, through the decrease in Ce(IV) absorption at 365 nm.[268] Extension of the concept of chemically amplified detection in HPLC to photochemical reactions is rather recent. Two reaction schemes serve as illustrations of photochemically amplified or *photocatalytic* reaction detection in HPLC.

Photochemical Amplification by Singlet Oxygen Sensitizers

Excited singlet states of molecular oxygen are produced when oxygen quenches the electronically excited states of organic molecules (see Chap. 1). Since most compounds that absorb light will be quenched by molecular oxygen, at least to some degree, detection schemes which utilize the enhanced reactivity of singlet oxygen have the potential to be applied to a wide variety of analytes. Photochemical amplification methods based on singlet oxygen trapping involve spiking an acceptor molecule into the HPLC mobile phase which reacts with the singlet oxygen. After reacting with singlet oxygen the spectroscopic characteristics of the acceptor molecule are altered. Changes in the spectra of the acceptor cause a change in its response—in either a positive or quench direction—to a conventional spectroscopic detector. Both UV absorption and fluorescence detectors have been used.[124, 269] To a first approximation, the analytes are not altered during the sequence of reactions. Each analyte molecule will go through the photoexcitation/oxygen quench cycle many times and thus cause the reaction of a large number of the acceptor molecules. In cases where the reaction of the acceptor produces a positive response to the detector, any native response by the analyte will add to the total signal. Systems based on singlet oxygen trapping, such as the photochemical amplifier for UV absorption detectors, are discussed in detail in the following chapter.

Photocatalytic Chemiluminescence Detection of Quinones (PCCL)

In the second example of photochemically amplified HPLC detection, the photocatalytic chemiluminescence detector (PCCL), surrogate molecules are produced photochemically by the analytes and subsequently trigger a chemiluminescent reaction.[158, 159] An advantage of this system is the extremely high selectivity that is achieved by coupling two reaction-detection methods in series.

As reaction (6-23) in the previous section indicates, in the presence of oxygen, photoreduced quinones will oxidize back to the parent compound with the concurrent production of H_2O_2. In fact, the photoreduction reaction sequence is stopped after the initial abstraction of the hydrogen, because the semiquinone radical rapidly reacts with oxygen [reaction (6-27a)].[215, 217, 270–273] α-hydroxyalkyl radicals also will react with oxygen [reaction (6-27b)]. The most important feature of this scheme is the fact that the analyte is regenerated by the oxidation reaction and as such can act as a catalyst by going through the reaction cycle many times. Catalytic yields of H_2O_2 depend on the length of the reactor, composition of the mobile phase and the substituents on the quinone sensitizer. Under conditions optimal for their separation, yields of H_2O_2 for the best sensitizers range from ~ 20 to 100 molecules of H_2O_2 per sensitizer molecule. The initial steps of the aerobic and anaerobic photochemical reaction detection schemes for quinones are identical [reactions (6-18) to (6-20)]. Reactions (6-28) to (6-31) are some of the productive reaction pathways possible when methanol is the HAD mobile-phase constituent[215, 270–272]:

$$Q + h\nu \rightarrow {}^1Q^* \tag{6-18}$$

$${}^1Q^* \rightarrow {}^3Q^* \tag{6-19}$$

$${}^3Q^* + CH_3OH \rightarrow \cdot CH_2OH + \cdot QH \tag{6-20}$$

$$\cdot QH + O_2 \rightarrow \cdot QOOH \tag{6-27a}$$

$$\cdot CH_2OH + O_2 \rightarrow \cdot OOCH_2OH \tag{6-27b}$$

$$2 \cdot QOOH \rightarrow 2Q + H_2O_2 + O_2 \tag{6-28a}$$

$$2 \cdot OOCH_2OH \rightarrow H_2O_2 + O_2 + 2CH_2O \tag{6-28b}$$

$$2 \cdot OOCH_2OH \rightarrow H_2O_2 + 2HCOOH \tag{6-28c}$$

$$\cdot QOOH + \cdot OOCH_2OH \rightarrow H_2O_2 + O_2 + CH_2O + Q \tag{6-28d}$$

$$\cdot QOOH \rightarrow Q + HOO\cdot \tag{6-29a}$$

$$\cdot OOCH_2OH \rightarrow CH_2O + HOO\cdot \tag{6-29b}$$

$$HOO\cdot + HOO\cdot \rightarrow H_2O_2 + O_2 \tag{6-30}$$

$$\cdot QOOH + \cdot CH_2OH \rightarrow Q + CH_2O + H_2O_2 \tag{6-31a}$$

$$\cdot QOOH + \cdot QH \rightarrow 2Q + H_2O_2 \tag{6-31b}$$

$$\cdot OOCH_2OH + \cdot CH_2OH \rightarrow 2CH_2O + H_2O_2 \tag{6-31c}$$

$$\cdot OOCH_2OH + \cdot QH \rightarrow Q + CH_2O + H_2O_2 \tag{6-31d}$$

Q = quinone analyte

$\cdot QH$ = semiquinone radical

Triplet deactivation mechanisms other than hydrogen abstraction are unimportant in this system because the concentration of the hydrogen-atom donor (HAD), methanol, is high. Quenching by hydrogen abstraction is extremely efficient under these conditions.[274] Similarly, the high concentration of methanol makes it the most likely HAD even though its oxidation products, such as formaldehyde, are better donors. Few compounds other than type 1 photooxygenation sensitizers respond in the PCCL. Type 1 sensitizers have triplet states that are capable of abstracting hydrogen atoms from compounds with weak carbon-hydrogen bonds.[259] Hence, type 1 photooxidations always involve secondary free radical reactions.

H_2O_2 is detected by peroxyoxalate chemiluminescence.[275–284] The peroxyoxalate reaction is very sensitive and relatively interference-free. For the detection of H_2O_2 an oxalate ester, *bis*-2,4,6-trichlorophenyl oxalate (TCPO), and a fluorophore are introduced prior to entering the detector cell.[35–39, 54, 55, 285–290] Alternatively, peroxyoxalate chemiluminescence can be used in detecting oxidizable fluorophores by adding H_2O_2

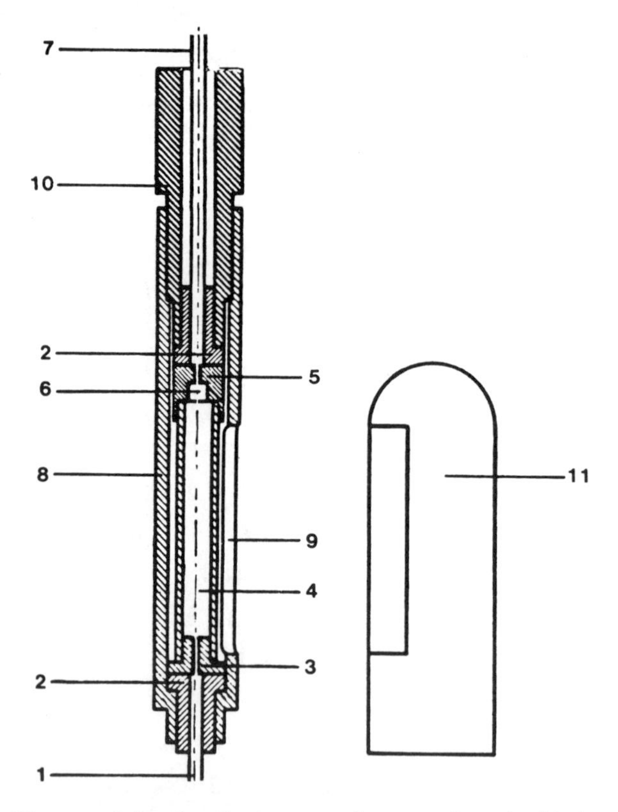

Figure 6-27 ■ Diagram of the chemiluminescence detector cell used with immobilized fluorophores: (1) PTFE inlet capillary. (2) Stainless-steel flange. (3) Fritless PTFE cap (male end). (4) Quartz tube, 3.0 cm long × 1.5 mm i.d. (5) PTFE outlet cap (female end). (6) Frit. (7) PTFE outlet capillary. (8) Metal casing of the cell holder. (9) Slot in holder casing to expose the cell to the PMT. (10) Threaded cell holder end fitting which seals the cell and holds it in position. (11) RCA IP-28 photomultiplier tube. Adapted from G. Gübitz, P. van Zoonen, C. Gooijer, N. H. Velthorst and R. W. Frei, *Anal. Chem.* **57**, 2073 (1985).

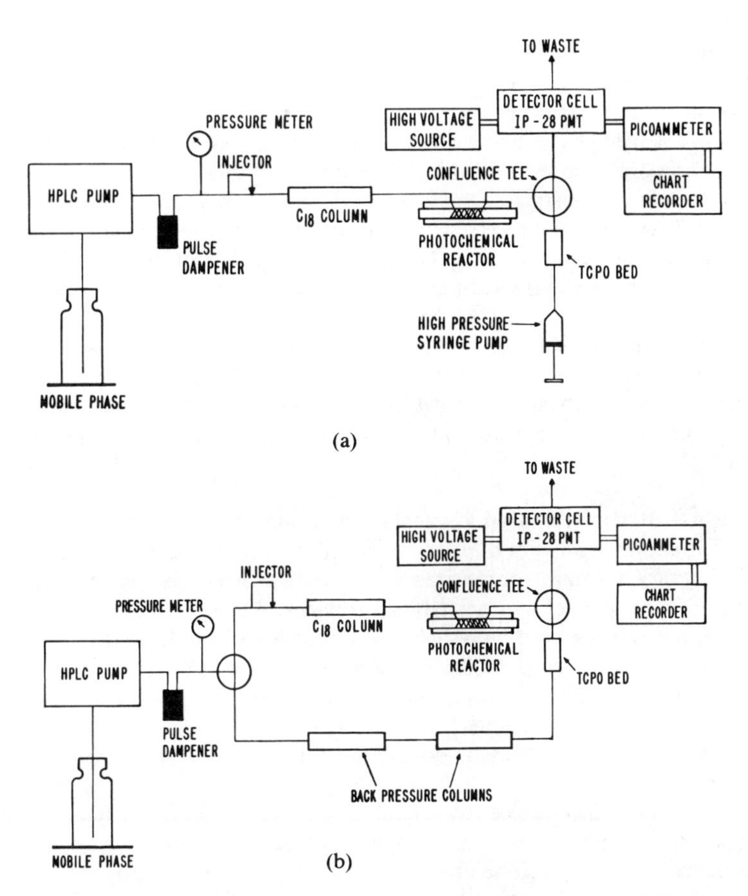

Figure 6-28 ■ Schematic diagrams of solid-phase reagent addition systems for the photocatalytic chemiluminescence detector (PCCL). (a) Dual-pump mode. (b) Split-flow mode. Both systems utilize the immobilized fluorophore detector cell depicted in Fig. 6-27. Adapted from J. R. Poulsen, J. W. Birks, P. van Zoonen, C. Gooijer, N. H. Velthorst and R. W. Frei, *Chromatographia* **21**, 587 (1986).

and TCPO to the column effluent.[178, 250, 291–302] The PCCL provided an excellent model system for investigating ways to simplify the detection of H_2O_2 by peroxyoxalate chemiluminescence in HPLC. Rather than adding the fluorophore in solution, 3-aminofluoranthene is immobilized on silica particles or glass beads and packed in the detector flow cell, a 1 mm i.d. quartz tube (Fig. 6-27).[37, 38] It also is possible to solubilize TCPO from a solid-phase reagent bed (Fig. 6-28a).[39] If the solvent composition of the eluent is suitable, its flow can be split so that a portion of it flows through a solid-phase reagent bed connected in parallel to the HPLC column (Fig. 6-28b).[39] In combination with the immobilized fluorophore, flow splitting allows peroxyoxalate detection of H_2O_2 without the use of any post-column reagent pumps. These simplified detection techniques are applicable to the detection of H_2O_2 produced in other types of post-column reactions.

One of the primary goals of reaction detection, namely enhanced selectivity relative to conventional methods, is demonstrated clearly by the application of PCCL to the detection of anthraquinone in a cardboard extract (Fig. 6-29).[158, 159] The presence of anthraquinone in cardboard results from its addition to pulping mixtures in order to enhance the delignification process.[303-311] Obviously, the low level of anthraquinone in this sample cannot be detected by UV absorption (251 nm), yet the PCCL chromatogram exhibits a well resolved peak. No prefractionation of the cardboard extract is necessary for PCCL detection, so the analysis time is greatly reduced in comparison to that required for the determination of anthraquinone with UV absorption detection. PCCL also is very sensitive. Detection limits for anthraquinone and most alkyl anthraquinones are less than 25 pg, with those of alkyl naphthoquinones less than 35 pg. This represents an improvement of up to a factor of 5 over optimized UV detection.[158, 159]

Photocatalytic Derivatization Reagents. Photocatalytic detection schemes also are of interest from the standpoint of pre-column derivatization reactions. Development of reagents that take advantage of sensitive photocatalytic detection schemes is a promising way of solving some of the detectability problems that are common in HPLC. The nitroaromatic reagents currently available for reductive HPLC-EC, HPLC-$h\nu$-EC and UV detection are well suited to the photochemical amplification methods based on singlet oxygen trapping.[124, 269] Nitroaromatic compounds are very good singlet oxygen sensitizers. Although these methods are not extremely selective when UV absorbing compounds are detected, it may be possible to synthesize a reagent tag that absorbs longer wavelength light.

The PCCL is another scheme that would be ideal for detecting analytes derivatized pre-column with a photosensitizing reagent. Anthraquinones have good chromatographic characteristics for reverse-phase HPLC separations, while being sensitively and selectively detected by PCCL.[158, 159] Quinone tags also would be useful in conjunction with the photoreduction fluorescence detection method.[123, 159] Furthermore, quinones are good singlet oxygen sensitizers in the absence of HAD compounds.

Another photocatalytic reaction currently is under investigation and may prove useful for HPLC detection. It involves the photooxidation of tris(2,2′-bypyridine)ruthenium(II) complexes ($Ru(bpy)_3^{2+}$) followed by electrochemical reduction.[312, 313] Excited states of $Ru(bpy)_3^{2+}$ transfer an electron to quenching molecules. In principle, it should be possible to use this reaction to detect either quenchers or $Ru(bpy)_3^{2+}$ itself. The best quenchers are cobalt(III) complexes.

A FIA system has been used to optimize the photochemistry and amperometric flow-cell design. The flow cell has windows to allow irradiation of the region near the electrodes by a HeCd continuous-wave laser (441.6 nm). A low-frequency chopper is used to discriminate against electrochemically active interferents. By adding a Co(III) complex to the flow stream, a detection limit of 1.1×10^{-4} mol (6.3 pg) is achieved for $Ru(bpy)_3^{2+}$.[312] This represents an average of 36 catalytic cycles per ruthenium complex.[313]

Coupling of this detection cell to HPLC has not been reported yet. Obviously, interest in detecting cobalt(III) complexes and $Ru(bpy)_3^{2+}$ is limited. In order to

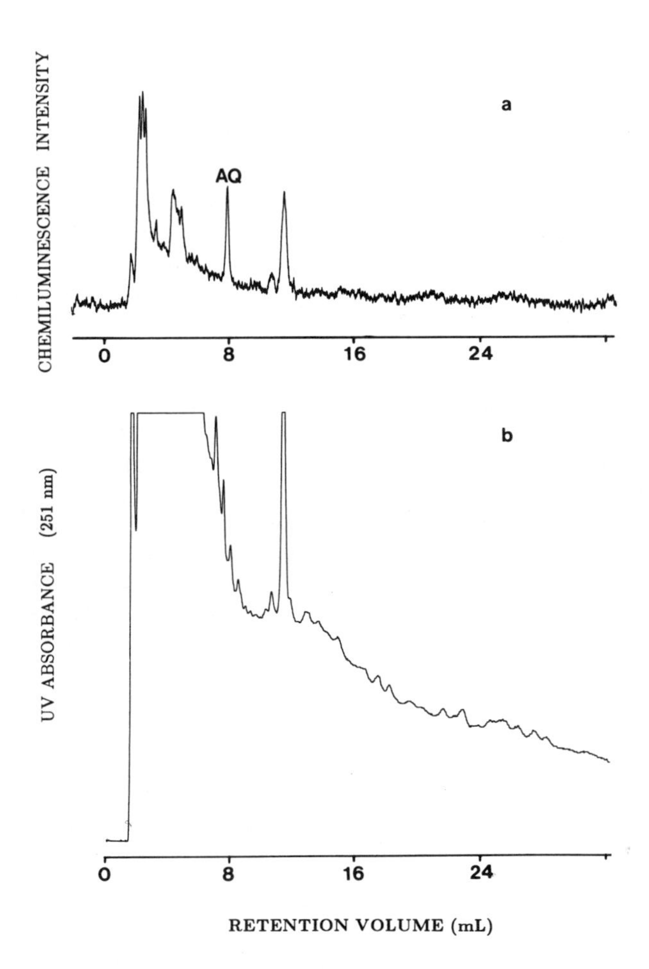

Figure 6-29 ■ Chromatograms of a cardboard extract. The soxhlet apparatus ran through 12 to 15 cycles. If the extraction were 100% efficient, the anthraquinone (AQ) concentration of the extract would correspond to ≈ 50 ppb in the original cardboard sample. No prefractionation or preconcentration of the sample was performed. HPLC conditions: Dupont Zorbax C_{18} column, 250 × 4.6 mm; mobile phase, methanol : water 85 : 15. (a) PCCL detection ($t_{rxn} = 155$ s). TCPO (0.83 g/l) and rubrene (23 mg/l) were delivered in acetone at 0.33 ml/min. TRIS buffer (pH = 7.9) in water : acetone : methanol 4 : 58 : 38 at 0.15 ml/min. The PMT was set at 1300 V. (The retention volume axis is offset to compensate for the 2.1 ml reactor volume.) (b) UV absorption detection at 251 nm. (The anthraquinone peak would be an order of magnitude smaller on this attenuation than that in the PCCL chromatogram.) From J.R. Poulsen, Ph.D. thesis, 1988.

enhance the applicability of this photoelectrochemical reaction to HPLC detection, reagents for attaching a $Ru(bpy)_3^{2+}$ moiety to analytes in a pre-column mode are being developed.[312, 313] A reagent of this type might provide a selective and sensitive tag for derivatizable analytes. The photoelectrochemical properties of the ruthenium complexes are not affected strongly by modification of the ligands.[313] One known drawback of the $Ru(bpy)_3^{2+}/Ru(bpy)_3^{3+}$ redox couple is the possibility for the $Ru(bpy)_3^{3+}$ to be reduced by other compounds before diffusing to the electrode. The addition of a "mediator" to the flow stream has been investigated as a means of preventing this form of interference.[313] The mediator is oxidized by $Ru(bpy)_3^{3+}$ and then reduced at the electrode. Because the mediator is a weaker oxidizing agent than the photoproduct ($Ru(bpy)_3^{3+}$), it is less likely to react with organic interferents present in the detector flow cell.

Some doubt remains as to whether a suitable reagent for pre-column derivatization in HPLC can be developed. The large size and ionic character of $Ru(bpy)_3^{2+}$ functionality may complicate the separation of compounds derivatized with it. In any case, catalytic photoelectrochemical reactions could be very useful for HPLC reaction detection. The $Ru(bpy)_3^{2+}/Ru(bpy)_3^{3+}$ redox couple has allowed the design parameters for photoelectrochemical detection cells and their associated electronics to be investigated. Even if the $Ru(bpy)_3^{2+}$ derivatives prove to be awkward chromatographically, other reactions of this type may be useful. Photoelectrochemical reactions also might be applied to the detection of nonabsorbing analytes, similar to the schemes discussed in the following section. For example, the ruthenium bipyridine complexes are quenched by a variety of Co(III) complexes. If a constant amount of photoexcited $Ru(bpy)_3^{2+}$ is present in the detector flow cell, the introduction of the Co(III) complexes would produce a response. A sensitized photoelectrochemical reaction triggered by compounds of wider analytical interest could be the basis of a sensitive and useful reaction detector.

Detection of Compounds without Chromophores

All of the detection schemes reviewed up to this point require the analyte to absorb light and undergo some type of photoreaction. While many of these schemes provide advantages in terms of sensitivity and/or selectivity relative to conventional detectors, the analytes will respond at some level to absorption detection. In the following reaction schemes, sensitizers that absorb light and undergo reactions with the analyte provide the basis for detection. The analytes are not required to absorb light to be detected. However, suitable sensitizers are rare because of photochemical background reactions. The detection schemes developed by Gandelman *et al.* provide an example of this approach.[122, 125, 314–316]

Abstraction of hydrogen from hydrogen-atom-donating (HAD) analytes—compounds with weak carbon-hydrogen bonds—by quinones in their lowest excited triplet state forms the basis of these detection schemes. Two schemes were developed for detecting HAD analytes, one utilizing aerobic photochemistry and the other occurring under anaerobic conditions.[122, 125, 314–316] In fact, the results from these systems inspired the quinone detection work based on the same photochemistry.[123, 158, 159] In each system, the quinones absorb UV light and populate an excited singlet state.

Regardless of the energy level of this singlet state, it decays to the lowest excited singlet (S_1) through internal conversion and subsequently intersystem crosses to the lowest excited triplete state (T_1). The separation between these two states is very small, on the order of 4 kcal/mol.[274]

Photooxygenation Chemiluminescence (POCL). In the first of the Gandelman and Birks detection schemes for HAD analytes, hydrogen peroxide is produced through the aerobic photoreaction of an anthraquinone sensitizer with a HAD analyte (e.g., ethanol) (Fig. 6-30).[314] After its photochemical production, the H_2O_2 is detected by either the luminol[314] or the peroxyoxalate chemiluminescent reaction.[316] A large photooxygenation background kept the detection limits in the 1 to 15 μg range for analytes such as aliphatic alcohols, ethers, aldehydes and saccharides.[314] It has been shown that anthraquinones in aqueous solution undergo a complicated reaction which has α- and β-hydroxy anthraquinones and H_2O_2 as products. These are formed in equal amounts. A photosolvate between the anthraquinone triplet and water is the most likely intermediate.[214, 317–324]

The photoreactor consisted of a PTFE OTR coiled around a Pyrex sleeve that fit over an 8 W fluorescent lamp. This lamp had a maximum emission at 366 nm. A photosensitizer, anthraquinone-2,6-disulfonate (5×10^{-5} M), and cobalt(II) ion (10^{-4} M) were added to the chromatographic eluent. HAD mobile phases must be avoided with the POCL; otherwise, a large photochemical background will result. Acetonitrile and water mixtures are suitable for both the POCL and photoreduction fluorescence (PRF) detection schemes. By spiking the photosensitizer and catalyst for the luminol reaction [Co(II)] into the mobile phase, only the luminol reagent itself had to be added with a post-column pump. Reverse-phase separations of these polar analytes actually were improved by the presence of the reagents in the eluent. After

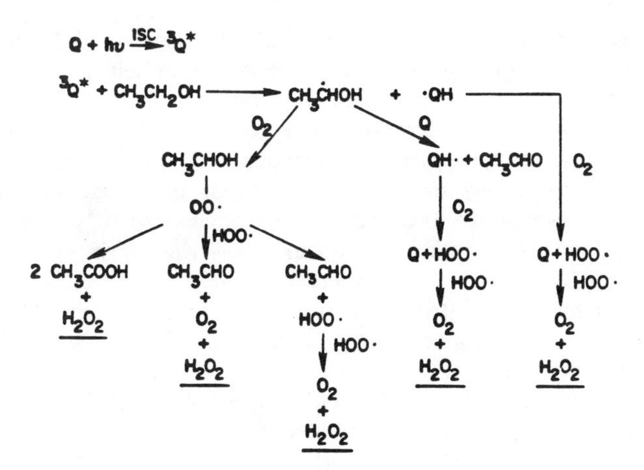

Figure 6-30 ■ Probable mechanism for the AQDS sensitized photooxygenation of ethanol in neutral aqueous solution. ISC is intersystem crossing, Q is anthraquinone-2,6-disulfonate and QH · is the semiquinone radical. From M. S. Gandelman and J. W. Birks, *J. Chromatogr.* **242**, 21 (1982).

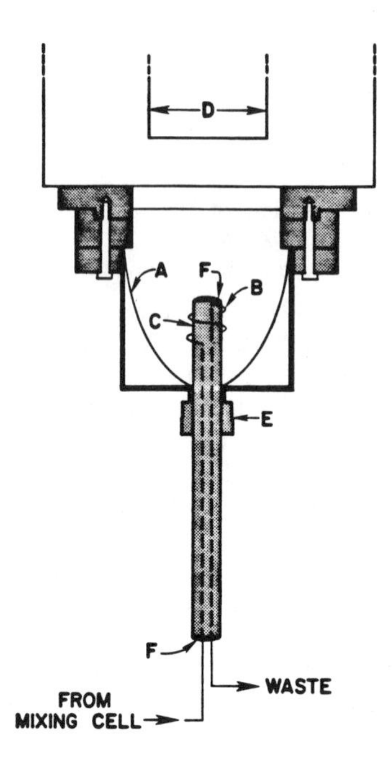

Figure 6-31 ▪ Schematic diagram of the chemiluminescence cell. (A) Ellipsoidal mirror. (B) PTFE coil. (C) 1 cm diameter metal tube. (D) PMT photocathode. (E) O-ring compression fitting. (F) RTV black rubber seal (light tight). From M. S. Gandelman and J. W. Birks, *J. Chromatogr.* **242**, 21 (1982).

exiting the photoreactor, the eluent stream and luminol reagent were mixed and flowed into a home-built chemiluminescence detection cell. This cell consisted of a PTFE coil located at the focus of an ellipsoidal mirror. A red sensitive photomultiplier tube was positioned at the other focus of the mirror (Fig. 6-31).[314]

Photoreduction Fluorescence Detection of Hydrogen-Atom Donors (PRF). By excluding oxygen, the photoreduction reaction can be exploited for HAD detection. Since the photoreduced anthraquinone sensitizers—dihydroxyanthracene derivatives—are highly fluorescent, they are readily detected.[122, 125, 315, 316] Quinones themselves are extremely nonfluorescent, with quantum yields so low that a sensitive system can be based on the change in fluorescence resulting from their photoreduction. An additional advantage of the PRF detector is the absence of post-column reagent addition apparatus (Fig. 6-32).[125] A quinone sensitizer such as 2-*tert*-butyl-9,10-anthraquinone (tBAQ) or 9,10-anthraquinone-2,6-disulfonate (AQDS) is spiked into the degassed mobile phase. tBAQ is the sensitizer of choice for the PRF, because the sensitivity is improved through reductions in the background, while its nonionic nature minimizes the effect on the chromatography.[122] Only the need for degassing the mobile phase and the addition of the photoreactor distinguish the HPLC apparatus from a conventional system utilizing fluorescence detection.

Three different photoreactor sources were investigated in combination with several OTRs and a novel DOTR. With a low-pressure, mercury pencil lamp and a coiled PTFE OTR, detection limits of 2 ng were achieved for several cardiac glycosides.[122] Saccharides had detection limits of about 500 ng with this 254 nm excitation source. An 8 W fluorescent lamp ($\lambda_{max} = 366$ nm) in conjunction with a 12.5 m "knitted"

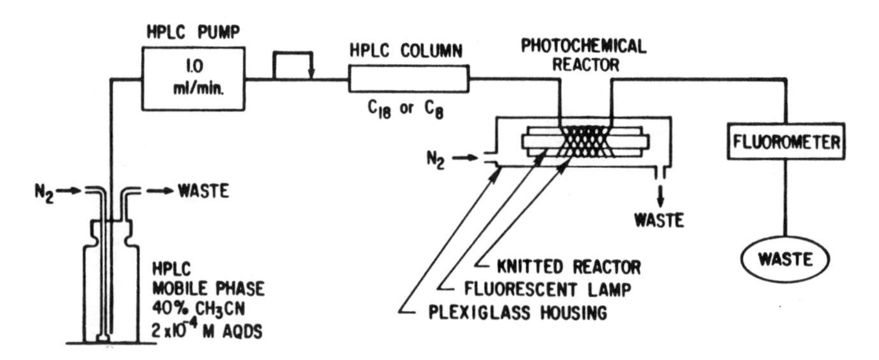

Figure 6-32 ■ Schematic diagram of the single-pump PRF detector. From M. S. Gandelman, J. W. Birks, U. A. Th. Brinkman and R. W. Frei, *J. Chromatogr.* **282**, 193 (1983).

DOTR improved the detection limits of the common food saccharides to 80 ng, while maintaining the high sensitivity toward the cardiac glycosides and other HAD analytes. A later paper describes the method for fabricating these DOTRs.[70] The optimal residence time in the 254 nm photoreactor, where the wavelength of the light emitted by the source coincides with stronger absorption bands of the sensitizers, is about three times less than that in the 366 nm photoreactor. A 200 W Xe-Hg arc lamp reduced the optimal reaction time, but sensitivity decreased as well (see also Fig. 6-6). Presumably, the sensitivity decline results from photodegradation reactions.[125] All three reactor housings had provisions for deoxygenation, which is necessary when anaerobic reactions are carried out in PTFE reactors. Oxygen will react with the dihyroxyanthracene produced from the sensitizer as well as the radical intermediates of the photoreduction reaction.

Photoreduction fluorescence detection (PRF) is ~ 2 to 3 orders of magnitude more sensitive than the oxygenated HAD detection system. However, it still is limited by a photochemical background of somewhat uncertain nature. Possible causes of the background photoreduction include reaction by the excited quinones with water (photosolvation), with impurities in the mobile-phase solvents or with ground-state quinone sensitizers.[122, 316] While the reaction sequence is the same as that for the photoreductive detection of quinones presented earlier [reactions (6-18) to (6-22)], the conditions are much different. In detecting HAD analytes, the sensitizer is in relatively high concentrations (10^{-3} to 10^{-5} M) and the hydrogen donor at very low concentrations ($\approx 10^{-6}$ to 10^{-7} M at the detection limit). For quinone detection the mobile-phase polarity modifier is the donor, so its concentration is in excess of 15 M. This contrast in conditions explains the difference in reaction time required for optimal sensitivity between the two systems. Nonradiative deactivation of the reactive triplet states ($T_1 \rightarrow S_0$) competes with hydrogen abstraction and decreases the efficiency of the photoreduction reaction in the PRF. Furthermore, the mechanism of quinone photoreduction becomes much more complex at these very low donor concentrations.[214, 317–324] For these reasons, PRF detection of HAD analytes requires a longer residence time in the photoreactor than that used in the detection of quinones (PRFQ).[123]

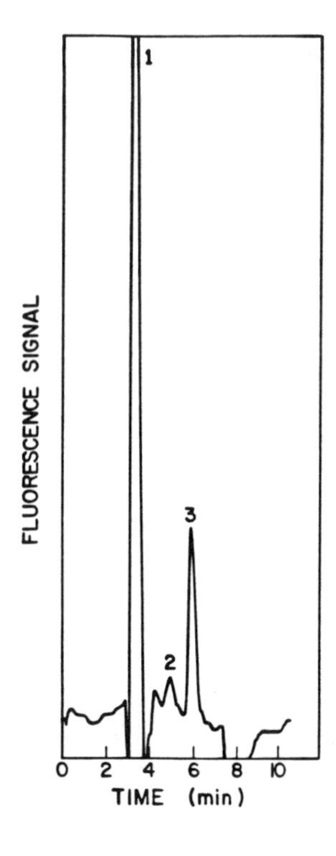

Figure 6-33 ■ Chromatogram of digoxin tablet extract with single-pump PRF detection. Conditions: C_{18} column; mobile phase, acetonitrile : water 40 : 60 with 2×10^{-4} M anthraquinone-2,6-disulfonate at 1.0 ml/min; residence time, 81 s in a DOTR irradiated by a fluorescent poster lamp; Schoeffel FS-970 fluorometer (see also Fig. 7-32). Peaks: 1, lactose; 2, impurity; 3, 200 ng of digoxin. From M. S. Gandelman, J. W. Birks, U. A. Th. Brinkman and R. W. Frei, *J. Chromatogr.* **282**, 193 (1983).

Reactivity of the quinone triplet states also depends on the solvent composition. As the polarity increases, the efficiency of the photoreduction reaction decreases (see Chap. 1). An explanation for this behavior is found in the effect of solvent polarity on the energy levels of the two lowest triplet states. Increased polarity causes the energy of the reactive triplet (n, π^*) to be raised while the unreactive triplet state (π, π^*) remains at about the same energy level or actually is lowered in energy. A transition from a lowest triplet state with primarily n, π^* character to one which is dominated by π, π^* characteristics occurs at high polarity. This limits flexibility in selecting the mobile phase while optimizing a separation. Fortunately, polar bonded phases usually allow polar analytes to be separated at sufficiently high organic modifier percentages. A chromatogram of a digoxin tablet extract demonstrates the ability to carry out useful separations with the PRF detector (Fig. 6-33).[125]

Detection through Quenched Photochemistry. Analytes that lack significant chromophores also can be detected indirectly if they are capable of quenching a steady-state signal produced by a photochemical reaction. An example of this type of reaction scheme is included in the chapter on singlet-oxygen-based detection for HPLC. Briefly, triplet quenchers such as trialkyl amines short-circuit the oxidation of fluorescent furan molecules by singlet oxygen.[269] Quenching of the singlet oxygen, or of the triplet state of the sensitizer that produces it, will cause this to happen. Either the furan derivative itself or an additional sensitizer produces this singlet oxygen. By inhibiting the photooxidation reaction which transforms the fluorescent furans to nonfluorescent products, a quencher produces a positive fluorescence peak.

Photochemical Studies with HPLC Photoreactors

A better understanding of the photochemical reaction products of many compounds is urgently needed. This information is not only desirable for better understanding of the reactions used for photochemical reaction detection, but also with regard to the light-initiated breakdown reactions of a wide variety of important compounds. Pharmaceutical companies expend great effort to study the decomposition pathways of their products on the shelf as well as through metabolism. Many of these nonmetabolic reactions are photochemically initiated. The experimental requirements for studying these reactions often can be satisfied by post-column photoreactors. Other compounds of environmental interest, such as pesticides, combustion products and refrigerants undergo light initiated reactions in the biosphere, and several of these anthropogenically generated contaminants are decomposed most effectively through photochemically activated pathways.

The principal advantages of using a photoreactor coupled to an HPLC system in photodegradation product studies are:

1. Under suitable chromatographic conditions the sample of the compound introduced into the photoreactor is pure.
2. An analyte is subjected to a reproducible reaction time with a uniform photon flux throughout the reaction.
3. It is possible to explore both aerobic and anaerobic photochemistry with the only modification of the system being a change in the gas used to purge the mobile phase and photoreactor. Depending on the design of the reactor housing, the photoreactor source may be varied to change the excitation wavelength.
4. It is possible to *heart-cut* and rapidly reinject the photoproduct solution into the same chromatographic system where the retention time of the parent compound is well known.

If the reaction needs to be investigated in a solvent system that is incompatible with separating the reactant from possible contaminants or in the extreme case, unable to elute it from a HPLC column, one can carry out the reaction in a *plug-injection* mode. Of course, the sample must be rigorously purified before injection.

After heart-cutting the analyte peak, which now consists of all the stable photoproducts formed in the post-column reactor, an aliquot of this solution is reinjected into the chromatograph. When concentrated solutions are used for the reaction, it may be possible to directly inject the photoproducts without any preconcentration. When the yield is low or the original solution dilute, the photoproducts can be concentrated by pooling several injections and evaporating the solvent. This may be necessary, because chromatographic peaks are diluted by dispersion in the HPLC column and photoreactor.

As an example of the application of such a system to a photochemical study, we investigated the photoproducts, specifically quinones, produced from PAHs during aerobic photoreactions. Our interest in these reactions stemmed from our observation of response to the PCCL detector (designed for the detection of quinones) to these PAHs. In order to distinguish between a response caused by organic peroxides or singlet oxygen and that due to quinones formed from PAHs, the reaction products of rather concentrated solutions were monitored. Multiple chromatograms of the products were run to compare the response of our reaction detectors to conventional systems.[158, 159] Quinones were identified by their retention times and relative responses to the various detectors.

Conclusions and Future Directions

Even as HPLC continues to grow in popularity the commercially available spectroscopic, and to a lesser extent, electrochemical detectors are maintaining their share of the market. However, column liquid chromatography still is a technique limited by detection capability. Given the resolution difference between HPLC and gas chromatography (GC), even a sensitive universal HPLC detector would not be able to function as effectively as the flame ionization detector does for GC. After more than a decade of being a "field growing in popularity" reaction-detection in HPLC remains just that. Photochemical reaction detection as a subset of this field is even more obscure. We hope that the examples reviewed here serve two purposes: (1) to illustrate both the selectivity and sensitivity which can be achieved through photochemical reaction detection and (2) to stimulate interest and new ideas with respect to widening the application of existing photoreaction schemes and the development of new ones.

From this review it is readily apparent that many of these systems originate from the area of medicinal chemistry. Extreme matrix problems, the low levels of both endogenous compounds and clinical agents, and still lower levels of their metabolic products provide the impetus to develop selective detection methods that also are extremely sensitive. After all, clinical subjects only are able to part with limited quantities of their biological fluids. We see continued efforts toward improving methods related to pharmaceuticals with emphasis on dedicated systems for specific, research-oriented applications.

One of the main reasons for the preferential use of conventional detectors is their simplicity and reliability of operation. For this reason the trend in reaction detectors of all types should continue toward the simpler, more reliable and less expensive pumpless reaction systems. Photochemically activated reactions fit in well with this trend. Miniaturization of HPLC systems, with the goal of conserving solvents, expensive reagents and, most importantly, use in sample-limited applications is compatible with on-line photochemical reaction detection. Especially in combination with laser induced photoreactions, photochemical reaction detection holds great promise for microcolumn HPLC applications.

In this sense, the evolution of photochemical reaction detectors for HPLC is following a pattern similar to peroxyoxalate chemiluminescence detection. First adapted to normal-bore HPLC systems for the detection of compounds (fluorophores and H_2O_2) that respond natively, the use of peroxyoxalate reaction systems has been broadened through the development of new derivatization reagents, simplified reagent introduction methods and coupling to photochemical reactions and immobilized enzyme reactors. Furthermore, this scheme is well suited to microcolumn HPLC because of the "chemical band-narrowing" effect.[325, 326] When this effect is observed, the effective dead volume of the detector cell is determined by the kinetics of the peroxyoxalate reaction rather than its physical dimensions. Perhaps given time, more derivatization reagents specifically designed for photochemical reaction detection will be developed, just as they are currently being developed for peroxyoxalate chemiluminescence detection. Such specific tags will prove to be extremely valuable in expanding the range of application of these selective detection methods. Given their adaptability to microcolumn HPLC—peroxyoxalate chemiluminescence through the chemical band-narrowing effect and photochemical reaction detection through miniaturized, pumpless reaction systems—pre-column derivatization tags which improve some combination of the extractability, separability and detectability of analytes by these methods would be very useful for sample-limited applications. The reagent for photochemical reaction detection of metal ions, developed by Shih and Carr, is a fine example of this approach in normal-bore HPLC.[176, 177] As more applications are reported, the likelihood of reaction-detection apparatus becoming commercially available also will increase.

Another feature of reaction detectors that is frequently criticized is the cost. Usually schemes are developed which utilize a conventional detector in addition to the post-column apparatus. Often the choice of this detector is made on the basis of availability. In many cases these systems appear to be much more expensive than what would be necessary to build a completely dedicated reaction detector. For example, a Kratos FS-970 fluorescence detector was used for the photoreduction fluorescence detector of Gandelman and Birks.[122] Each analyte is detected as the dihydroxyanthracene derived from the photosensitizer; thus the PRF, though it responds to a wide variety of analytes, does not require a general purpose fluorometer. It should be possible to utilize a simple filter fluorometer in which the 254 nm line of a low-pressure mercury lamp serves as the excitation source. Dihydroxyanthracenes have strong absorptions at this wavelength. In this way the cost of the monochromator and high-pressure excitation lamp would be eliminated.

In fact, any photoactivated release tag that releases the same detectable molecule from each derivatized analyte should be amenable to this approach. The tagging reagent for metal ions discussed earlier is another example. *n*-butylnaphthylmethylamine is released regardless of the identity of the metal ion in the complex.[177]

Another detection method that is frequently guilty of using unnecessarily expensive equipment is peroxyoxalate chemiluminescence. Once again a Kratos FS-970 fluorometer often is utilized in a role which does not require its full capabilities. With the excitation source turned off and emission filters removed, it is essentially an expensive photomultiplier tube housing. While convenience may justify using such a detector in this manner while researching and developing new detection methods, it would be wasteful of laboratory resources in routine operation. Several authors have designed flow cells and PMT housings suitable for chemiluminescence reaction detectors.[35–40, 287, 290, 314, 316, 327–332] However, the flow cell in the Kratos detector remains the best commercially available cell for chemiluminescence applications because it is very efficient in the collection of the emitted light.

To date, the only photochemical schemes that have utilized peroxyoxalate chemiluminescence detection are the POCL[316] and PCCL.[38, 39, 158, 159] Both of these reaction schemes generate H_2O_2 as a detectable surrogate. Work with the PCCL detection scheme has demonstrated the feasibility of pumpless addition of oxalate esters from solid-phase reagent beds and the use of immobilized fluorophores in HPLC.[37–39] These systems are much simpler to operate. Despite the high sensitivity of peroxyoxalate chemiluminescence detection toward oxidizable fluorophores such as amino-PAHs, the development of photochemically activated tags which release these compounds and/or photochemical reactions which generate them on-line remain to be demonstrated.

An additional advantage of lower cost, dedicated reaction detectors is the potential for use in tandem with conventional detectors. Multiple detection in either a sequential or split-flow mode yields additional qualitative information. This information can help prevent errors in quantifying analytes caused by coelution with matrix constituents, as well as aiding in the confirmation of peak identities through relative responses to each detector.

By paying careful attention to system design and reducing cost where possible, the use of reaction-detection schemes almost certainly will increase. This is particularly important in systems dedicated to specific applications and to those being adapted for commercialization. Given the impressive capabilities of both chemiluminescence and photochemical reaction detection, it seems that off-the-shelf photoreactors and chemiluminescence detection apparatus designed for HPLC will be available in the future. Perhaps more than any other development, the availability of such commercial systems would result in more universal acceptance and utilization of these methods.

References

1. Frei, R. W. In R. W. Frei and J. F. Lawrence, eds., *Chemical Derivatization in Analytical Chemistry*, Plenum, New York, 1981, Vol. 1, pp. 211–340.
2. Krull, I. S., ed., *Reaction Detection in Liquid Chromatography*, *Chromatographic Science Series*, Dekker, New York, 1986, Vol. 34.
3. Frei, R. W., Jansen, H., and Brinkman, U. A. Th. *Anal. Chem.* **57**, 1529A (1985).
4. Frei, R. W. *Chromatographia* **15**, 161 (1982).
5. Frei, R. W. *J. Chromatogr.* **165**, 75 (1979).
6. Frei, R. W., and Scholten, A. H. M. T. *J. Chromatogr. Sci.* **17**, 152 (1979).
7. Ettre, L. S. *J. Chromatogr. Sci.* **16**, 396 (1978).
8. Stewart, J. T. *Trends Anal. Chem.* **1**, 170 (1982).
9. Fink, D. W. *Trends Anal. Chem.* **1**, 254 (1982).
10. Frei, R. W., Michel, L., and Santi, W. *J. Chromatogr.* **142**, 261 (1977).
11. Birks, J. W., and Frei, R. W. *Trends Anal. Chem.* **1**, 361 (1982).
12. Krull, I. S. In I. S. Krull, ed., *Reaction Detection in Liquid Chromatography*, *Chromatographic Science Series*, Dekker, New York, 1986, Vol. 34, pp. 303–352.
13. Birks, J. W., Poulsen, J. R., and Shellum, C. L. In D. Eastwood and L. J. Cline Love, Eds. *Progress in Analytical Luminescence*, American Society for Testing and Materials, Philadelphia, 1987.
14. Winefordner, J. D., and Long, G. L. *Anal. Chem.* **55**, 712A (1983).
15. Parsons, M. L. *J. Chem. Ed.* **46**, 290 (1969).
16. Foley, J. P., and Dorsey, J. G. *Chromatographia* **18**, 503 (1984).
17. Ebel, S., Kühnert, H. and Mück, W. *Chromatographia* **23**, 934 (1987).
18. Mellinger, T. J., and Keeler, C. E. *Anal. Chem.* **35**, 554 (1963).
19. Mellinger, T. J., and Keeler, C. E. *Anal. Chem.* **36**, 1840 (1964).
20. Ragland, J. B., and Kinross-Wright, V. J. *Anal. Chem.* **36**, 1356 (1964).

21. Ragland, J. B., Kinross-Wright, V. J., and Ragland, R. S. *Anal. Biochem.* **12**, 60 (1965).
22. Gurka, D. F., Kolinski, R. E., Myrick, J. W., and Wells, C. E. *J. Pharm. Sci.* **69**, 1069 (1980).
23. White, V. R., Frings, C. S., Villafranca, J. E., and Fitzgerald, J. M. *Anal. Chem.* **48**, 1314 (1976).
24. Muusze, R. G., and Huber, J. F. K. *J. Chromatogr. Sci.* **12**, 779 (1974).
25. Scholten, A. H. M. T., Brinkman, U. A. Th., and Frei, R. W. *Anal. Chim. Acta* **114**, 137 (1980).
26. Scholten, A. H. M. T., Welling, P. L. M., Brinkman, U. A. Th., and Frei, R. W. *J. Chromatogr.* **199**, 239 (1980).
27. Brinkman, U. A. Th., Welling, P. L. M., De Vries, G., Scholten, A. H. M. T., and Frei, R. W. *J. Chromatogr.* **217**, 463 (1981).
28. Gfeller, J. C., Frey, G., Huen, J. M., and Thevenin, J. P. *J.H.R.C. and C.C.* **1**, 213 (1978).
29. Frei, R. W., Lawrence, J. F., Brinkman, U. A. Th., and Honigberg, I. *J.H.R.C. and C.C.* **2**, 11 (1979).
30. Van Buuren, C., Lawrence, J. F., Brinkman, U. A. Th., Honigberg, I. L., and Frei, R. W. *Anal. Chem.* **52**, 700 (1980).
31. Lawrence, J. F., Brinkman, U. A. Th., and Frei, R. W. *J. Chromatogr.* **171**, 73 (1979).
32. Lawrence, J. F., Brinkman, U. A. Th., and Frei, R. W. *J. Chromatogr.* **185**, 473 (1979).
33. Werkhoven-Goewie, C. E., Brinkman, U. A. Th., and Frei, R. W. *Anal. Chem. Acta* **114**, 147 (1980).
34. Jansen, H., Brinkman, U. A. Th., and Frei, R. W. *J. Chromatogr. Sci.* **23**, 279 (1985).
35. Van Zoonen, P., Kamminga, D. A., Gooijer, C., Velthorst, N. H., and Frei, R. W. *Anal. Chim. Acta* **167**, 249 (1985).
36. Van Zoonen, P., Kamminga, D. A., Gooijer, C., Velthorst, N. H., Frei, R. W., and Gübitz, G. *Anal. Chim. Acta* **174**, 151 (1985).
37. Gübitz, G., van Zoonen, P., Gooijer, C., Velthorst, N. H., and Frei, R. W. *Anal. Chem.* **57**, 2071 (1985).
38. Poulsen, J. R., Birks, J. W., Gübitz, G., van Zoonen, P., Gooijer, C., Velthorst N. H., and Frei, R. W. *J. Chromatogr.* **360**, 371 (1986).
39. Poulsen, J. R., Birks, J. W., van Zoonen, P., Gooijer, C., Velthorst, N. H., and Frei, R. W. *Chromatographia* **21**, 587 (1986).
40. Van Zoonen, P., Kamminga, D. A., Gooijer, C., Velthorst, N. H., and Frei, R. W. *J. Liq. Chromatogr.* **10**, 819 (1987).
41. Studebaker, J. F. *J. Chromatogr.* **185**, 497 (1979).
42. Studebaker, J. F., Slocum, S. A., and Lewis, E. L. *Anal. Chem.* **50**, 1500 (1978).
43. Krull, I. S., Xie, K.-H., Colgan, S., Neue, U., Izod, T., King, R., and Bidlingmeyer, B. *J. Liq. Chromatogr.* **6**, 605 (1983).
44. Krull, I. S., Colgan, S., Xie, K.-H., Neue, U., King, R., and Bidlingmeyer, B. *J. Liq. Chromatogr.* **6**, 1015 (1983).
45. Jansen, J., Frei, R. W., Brinkman, U. A. Th., Deelder, R. S., and Snellings, R. P. J. *J. Chromatogr.* **325**, 255 (1985).
46. Jansen, H., Brinkman, U. A. Th., and Frei, R. W. *Chromatographia* **20**, 453 (1985).
47. Colgan, S. T., and Krull, I. S. In I. S. Krull, ed., *Reaction Detection in Liquid Chromatography*, *Chromatographic Science Series*, Dekker, New York, 1986, Vol. 34, pp. 227–258.
48. Boppana, V. K., Fong, K.-L. L., Ziemniak, J. A., and Lynn, R. K. *J. Chromatogr.* **353**, 231 (1986).
49. Dalgaard, L. *Trends Anal. Chem.* **5**, 185 (1986).
50. Bowers, L. D. In I. S. Krull, ed., *Reaction Detection in Liquid Chromatography*, *Chromatographic Science Series*, Dekker, New York, 1986, Vol. 34, pp. 195–225.
51. Nondek, L. *Anal. Chem.* **56**, 1192 (1984).

52. Nondek, L., Brinkman, U. A. Th., and Frei, R. W. *Anal. Chem.* **55**, 1466 (1983).
53. Vrátný, P., Frei, R. W., Brinkman, U. A. Th., and Nielen, M. W. F. *J. Chromatogr.* **295**, 355 (1984).
54. Bostick, D. T., and Hercules, D. M. *Anal. Chem.* **47**, 447 (1975).
55. Williams, III, D. C., Huff, G. R., and Seitz, W. R. *Anal. Chem.* **48**, 1003 (1976).
56. Dalgaard, L., Nordholm, L., and Brimer, L. *J. Chromatogr.* **265**, 183 (1983).
57. Honda, K., Miyaguchi, K., Nishino, H., Tanaka, H., Yao, T., and Imai, K. *Anal. Biochem.* **153**, 50 (1986).
58. Schlabach, T. D., and Weinberger, R. In I. S. Krull, ed., *Reaction Detection in Liquid Chromatography*, *Chromatographic Science Series*, Dekker, New York, 1986, Vol. 34, pp. 63–127.
59. Hupe, K.-P., and Frei, R. W., *Chromatographia* **15**, 401 (1982).
60. Mendenhall, D. W., Kobayashi, H., Shih, F. M. L., Sternson, L. A., Higuchi, T., and Fabian, C. *Clin. Chem.* **24**, 1518 (1978).
61. Golander, Y., and Sternson, L. A. *J. Chromatogr.* **181**, 41 (1980).
62. Frith, R. G., and Phillipou, G. *J. Chromatogr.* **367**, 260 (1986).
63. Popovich, D. J., Dixon, J. B., and Ehrlich, B. J. *J. Chromatogr. Sci.* **17**, 643 (1979).
64. LaCourse, W. R., and Krull, I. S. *Anal. Chem.* **57**, 1810 (1985).
65. LaCourse, W. R., and Krull, I. S. *Anal. Chem.* **59**, 49 (1987).
66. LaCourse, W. R., and Krull, I. S. *Trends Anal. Chem.* **4**, 118 (1985).
67. S. L. Murov, ed., *Handbook of Photochemistry*, Dekker, New York, 1973, pp. 3–7.
68. Turro, N. J. *Modern Molecular Photochemistry*, Benjamin, New York, 1981, p. 374.
69. LaCourse, W. R., Selavka, C. M., and Krull, I. S. *Anal. Chem.* **59**, 1366 (1987).
70. Poulsen, J. R., Birks, K. S., Gandelman, M. S., and Birks, J. W. *Chromatographia* **22**, 231 (1986).
71. Snyder, L. R. *J. Chromatogr.* **125**, 287 (1976).
72. Scholten, A. H. M. T., Brinkman, U. A. Th., and Frei, R. W. *J. Chromatogr.* **205**, 229 (1981).
73. Scholten, A. H. M. T., Brinkman, U. A. Th., and Frei, R. W. *Anal. Chem.* **54**, 1932 (1982).
74. Apffel, A., Brinkman, U. A. Th., and Frei, R. W. *Chromatographia* **17**, 125 (1983).
75. Apffel, A., Brinkman, U. A. Th., and Frei, R. W. *Chromatographia* **18**, 5 (1984).
76. Apffel, A., Brinkman, U. A. Th., and Frei, R. W. *J. Chromatogr.* **312**, 153 (1984).
77. Kucera, P., and Umagat, H. *J. Chromatogr.* **255**, 563 (1983).
78. Lillig, B., and Engelhardt, H. In I. S. Krull, ed., *Reaction Detection in Liquid Chromatography*, *Chromatographic Science Series*, Dekker, New York, 1986, Vol. 34, pp. 1–61.
79. Deelder, R. S., Kuijpers, A. T. J. M., and van den Berg, J. H. M. *J. Chromatogr.* **255**, 545 (1983).
80. Huber, J. F. K., Jonker, K. M., and Poppe, H. *Anal. Chem.* **52**, 2 (1980).
81. Van den Berg, J. H. M., Deelder, R. S., and Egberink, H. G. M. *Anal. Chim. Acta* **114**, 91 (1980).
82. Deelder, R. S., Kroll, M. G. F., Beesen, A. J. B., and van den Berg, J. H. M. *J. Chromatogr.* **149**, 669 (1978).
83. Horvath, C. G., Preiss, B. A., and Lipsky, S. R. *Anal. Chem.* **39**, 1422 (1967).
84. Hofmann, K., and Halász, I. *J. Chromatogr.* **173**, 211 (1979).
85. Hofmann, K., and Halász, I. *J. Chromatogr.* **199**, 3 (1980).
86. Golay, M. J. E. *J. Chromatogr.* **186**, 341 (1979).
87. Atwood, J. G., and Golay, M. J. E. *J. Chromatogr.* **186**, 353 (1979).
88. Atwood, J. G., and Golay, M. J. E. *J. Chromatogr.* **218**, 97 (1981).
89. Tijssen, R. *Anal. Chim. Acta* **114**, 71 (1980).
90. Snyder, L. R., and Adler, H. J. *Anal. Chem.* **48**, 1017 (1976).
91. Snyder, L. R., and Adler, H. J. *Anal. Chem.* **48**, 1022 (1976).
92. Snyder, L. R. *Anal. Chim. Acta* **114**, 3 (1980).

93. Deelder, R. S., and Hendricks, P. J. H. *J. Chromatogr.* **83**, 343 (1973).
94. Ertingshausen, G., Adler, H. J., and Reichler, A. S. *J. Chromatogr.* **42**, 355 (1969).
95. Deelder, R. S., Kroll, M. G. F., and van den Berg, J. H. M. *J. Chromatogr.* **125**, 307 (1976).
96. Sternberg, J. C. In J. C. Giddings and R. A. Keller, eds., *Advances in Chromatography*, Dekker, New York, 1966, Vol. 2, pp. 205–270.
97. Kirkland, J. J., Yau, W. W., Stoklosa, H. J., and Dilks, Jr., C. H. *J. Chromatogr. Sci.* **15**, 303 (1977).
98. Hupe, K.-P., Jonker, R. J., and Rozing, G. *J. Chromatogr.* **285**, 253 (1984).
99. Foley, J. P., and Dorsey, J. G. *Anal. Chem.* **55**, 730 (1983).
100. Haddad, P. R., Low, G. K.-C., and Heckenburg, A. L. *Chromatographia* **18**, 417 (1984).
101. Shih, Y.-T., and Carr, P. W. *Anal. Chim. Acta* **167** 137 (1985).
102. Uihlein, M., and Schwab, E. *Chromatographia* **15**, 140 (1982).
103. Engelhardt, H., and Neue, U. D. *Chromatographia* **15**, 403 (1982).
104. Engelhardt, H., and Klinkner, R. *Fres. Z. Anal. Chem.* **319**, 277 (1984).
105. Engelhardt, H., and Lillig, B. *J.H.R.C. and C.C.* **8**, 531 (1985).
106. Selavka, C. M., Jiao, K.-S., and Krull, I. S. *Anal. Chem.* **59**, 2221 (1987).
107. Davis, J. C., and Peterson, D. P. *Anal. Chem.* **57**, 768 (1985).
108. Haginaka, J., and Wakai, J. *J. Chromatogr.* **396**, 297 (1987).
109. Haginaka, J., Wakai, J., and Yasuda, H. *Anal. Chem.* **59**, 324 (1987).
110. Nord, L., and Karlberg, B. *Anal. Chim. Acta* **118**, 285 (1980).
111. Ogata, K., Taguchi, K., and Imanari, T. *Anal. Chem.* **54**, 2127 (1982).
112. Fossey. L., and Cantwell, F. F. *Anal. Chem.* **54**, 1693 (1982).
113. Bäckström, K., Danielsson, L.-G., and Nord, L. *Anal. Chim. Acta* **169**, 43 (1985).
114. Habig, R. L., Schlein, B. W., Walters, L., and Thiers, R. E. *Clin. Chem.* **15**, 1045 (1969).
115. Nord, L., and Karlberg, B. *Anal. Chim. Acta* **164**, 233 (1984).
116. Gübitz, G., Aischinger, I., and Birks, J. W. Unpublished work.
117. Jonker, K. M., Poppe, H., and Huber, J. F. K. *Chromatographia* **11**, 123 (1978).
118. Reijn, J. M., van der Linden, W. E., and Poppe, H. *Anal. Chim. Acta* **123**, 229 (1981).
119. Reijn, J. M., Poppe, H., and van der Linden, W. E. *Anal. Chem.* **56**, 943 (1984).
120. MacCrehan, W. A., Durst, R. A., and Bellama, J. M. *Anal. Lett.* **10**, 1175 (1977).
121. Wightman, R. M., Paik, E. C., Borman, S., and Dayton, M. A. *Anal. Chem.* **50**, 1410 (1978).
122. Gandelman, M. S., and Birks, J. W. *Anal. Chim. Acta* **155**, 159 (1983).
123. Poulsen, J. R., and Birks, J. W. Unpublished manuscript.
124. Shellum, C. L., and Birks, J. W. *Anal. Chem.* **59**, 1834 (1987).
125. Gandelman, M. S., Birks, J. W., Brinkman, U. A. Th., and Frei, R. W. *J. Chromatogr.* **282**, 193 (1983).
126. Lefevere, M. F., Frei, R. W., Scholten, A. H. M. T., and Brinkman, U. A. Th. *Chromatographia* **15**, 459 (1982).
127. Murov, S. L., ed., *Handbook of Photochemistry*, Dekker, New York, 1973, pp. 111–115.
128. Winefordner, J. D., ed., *Trace Analysis: Spectroscopic Methods for Elements*, *Chemical Analysis Series*, Wiley, New York, 1976, Vol. 46, pp. 233–235.
129. Green, R. B. In E. S. Yeung, ed., *Detectors for Liquid Chromatography*, *Chemical Analysis Series*, Wiley, New York, 1986, Vol. 89, pp. 29–63.
130. Product bulletins, #109B, #101-H/OL, #110B and #125A, Ultra-Violet Products, Inc., P.O. Box 1501, San Gabriel, CA 91778.
131. Product literature, #17620, #19208, #24559, #26421, and #24687, BHK, Inc., 1000 S. Magnolia Ave., Monrovia, CA 91016.
132. Yeung, E. S. In E. S. Yeung, ed., *Detectors for Liquid Chromatography*, *Chemical Analysis Series*, Wiley, New York, 1986, Vol. 89, pp. 1–28.
133. Shelly, D. C., and Warner, I. M. In T. M. Vickrey, ed., *Liquid Chromatography Detectors*, *Chromatographic Science Series*, Dekker, New York, 1983, Vol. 23, pp. 87–123.

134. Sepaniak, M. J., and Kettler, C. N. In E. S. Yeung, ed., *Detectors for Liquid Chromatography*, *Chemical Analysis Series*, Wiley, New York, 1986, Vol. 89, pp. 148–203.
135. Yeung, E. S. In E. S. Yeung, ed., *Detectors for Liquid Chromatography*, *Chemical Analysis Series*, Wiley, New York, 1986, Vol. 89, pp. 204–228.
136. Morris, M. D. In E. S. Yeung, ed., *Detectors for Liquid Chromatography*, *Chemical Analysis Series*, Wiley, New York, 1986, Vol. 89, pp. 105–147.
137. Voigtman, E., and Winefordner, J. D. *J. Liq. Chromatogr.* **5**, 2113 (1982).
138. Iwaoka, W., and Tannenbaum, S. R. In E. A. Walker, P. Bogovski, and L. Griciute, eds., *Inter. Agency for Res. Cancer*, *Sci. Publ.* **14**, 51 (1976).
139. Daiber, D., and Preussmann, R. *Fres. Z. Anal. Chem.* **206**, 344 (1964).
140. Fan, T.-Y., and Tannenbaum, S. R. *J. Agric. Food Chem.* **19**, 1267 (1971).
141. Shuker, D. E. G., and Tannenbaum, S. R. *Anal. Chem.* **55**, 2152 (1983).
142. Placek-Llanes, B. G., and Tannenbaum, S. R. *Carcinogenesis* **3**, 1379 (1982).
143. Singer, G. M., Singer, S. S., and Schmidt, D. G. *J. Chromatogr.* **133**, 59 (1977).
144. Steinberg, M., Leist, Y., Goldschmidt, P., and Tassa, M. *J. Forensic Sci.* **29**, 464 (1984).
145. Snider, B. G., and Johnson, D. C. *Anal. Chim. Acta* **106**, 1 (1979).
146. Ciccioli, P., Tappa, R., and Guiducci, A. *Anal. Chem.* **53**, 1309 (1981).
147. Batley, G. E. *Anal. Chem.* **56**, 2261 (1984).
148. Werkhoven-Goewie, C. E., Boon, W. M., Praat, A. J. J., Frei, R. W., Brinkman, U. A. Th., and Little, C. J. *Chromatographia* **16**, 53 (1983).
149. McKinley, W. A. *J. Anal. Tox.* **5**, 209 (1981).
150. Jasinski, J. S. *Anal. Chem.* **56**, 2214 (1984).
151. Büttler, B., and Hörmann, W. D. *J. Agric. Food Chem.* **29**, 257 (1981).
152. Walters, S. M. *J. Chromatogr.* **259**, 227 (1983).
153. Omura, K., and Matsuura, T. *Tetrahedron* **27**, 3101 (1971).
154. Wells, C. H. J. *Introduction to Molecular Photochemistry*, Chapman and Hall, London, 1972, pp. 121–122.
155. Werkhoven-Goewie, C. E., Brinkman, U. A. Th., and Frei, R. W. *Anal. Chem.* **53**, 2072 (1981).
156. Wolkoff, A. W., and Larose, R. H. *J. Chromatogr.* **99**, 731 (1974).
157. Katz, S., and Pitt, Jr., W. W. *Anal. Lett.* **5**, 177 (1972).
158. Poulsen, J. R., and Birks, J. W. Unpublished manuscript.
159. Poulsen, J. R. Ph.D. Thesis, 1988.
160. Šalamoun, J., and František, J. *J. Chromatogr.* **378**, 173 (1987).
161. Šalamoun, J., Smrž, M., Kiss, F., and Šalamounová, A. *J. Chromatogr.* **419**, 213 (1987).
162. Krull, I. S., Ding, X.-D., Selavka, C., Bratin, K., and Forcier, G. *J. Forensic Sci.* **29**, 449 (1984).
163. Selavka, C. M., Krull, I. S., and Lurie, I. S. *J. Chromatogr. Sci.* **23**, 499 (1985).
164. Ding, X.-D., and Krull, I. S. *J. Agric. Food Chem.* **32**, 622 (1984).
165. Selavka, C. M., Nelson, R. J., Krull, I. S., and Bratin, K. *J. Pharm. Biomed. Anal.* **4**, 83 (1986).
166. Selavka, C. M., and Krull, I. S. *Anal. Chem.* **59**, 2699 (1987).
167. Selavka, C. M., and Krull, I. S. *Anal. Chem.* **59**, 2704 (1987).
168. Colgan, S. T., Krull, I. S., Neue, U., Newhart, A., Dorschel, C., Stacey, C., and Bidlingmeyer, B. *J. Chromatogr.* **333**, 349 (1985).
169. Krull, I. S., Selavka, C., Duda, C., and Jacobs, W. *J. Liq. Chromatogr.* **8**, 2845 (1985) and references therein.
170. Priebe, S. R., and Howell, J. A. *Anal. Lett.* **16**, 1219 (1983).
171. Priebe, S. R., and Howell, J. A. *J. Chromatogr.* **324**, 53 (1985).
172. Luchtefeld, R. G. *J. Chromatogr. Sci.* **23**, 516 (1985).
173. Miles, C. J., and Moye, H. A. *Anal. Chem.* **60**, 220 (1988).

174. Moye, H. A., Scherer, S. J., and St. John, P. A. *Anal. Lett.* **10**, 1049 (1977).
175. Krause, R. T. *J. Chromatogr.* **185**, 615 (1979).
176. Shih, Y.-T., and Carr, P. W. *Anal. Chim. Acta* **142**, 55 (1982).
177. Shih, Y.-T., and Carr, P. W. *Anal. Chim. Acta* **159**, 211 (1984).
178. Sigvardson, K. W., Kennish, J. M., and Birks, J. W. *Anal. Chem.* **56**, 1096 (1984).
179. Schmermund, J. T., and Locke, D. C. *Anal. Lett.* **8**, 611 (1975).
180. Driscoll, J. N., Conron, D. W., Ferioli, P., Krull, I. S., and Xie, X.-H. *J. Chromatogr.* **302**, 43 (1984).
181. Locke, D. C., Dhingra, B. S., and Baker, A. D. *Anal. Chem.* **54**, 447 (1982).
182. Voigtman, E., Jurgensen, A., and Winefordner, J. D. *Anal. Chem.* **53**, 1921 (1981).
183. Voigtman, E., and Winefordner, J. D. *Anal. Chem.* **54**, 1834 (1982).
184. Yamada, S., Hino, A., Kano, K., and Ogawa, T. *Anal. Chem.* **55**, 1914 (1983).
185. Yamada, S., Hino, A., and Ogawa, T. *Anal. Chim. Acta* **156**, 273 (1984).
186. Mallory, F. B., Wood, C. S., and Gordon, J. T. *J. Am. Chem. Soc.* **86**, 3094 (1964).
187. Mallory, F. B., Wood, C. S., Gordon, J. T., Lindquist, L. C., and Savitz, M. L. *J. Am. Chem. Soc.* **84**, 4361 (1962).
188. Mallory, F. B., Gordon, J. T., and Wood, C. S. *J. Am. Chem. Soc.* **85**, 828 (1963).
189. Moore, W. M., Morgan, D. D., and Stermitz, F. R. *J. Am. Chem. Soc.* **85**, 829 (1963).
190. Wood, C. S., and Mallory, F. B. *J. Org. Chem.* **29**, 3373 (1964).
191. Zweig, A. *Pure Appl. Chem.* **33**, 389 (1973).
192. Goodyear, J. M., and Jenkinson, N. R. *Anal. Chem.* **32**, 1203 (1960).
193. Goodyear, J. M., and Jenkinson, N. R. *Anal. Chem.* **33**, 853 (1961).
194. Rhys Williams, A. T., Winfield, S. A., and Belloli, R. C. *J. Chromatogr.* **235**, 461 (1982).
195. Harman, P. J., Blackman, G. L., and Phillipou, G. *J. Chromatogr.* **225**, 131 (1981).
196. Adam, H. K., Gay, M. A., and Moore, R. H. *J. Endocr.* **84**, 35 (1980).
197. Brown, R. R., Bain, R., and Jordan, V. C. *J. Chromatogr.* **272**, 351 (1983).
198. Nieder, M., and Jaeger, H. *J. Chromatogr.* **413**, 207 (1987).
199. Dyck, R. H., and McClure, D. S. *J. Chem. Phys.* **36**, 2326 (1962).
200. Twitchett, P. J., Williams, P. L., and Moffat, A. C. *J. Chromatogr.* **149**, 683 (1978).
201. Bowd, A., Byrom, P., Hudson, J. B., and Turnbull, J. H. *Talanta* **18**, 697 (1971).
202. Bowd, A., Swann, D. A., and Turnbull, J. H. *J. Chem. Soc., Chem. Comm.*, 797 (1975).
203. Scholten, A. H. M. T., and Frei, R. W. *J. Chromatogr.* **176**, 349 (1979).
204. Stoll, A., and Schlientz, W. *Helv. Chim. Acta* **38**, 585 (1958).
205. Gillespie, A. M. *Anal. Lett.* **2**, 609 (1969).
206. Stewart, J. T., Honigberg, I. L., Tsai, A. Y., and Hajdú, P. *J. Pharm. Sci.* **68**, 494 (1979).
207. Strojny, N., and de Silva, J. A. F. *Anal. Chem.* **52**, 1554 (1980).
208. Koechlin, B. A., and D'Arconte, L. *Anal. Biochem.* **5**, 195 (1963).
209. Schwartz, M. A., and Postma, E. *J. Pharm. Sci.* **55**, 1358 (1966).
210. Wilkinson, F. *J. Phys. Chem.* **66**, 2569 (1962).
211. Tickle, A., and Wilkinson, F. *Trans. Faraday Soc.* **61**, 1981 (1965).
212. Bruce, J. M. In S. Patai, ed., *The Chemistry of Quinoid Compounds*, *Part 1*, Wiley, New York, 1974, pp. 494–507.
213. Turro, N. J. *Modern Molecular Photochemistry*, Benjamin, New York, 1981, pp. 362–392.
214. Harriman, A., and Mills, A. *Photochem. Photobiol.* **33**, 619 (1981).
215. Carlson, S. A., and Hercules, D. M. *Photochem. Photobiol.* **17**, 123 (1973).
216. Hamanoue, K., Nakayama, T., Tanaka, A., Kajiwara, Y., and Teranishi, H. *J. Photochem.* **34**, 73 (1986).
217. Hulme, B. E., Land, E. J., and Phillips. G. O. *J. Chem. Soc., Faraday Trans. I* **68**, 1992 (1972).
218. Hulme, B. E., Land, E. J., and Phillips, G. O. *J. Chem. Soc., Faraday Trans. I* **68**, 2003 (1972).

219. Bridge, N. K., and Porter, G. *Proc. Roy. Soc. (London), Ser. A* **244**, 259 (1958).
220. Hercules, D. M., and Surash, J. J. *Spectrochim. Acta* **19**, 788 (1963).
221. Hamanoue, K., Yokoyama, K., Miyake, T., Kasuya, T., Nakayama, T., and Teranishi, H. *Chem. Lett., Chem. Soc. Japan*, 1967 (1982).
222. Jansson, O. *Acta Chem. Scand.* **24**, 2839 (1970).
223. Aaron, J. J., Villafranca, J. E., White, V. R., and Fitzgerald, J. M. *Appl. Spectroscopy* **30**, 159 (1976).
224. Aaron, J. J. *Methods Enzymol.* **67**, 140 (1980).
225. Langenberg, J. P., and Tjaden, U. R. *J. Chromatogr.* **305**, 61 (1984).
226. Langenberg, J. P., and Tjaden, U. R. *J. Chromatogr.* **289**, 377 (1984).
227. Kusube, K., Abe, K., Hiroshima, O., Ishiguro, Y., Ishikawa, S., and Hoshida, H. *Chem. Pharm. Bull.* **32**, 179 (1984).
228. Lambert, W. E., De Leenheer, A. P., and Lefevere, M. F. *J. Chromatogr. Sci.* **24**, 76 (1986).
229. Lambert, W. E., De Leenheer, A. P., and Baert, E. J. *Anal. Biochem.* **158**, 257 (1986).
230. Brunt, K., Bruins, C. H. P., and Doornbos, D. A. *Anal. Chim. Acta* **125**, 85 (1981).
231. Ikenoya, S., Abe, K., Tsuda, T., Yamano, Y., Hiroshima, O., Ohmae, M., and Kawabe, K. *Chem. Pharm. Bull.* **27**, 1237 (1979).
232. Ueno, T., and Suttie, J. W. *Anal. Biochem.* **133**, 62 (1983).
233. Hart, J. P., Shearer, M. J., and McCarthy, P. T. *Analyst (London)* **110**, 1181 (1985).
234. Hart, J. P. *Trends Anal. Chem.* **5**, 20 (1986).
235. Ettlinger, M. G. *J. Am. Chem. Soc.* **72**, 3666 (1950).
236. Green, J. P. *Nature* **174**, 369 (1954).
237. Green, J. P., and Dam, H. *Acta Chem. Scand.* **8**, 1341 (1954).
238. Leary, G., and Porter, G. *J. Chem. Soc. A* 2273 (1970).
239. Snyder, C. D., and Rapoport, H. *J. Am. Chem. Soc.* **91**, 731 (1969).
240. Mee, J. M. L., Brooks, C. C., and Yanagihara, K. H. *J. Chromatogr.* **110**, 178 (1975).
241. Vire, J. C., Patriarche, G. J., and Christian, G. D. *Anal. Chem.* **51**, 752 (1979).
242. Wilson, R. M., Walsh, T. F., and Gee, S. K. *Tetrahedron Lett.* **21**, 3459 (1980).
243. Nakata, T., and Tsuchida, E. *Methods Enzymol.* **67**, 148 (1980).
244. Fujisawa, S., Kawabata, S., and Yamamoto, R. *J. Pharm. Soc. Jpn.* **87**, 1451 (1967).
245. Lefevere, M. F., De Leenheer, A. P., and Claeys, A. E. *J. Chromatogr.* **186**, 749 (1979).
246. Zonta, F., and Stancher, B. *J. Chromatogr.* **329**, 257 (1985).
247. Chapman, O. L., Heckert, D. C., Reasoner, J. W., and Thackaberry, S. P. *J. Am. Chem. Soc.* **88**, 5550 (1966).
248. Barltrop, J. A., and Bunce, N. J. *J. Chem. Soc., Sec. C*, 1467 (1968).
249. Cu, A., and Testa, A. C. *J. Am. Chem. Soc.* **96**, 1963 (1974).
250. Sigvardson, K. W., and Birks, J. W. *J. Chromatogr.* **316**, 507 (1984).
251. MacCrehan, W. A., and May, W. E. *Anal. Chem.* **56**, 625 (1984).
252. MacCrehan, W. A., and May, W. E. In A. J. Dennis and M. Cooke, eds., *Proc. 8th Inter. Symp. on PAHs*, Batelle, Columbus, OH, 1984, pp. 857–869.
253. Tejada, S. B., Zweidinger, R. B., and Sigsby, J. E. *Anal. Chem.* **58**, 1827 (1986).
254. MacCrehan, W. A., May, W. E., and Yang, S. D. *Anal. Chem.* **60**, 194 (1988).
255. MacCrehan, W. A., Yang, S. D., and Benner, Jr., B. A. *Anal. Chem.* **60**, 284 (1988).
256. Wright, G. E., and Tang, Y. T. *J. Pharm. Sci.* **61**, 299 (1972).
257. Lang, J. R., Honigberg, I. L., and Stewart, J. T. *J. Chromatogr.* **252**, 288 (1982).
258. Lang, J. R., Stewart, J. T., and Honigberg, I. L. *J. Chromatogr.* **264**, 144 (1983).
259. Wells, C. H. J. *Introduction to Molecular Photochemistry*, Chapman and Hall, London, 1972, pp. 105–107.
260. Davidson, A. G. *J. Pharm. Pharmac.* **28**, 795 (1976).
261. Davidson, A. G. *J. Pharm. Pharmac.* **30**, 410 (1978).
262. Basińska, H., and Nowakowski, K. *Acta Polon. Pharm.* **29**, 463 (1972).

263. Ramappa, P. G., Gowda, H. S., and Nayak, A. N. *Analyst* **105**, 663 (1980).
264. Breyer, U. *Biochem. Pharmac.* **18**, 777 (1969).
265. Essien, E. E., Cowan, D. A., and Beckett, A. H. *J. Pharm. Pharmac.* **27**, 334 (1975).
266. Dekirmenjian, H., Javaid, J. I., Duslak, B., and Davis, J. M. *J. Chromatogr.* **160**, 291 (1978).
267. Cooper, S. F., and LaPierre, Y. D. *J. Chromatogr.* **222**, 291 (1981).
268. Nachtmann, F., Knapp, G., and Spitzy, H. *J. Chromatogr.* **149**, 693 (1978).
269. Shellum, C. L. Ph.D. Thesis, 1988.
270. Bolland, J. L., and Cooper, H. R. *Nature* **172**, 413 (1953).
271. Bolland, J. L., and Cooper, H. R. *Proc. Roy. Soc.* (*London*), *Ser. A* **225**, 405 (1954).
272. Wells, C. F. *Trans. Faraday Soc.* **57**, 1703 (1961).
273. Wells, C. F. *Trans. Faraday Soc.* **57**, 1719 (1961).
274. Carlson, S. A., and Hercules, D. M. *J. Am. Chem. Soc.* **93**, 5611 (1971).
275. Rauhut, M. M., Sheehan, D., Clarke, R. A., and Semsel, A. M. *Photochem. Photobiol.* **4**, 1097 (1965).
276. Rauhut, M. M., Bollyky, L. J., Roberts, B. G., Loy, M., Whitman, R. H., Iannotta, A. V., Semsel, A. M., and Clarke, R. A. *J. Am. Chem. Soc.* **89**, 6515 (1967).
277. Rauhut, M. M. *Acc. Chem. Res.* **2**, 80 (1969).
278. Rauhut, M. M., Roberts, B. G., Maulding, D. R., Bergmark, W., and Coleman, R. *J. Org. Chem.* **40**, 330 (1975).
279. Rauhut, M. M., and Semsel, A. M. Office of Naval Research, Final Report #N00014-73-C-0343, 1974, pp. 1–41.
280. McCapra, F. *Prog. Org. Chem.* **8**, 231 (1971).
281. McCapra, F., Perring, K., Hart, R. J., and Hann, R. A. *Tetrahedron Lett.* **22**, 5087 (1981).
282. Lechtken, P., and Turro, N. J. *Mol. Photochem.* **6**, 95 (1974).
283. Mohan, A. G., and Turro, N. J. *J. Chem. Ed.* **51**, 528 (1974).
284. Alvarez, F. J., Parekh, N. J., Matuszewski, B., Givens, R. S., Higuchi, T., and Schowen, R. L. *J. Am. Chem. Soc.* **108**, 6435 (1986).
285. Seitz, W. R., and Neary, M. P. *Anal. Chem.* **46**, 188A (1974).
286. Seitz, W. R. *Methods Enzymol.* **57**, 445 (1978).
287. Seitz, W. R. *CRC Crit. Rev. Anal. Chem.* **13**, 1 (1981).
288. Sherman, P. A., Holzbecher, J., and Ryan, D. E. *Anal. Chim. Acta* **97**, 21 (1978).
289. Freeman, T. M., and Seitz, W. R. *Anal. Chem.* **50**, 1242 (1978).
290. Scott, G., Seitz, W. R., and Ambrose, J. *Anal. Chim. Acta* **115**, 221 (1980).
291. Curtis, T. G., and Seitz, W. R. *J. Chromatogr.* **134**, 343 (1977).
292. Curtis, T. G., and Seitz, W. R. *J. Chromatogr.* **134**, 513 (1977).
293. Kobayashi, S.-I., and Imai, K. *Anal. Chem.* **52**, 424 (1980).
294. Sigvardson, K. W., and Birks, J. W. *Anal. Chem.* **55**, 432 (1983).
295. Miyaguchi, K., Honda, K., and Imai, K. *J. Chromatogr.* **303**, 173 (1984).
296. Miyaguchi, K., Honda, K., and Imai, K. *J. Chromatogr.* **316**, 501 (1984).
297. Grayeski, M. L., and Weber, A. J. *Anal. Lett.* **17**, 1539 (1984).
298. Weinberger, R., Mannan, C. A., Cerchio, M., and Grayeski, M. L. *J. Chromatogr.* **288**, 445 (1984).
299. Mellbin, G., and Smith, B. E. F. *J. Chromatogr.* **312**, 203 (1984).
300. Weinberger, R. *J. Chromatogr.* **314**, 155 (1984).
301. Imai, K., and Weinberger, R. *Trends Anal. Chem.* **4**, 170 (1985).
302. Imai, K. *Methods Enzymol.* **133**, 435 (1986).
303. Bach, B., and Fiehn, G. *Zellstoff u. Papier* **21**, 111 (1972).
304. Holton, H. H., and Chapman, F. L. *Tappi* **60**, 121 (1977).
305. Košiková, B., Janson, J., Pekkala, O., and Sågfors, P.-E. *Pap. Puu* **62**, 229 (1980).
306. Pekkala, O. *Pap. Puu* **64**, 735 (1982).

307. Ingruber, O. V. *New Process Alternatives in the Forest Products Industries*, *AIChE Symp. Ser.* **76**, 196 (1980).
308. Renard, J. J., and Phillips, R. B. *New Process Alternatives in the Forest Products Industries*, *AIChE Symp. Ser.* **76**, 182 (1980).
309. Haggin, J. *Chem. Engin. News* **62**, No. 42, 20 (1984).
310. *Chem. Engin. News* **64**, No. 21, 19 (1986).
311. Kiba, N., Takamatsu, M., and Furusawa, M. *J. Chromatogr.* **328**, 309 (1985).
312. Weber, S. G., Morgan, D. M., and Elbicki, J. M. *Clin. Chem.* **29**, 1665 (1983).
313. Elbicki, J. M., Morgan, D. M., and Weber, S. G. *Anal. Chem.* **57**, 1746 (1985).
314. Gandelman, M. S., and Birks, J. W. *J. Chromatogr.* **242**, 21 (1982).
315. Gandelman, M. S., and Birks, J. W. *Anal. Chem.* **54**, 2131 (1982).
316. Gandelman, M. S. Ph.D. Thesis, 1983.
317. Loeff, I., Treinin, A., and Linschitz, H. *J. Phys. Chem.* **87**, 2536 (1983).
318. Roy, A., Bhattacharya, D., and Aditya, S. *J. Indian Chem. Soc.* **59**, 585 (1982).
319. Clark, K. P., and Stonehill, H. I. *J. Chem. Soc., Faraday Trans. I* **68**, 577 (1972).
320. Clark, K. P., and Stonehill, H. I. *J. Chem. Soc., Faraday Trans. I* **68**, 1676 (1972).
321. Clark, K. P., and Stonehill, H. I. *J. Chem. Soc., Faraday Trans. I* **73**, 722 (1977).
322. Burchill, C. E., Smith, D. M., and Charlton, J. L. *Can. J. Chem.* **54**, 505 (1976).
323. Charlton, J. L., Smerchanski, R. G., and Burchill, C. E. *Can. J. Chem.* **54**, 512 (1976).
324. Stonehill, H. I., and Clark, K. P. *Can. J. Chem.* **54**, 516 (1976).
325. De Jong, G. J., Lammers, N., Spruit, F. J., Brinkman, U. A. Th., and Frei, R. W. *Chromatographia* **18**, 129 (1984).
326. De Jong, G. J., Lammers, N., Spruit, F. J., Frei, R. W., and Brinkman, U. A. Th. *J. Chromatogr.* **358**, 249 (1986).
327. Birks, J. W., and Kuge, M. C. *Anal. Chem.* **52**, 897 (1980).
328. Shoemaker, B., and Birks, J. W. *J. Chromatogr.* **209**, 251 (1981).
329. Pilosof, D., and Nieman, T. A. *Anal. Chem.* **54**, 1698 (1982).
330. De Jong, G. J., Lammers, N., Spruit, F. J., Dewaele, C., and Verzele, M. *Anal. Chem.* **59**, 1458 (1987).
331. Van Zoonen, P., Kamminga, D. A., Gooijer, C., Velthorst, N. H., Frei, R. W., and Gübitz, G. *Anal. Chem.* **58**, 1245 (1986).
332. Mellbin, G. *J. Liq. Chromatogr.* **6**, 1603 (1983).

7

Photochemical Reaction Detection Based On Singlet Oxygen Sensitization

Curtis L. Shellum and John W. Birks

Department of Chemistry and Cooperative Institute for Research in Environmental Sciences (CIRES), University of Colorado, Boulder, Colorado

The presence of singlet oxygen as an intermediate defines the mechanism of photooxidation termed type II photooxidation. This chapter discusses the richness of possible detection schemes arising from this process. An overview of many methods will be presented, with detail given to the most successful ones. Detection schemes that have not yet been experimentally verified will also be presented for the sake of interest and completeness.

The use of a singlet oxygen mediated process for detection can yield improvements in sensitivity of between 1 and 2 orders of magnitude over UV absorption. This improvement lies in the fact that the reaction sequence is photocatalytic, as will be discussed in some detail. Detection will be discussed with respect to both liquid and gas chromatography, although the work to date has been done in liquid chromatography.

Theory

Photosensitized Oxidation

The reaction sequence presented below defines type II photooxidation. It involves the promotion of oxygen to its excited singlet via the energy transfer quenching of organic molecules in their singlet or triplet excited states. This is a common and well studied process. Reactions (7-1) through (7-7) can be collectively referred to as photosensitization of singlet oxygen, while (7-8) and (7-9) describe the alternate fates of singlet oxygen—either radiationless decay back to the ground state [reaction (7-8)] or reaction with an acceptor [reaction (7-9)].

Type II Photooxidation Sequence:

$S + h\nu \rightarrow {}^1S^*$	(light absorption)	(7-1)
${}^1S^* \rightarrow S + h\nu$	(fluorescence)	(7-2)
${}^1S^* \rightarrow S$	(internal conversion)	(7-3)
${}^1S^* \rightarrow {}^3S^*$	(intersystem crossing)	(7-4)
${}^3S^* \rightarrow S$	(intersystem crossing)	(7-5)
${}^1S^* + {}^3O_2 \rightarrow S + {}^1O_2^*$	(energy transfer)	(7-6)
${}^3S^* + {}^3O_2 \rightarrow S + {}^1O_2^*$	(energy transfer)	(7-7)
${}^1O_2^* \rightarrow {}^3O_2$	(quenching)	(7-8)
$A + {}^1O_2^* \rightarrow AO_2$	(reaction)	(7-9)

where S represents the singlet oxygen sensitizer, A the singlet oxygen acceptor and AO_2 the oxidized acceptor. S, ${}^1S^*$ and ${}^3S^*$ denote the ground, first excited singlet and lowest triplet states of the sensitizer, respectively. ${}^3O_2$ and ${}^1O_2^*$ represent the ground and singlet excited states of molecular oxygen.

This photooxidation mechanism is differentiated from type I photooxidation by the fact that the excited triplet states in the type I mechanism are capable of hydrogen abstraction from compounds with weakly bonded hydrogen atoms. The ensuing oxidation reaction is discussed in the previous chapter and also provides a basis for sensitive detection in HPLC.

Singlet oxygen research dates back to the 1920s and has undergone an interesting history of controversy and reinterpretation. The mechanism of photosensitized singlet oxygen production outlined above was actually proposed by Kautsky and de Bruijn[1] in 1931 but was not well accepted until the 1960s. The history has been carefully traced by Kasha and Khan,[2,3] Kearns[4] and Schaap.[5] Several books are devoted entirely to the subject of singlet oxygen.[5–7]

With the exception of the section on detection of singlet oxygen quenchers, the analytes discussed in this chapter will be those compounds capable of acting as singlet oxygen sensitizers. They exit the chromatographic column and enter the photochemical reactor where light drives the reaction sequence (7-1) through (7-9). If S is a singlet oxygen sensitizer it will induce singlet oxygen production and indirectly cause the oxidation of compound A, the singlet oxygen acceptor. Detection is brought about by detecting a decrease in the concentration of the acceptor or the appearance of the photooxidized product AO_2. The singlet oxygen sensitizer can catalytically oxidize the acceptor throughout its residence time in the photochemical reactor, i.e., the ground state S, which initiates the sequence, is always available for reexcitation after singlet oxygen production in reaction (7-6) or (7-7).

It is useful to examine each of the preceding reactions, especially from the perspective of how a particular compound will respond in the detection process. Some of the photophysical processes lead to detection, while other processes compete with those leading to detection.

In reaction (7-1), the analyte absorbs light and is promoted to its excited singlet state. The first-order rate constant for this process is given by

$$k_1 = \int I_\lambda \sigma_\lambda \, d\lambda \qquad (7\text{-}10)$$

where sigma is the absorption cross section of the analyte at wavelength λ, in units of centimeters squared per molecule ($2303\epsilon_\lambda / N_A$ where ϵ_λ is the molar absorption coefficient and N_A is Avogadro's number) and I_λ is the photon flux in units of photons per centimeter squared per second at wavelength λ. An increase in lamp intensity I will directly increase k_1 and will enhance the detection process, as will be discussed. The molar extinction coefficient of the analyte will also directly correlate with detection sensitivity.

Following absorption, the excited singlet state can follow one of many routes to deactivation, dependent both on the solvent conditions and the photophysical nature of the analyte. These routes are reactions (7-2) to (7-4) and (7-6) and are the processes of fluorescence, internal conversion, intersystem crossing and energy transfer, respectively.

Reactions (7-2) and (7-3) do not lead to singlet oxygen formation, while both (7-4) and (7-6) do. The apportionment between these reactions will therefore affect the quantum efficiency for singlet oxygen production. Although fluorescence in reaction (7-2) does not lead to singlet oxygen production, it should be noted that even highly fluorescent molecules (for example, anthracene with a $\Phi_{ISC} = 0.3$) will generally produce a significant amount of singlet oxygen, as will be observed later in this chapter. As for reaction (7-3), its efficiency decreases with increasing molecular rigidity and can usually be neglected for rigid aromatic hydrocarbons.[8]

Reaction (7-4) appears to be spin forbidden, but because of factors inducing spin-orbit coupling of ^1S to ^3S and other energetic considerations, relaxation of the selection rule occurs. These are discussed in detail elsewhere.[9] Examples of the important factors include the similarity of electronic configurations of the states undergoing intersystem crossing, the effect of "heavy atoms," the presence of paramagnetic oxygen and the size of the energy gap between the singlet and triplet states. As an example, molecules with an n, π^* singlet excited state and a close lying π, π^* triplet excited state (certain carbonyl compounds such as benzophenone) tend to have much greater spin-orbit coupling relative to the corresponding $^1(\pi, \pi^*)$ to $^3(\pi, \pi^*)$ transitions, and the quantum efficiency for intersystem crossing is much greater. The fluorescence quantum yields for the two types of molecules (n, π^* or π, π^* singlet states) are opposite in relative magnitudes for the same reason. Table 7-1 contains a list of several compounds with their excited state configurations and respective fluorescence and intersystem crossing quantum yields. It can be seen that these clearly support the preceding statements. In general, with other factors equal, compounds with the highest quantum yields for intersystem crossing will be the most sensitively detected. This will be illustrated by example later.

Reaction (7-6) is another potential means of deactivation of the excited singlet state. This is normally an unimportant pathway, as shown by experimental evidence and as predicted by Franck–Condon factors.[10] However, there is experimental evidence for several specific compounds that can sensitize singlet oxygen via their singlet state.

Table 7-1 ▪ Quantum Yields for Fluorescence and Intersystem Crossing[a]

Molecule (configuration of S_1)	ϕ_F	ϕ_{ISC}
Benzene (π, π^*)	0.05	0.25
1,4-dimethylbenzene (π, π^*)	0.35	0.65
Naphthalene (π, π^*)	0.20	0.80
Anthracene (π, π^*)	0.70	0.30
Tetracene (π, π^*)	0.15	0.65
Pentacene (π, π^*)	0.10	0.15
Acetone (n, π^*)	0.001	~ 1.0
Biacetyl (n, π^*)	0.002	~ 1.0
Benzophenone (n, π^*)	0.000	~ 1.0

[a] From F. Wilkinson, in J. B. Birks, eds., *Organic Molecular Photophysics*, Wiley, New York, 1975, p. 95.

These include 9-methylanthracene, 9-phenylanthracene, 9,10-dimethylanthracene, perylene, pyrene and naphthalene.[11,12] These are all highly fluorescent and in some of these compounds the efficiency of energy transfer from the singlet state to oxygen is almost equal to that from the triplet state, but singlet oxygen generation from the triplet state is always the favored process of the two.

Once the triplet excited state is produced in reaction (7-4), it can either intersystem cross to the ground state [reaction (7-5)] or be quenched by oxygen in reaction (7-7). For solutions saturated with oxygen, reaction (7-7) is typically 100 to 1000 times faster than reaction (7-5), as predicted by Franck–Condon factors.[13] Reaction (7-5) can thus be neglected for all practical purposes. Phosphorescence emission from the excited triplet is also much slower than quenching by oxygen and has thus been left out of the reaction sequence; phosphorescent lifetimes are 3 to 6 orders of magnitude longer.[13]

Quenching by oxygen in reaction (7-7) occurs at a diffusion controlled rate constant for triplets with an energy of ≥ 22.5 kcal mol^{-1}, the energy of the $^1\Delta_g$ state of oxygen, and triplets with energies in excess of 37.7 kcal mol^{-1} can produce the $^1\Sigma_g^+$ state of oxygen.[13–18] The $^1\Sigma_g^+$ state undergoes a spin-allowed radiationless decay within 10^{-10} s to the metastable $^1\Delta_g$ state.[19] Much work has been done to describe the production of $^1\Delta_g$ and $^1\Sigma_g^+$ as a function of the triplet state energy of the sensitizer, but this has been difficult because of the short-lived nature of the $^1\Sigma_g^+$ state.[20] Theoretically, the ratio $^1\Sigma_g^+ : {}^1\Delta_g$ should approach unity at donor triplet energies > 42 kcal mol^{-1}.[21] This, however, has not yet been experimentally validated.[15]

The rate constant for reaction (7-7), although diffusion controlled, is statistically only 1/9 that of k_d (pure diffusion controlled rate). This is because the precursor of the final quenching products is a collision complex of the triplet sensitizer with ground-state oxygen that is formed with either singlet, triplet or quintet multiplicity.[4] Only the singlet complex can yield $^1O_2$ directly, and thus only 1/9 of the collisions will yield $^1O_2$. Notice that at this point in the reaction sequence the sensitizer is returned to its ground state and is reavailable for absorption and the potential generation of another singlet oxygen molecule. Singlet oxygen will be continually produced as long as

Table 7-2 ▪ Quantum Yields of Singlet Oxygen Formation in Photosensitized Oxygenation Reactions of 2,5-Dimethylfuran (DMF) and Tetramethylethylene (TME) in Methanol at 20°C[a]

Sensitizer	Triplet Energy E_T (kcal/mol)	ϕ_{ISC}	$\phi_{^1O_2}$ (λ_{ex} = 283–372 nm) DMF	TME
Benzaldehyde	72.0	—	0.64–0.53	0.64–0.57
Acetophenone	73.6	0.99	0.59–0.56	0.17–0.16
Benzophenone	68.5	0.99	0.53–0.46	0.25–0.23
Fluorene	67.6	0.31	0.11–0.08	0.10–0.08
Quinoline	62.0	0.16	0.10–0.08	0.11–0.09
Naphthalene	60.9	0.39	0.15–0.12	0.15–0.13
Fluorenone	53.3	0.93	0.07–0.06	0.30–0.02
Pyrene	48.7	0.18	0.62–0.57	0.65–0.60
Rose bengal	39.5–42.2	0.8	0.83–0.80	0.82–0.78

[a]From K. Gollnick, T. Franken, G. Schade, and G. Dörhöfer, *Ann. N.Y. Acad. Sci.* **171**, 89 (1970).

light is present and as long as the sensitizer is not oxidized to a nonabsorbing compound. This is the basis of the catalytic nature of this reaction sequence.

The degree of singlet oxygen production can be defined in terms of quantum yield (probability of singlet oxygen formation per photon absorbed by the sensitizer). For UV-visible absorbing organic compounds with triplet state energies ≥ 22.5 kcal mol^{-1}, the quantum efficiency is almost always nonzero and can range up to 1.0. Table 7-2 gives some example quantum yields. Note that even highly fluorescent compounds such as fluorene, naphthalene and pyrene have significant singlet oxygen production. Reaction (7-6), singlet oxygen sensitization from the excited singlet state, must be a significant contributor in the case of pyrene.

As outlined in reactions (7-8) and (7-9), the singlet oxygen can either be quenched back to the ground state or react with a singlet oxygen acceptor. The quenching in reaction (7-8) can be by the solvent or by another quencher if present. Radiative emission is highly improbable, as the radiative lifetime for the $^1\Delta_g$ state is 45 min in a vacuum.[20] Quenching by nonsolvent molecules will be discussed in the section entitled "Detection of Singlet Oxygen Quenchers." Solvent quenching is worth noting here, however. There are enormous variations in the singlet oxygen lifetime from solvent to solvent (microseconds to milliseconds). The lifetime τ can be expressed as the inverse of the first-order rate constant for reaction (7-8) ($\tau = 1/k_8$). A list of solvents and the corresponding $^1\Delta_g$ lifetimes are given in Table 7-3 and in other reports.[22] The shortest lifetimes (2 μs) are observed in aqueous solution or alcohols, and the longest lifetimes (1000 μs) are seen in perfluorinated hydrocarbons. Three common reverse-phase HPLC solvents, water, methanol and acetonitrile, induce lifetimes of 2, 7 and 30 μs, respectively. Note that deuteration increases the lifetime by approximately an order of magnitude.

Many solvent characteristics have been examined in order to determine a mechanism for this solvent effect. These have included solvent polarity, viscosity, polarizabil-

Table 7-3 ▪ Lifetimes of Singlet Oxygen in Various Solvents[a]

Solvent	τ (μs)	Solvent	τ (μs)
H_2O	2	CH_3CN	30
CH_3OH	7	D_2O/CD_3OD (1 : 1)	35
D_2O/CH_3OH (1 : 1)	11	$CHCl_3$	60
C_2H_5OH	12	CS_2	200
C_6H_{12}	17	$CDCl_3$	300
D_2O	20	C_6F_6	600
C_6H_6	24	CCl_4	700
$(CH_3)_2CO$	26	$CFCl_3$	1000

[a]From Kearns, D. R., in H. H. Wasserman and R. W. Murray, eds., *Singlet Oxygen*, Academic, New York, 1979, Chap. 4, p. 120 and K. Gollnick and H. J. Kuhn, in H. H. Wasserman and R. W. Murray, eds., *Singlet Oxygen* Academic, New York, 1979, Chap. 8, p. 290.

ity, ionization potential, absorption characteristics and oxygen solubility. Of these, the only important factor that has emerged is the infrared absorption spectrum of the solvent. The intensity of absorption near 7880 and 6280 cm^{-1}, resonant with the $^1\Delta_g \rightarrow {}^3\Sigma_g^-$ (0,0) and (0,1) transitions of oxygen, dependably predict what the relative lifetime will be. Energy transfer quenching occurs which involves the conversion of electronic excitation energy of oxygen into vibrational excitation of the solvent. For small molecules such as diatomic oxygen which have few internal modes of vibration (one for O_2), the energy spacing of the solvent's vibrational levels plays the dominant role in radiationless deactivation. More details are discussed elsewhere.[23]

The final reaction in this sequence is reaction (7-9), the oxidation by singlet oxygen of an organic singlet oxygen acceptor. An acceptor is a compound that is readily oxidized by singlet oxygen. If such a compound is present in the solution, reaction (7-9) can compete with the radiationless decay of reaction (7-8). The degree of reaction with singlet oxygen will depend on the concentration of the acceptor, the type of solvent and the oxidation rate constant for the particular compound (k_9).

There is an immense variety of compounds that react to a greater or lesser degree with singlet oxygen. Large tabulations of acceptors and their rate constants for reaction can be found.[22, 24] Table 7-4 contains a collection of some of the acceptors that will be mentioned or discussed in this chapter.

The mechanisms of reaction fall into two major categories depending on the type of organic acceptor. These are (1) compounds that contain the structural element of *cis*-1,3-dienes (e.g., cyclic 1,3-dienes, aromatics such as anthracene and heterocyclic compounds such as furans) and (2) olefins containing allylic hydrogen atoms. The first class of compounds undergoes 1,4-cycloaddition reactions analogous to the Diels–Alder reaction in which intermediates such as hydroperoxides, dioxetanes or endoperoxides are formed and gives rise to the observed products.[15] There is an especially great diversity of products among the heterocycles. This results not from the initial oxygen addition, but from the varied processes involved in the breakdown of the intermediates.

Table 7-4 ▪ Reaction Rate Constants for Selected Singlet Oxygen Acceptors

Compound	k_r ($M^{-1} s^{-1}$)	Ref.
1,3-Diphenylisobenzofuran (DPBF)	9.1×10^8	*a*
2,5-Dimethylfuran (DMF)	3.6×10^8	*a*
2,5-Diphenylfuran (DPF)	1.1×10^8	*a*
2-Methylfuran (MF)	9.1×10^7	*a*
Tetramethylethylene	3.0×10^7	*b*
N,N-dimethylisobutenylamine	4.9×10^8	*c*
2,3-Diphenyl-1,4-dioxene	1.6×10^7	*d*

[a] R. H. Young, K. Wehrly and R. L. Martin, *J. Am. Chem. Soc.* **93**, 5774 (1971).
[b] F. Wilkinson and J. G. Brummer, *J. Phys. Chem. Ref. Data* **10**, 809 (1981).
[c] C. S. Foote, A. A. Dzakpasu, and J. W. Lin, *Tetrahedron Lett.* **14**, 1247 (1975).
[d] K. A. Zaklika, B. Kaskar and A. P. Schaap, *J. Am. Chem. Soc.* **102**, 386 (1980).

In the breakdown reactions, the effects of solvent, temperature and substituents all have an important role in determining the type of product(s) formed.[25]

The second class of compounds forms allylic hydroperoxides where the double bond has shifted into the allylic position, analogous to the "ene" reaction.[26] These reactions are not significantly solvent dependent.

The choice of an acceptor depends on many factors such as its reaction rate constant (k_9) in the particular solvent used, its absorption properties and its spectral or other detectable changes in physical properties upon oxidation. These will be discussed in later sections with respect to specific examples. All the compounds utilized thus far have been heterocycles. Most notable have been various furans which fall into class 1. Furans as well as other acceptors given in Table 7-4 have been used extensively in the study of singlet oxygen. The vast amount of groundwork on these reactions has made it straightforward to adapt this photooxidation sequence to detection.

Use of the Photooxidation Reaction for Detection. The type II photooxidation sequence has been discussed where the excitation of a singlet oxygen sensitizer (the analyte in this case) can result in the oxidation of an organic singlet oxygen acceptor. This can be restated and put into the context of chromatographic detection using a post-column photochemical reactor: The chromatographic effluent (spiked with a singlet oxygen acceptor) exits the separation column and enters the photochemical reactor. Through the course of irradiation time in the reactor the compounds which are singlet oxygen sensitizers [reaction (7-6) or (7-7)] will continuously catalyze the oxidation of the singlet oxygen acceptor [reaction (7-9)]. Many acceptor molecules can be oxidized for each sensitizer molecule because the sensitizer is reavailable to absorb another photon [reaction (7-1)] each time its excited state is quenched to form singlet oxygen. The effluent of the reactor then flows into a detector which responds to either a decrease in acceptor concentration or an appearance of the oxidation product. Absorption, fluorescence or other suitable method can be used for detection, depending on the characteristics of the singlet oxygen acceptor.

Kinetics of Sensitized Photooxidation

The reaction kinetics leading to oxidation of the acceptor are influenced by many factors including the nature of the sensitizer, the type of solvent, the excitation spectrum of the lamp, the absorption spectra of the sensitizer and acceptor and, of course, the reaction rates and concentrations of all the species involved. There are many factors that complicate things. One example is the case where the sensitizer itself can be oxidized to an extent by singlet oxygen. In this case, the rate of acceptor photooxidation would continually change throughout the residence time in the reactor, because the concentration of the original sensitizer is continually changing. Another complication could arise for certain quinone analytes capable of hydrogen abstraction (type I photooxidation). These could sensitize the formation of hydrogen peroxide, which is known to photodissociate at 254 nm to produce OH radicals.[27] These are quite reactive and could induce further chemistry to occur. Perhaps the greatest complication arises, however, when the acceptor has some absorbance at the wavelength of excitation. Here, the acceptor sensitizes the formation of singlet oxygen, and this leads to undesirable autooxidation of the acceptor.

Another important factor when considering the photooxidation kinetics is the branching ratio of singlet oxygen into reactions (7-8) and (7-9). Whether singlet oxygen will be quenched back to the ground state or react with the acceptor depends on the product $k_9[A]$ vs. k_8, the reciprocal of the singlet oxygen lifetime τ. Conditions can vary from those where reaction (7-8) $\gg$ (7-9) to those where reaction (7-9) $\gg$ (7-8). The ratio can even go from one regime into the other during the residence time in the photochemical reactor if significant autooxidation of the acceptor is occurring. The ratio $k_8/k_9[A]$, when equal to unity, has been defined as β. Thus, β is equal to the concentration of acceptor which will result in a 50% trapping of singlet oxygen; 1/2 the singlet oxygen will be trapped by the acceptor and 1/2 will be quenched back to the ground state. This index of reactivity has been used extensively in past singlet oxygen literature because it used to be quite difficult to sort out the individual values of k_8 and k_9. Techniques for measuring individual rate constants for these reactions are discussed in many reports,[22, 28-30] and it is interesting to see how the advent of better spectroscopic techniques have influenced the ability to gain such information. At this time, it is possible to estimate the branching ratio for a particular system, because there are extensive tabulations of the rate constants k_8 and k_9 in a variety of solvents.

In this section, rate equations will be discussed for several of the possible conditions of photooxidation. These are intended as theoretical guidelines in predicting the behavior of the detection systems that will be discussed. In the following equations we will make the assumption that $O_2(^1\Delta_g)$ is at a steady-state concentration throughout the residence time of a sensitizer in the photochemical reactor. This assumes that the sensitizer concentration is not changing in the reactor (due to photooxidation) and that the lifetime of singlet oxygen is much shorter than the residence time in the reactor. The latter is definitely true, since the lifetime of $O_2(^1\Delta_g)$ is measured in microseconds while the reactor residence time is typically 1 to 3 min. The earlier assumption may not be correct for all analytes. Also, reaction (7-5) is assumed to be slow relative to reaction (7-7) (it is 100 to 1000 times slower for solutions saturated with O_2). Finally, reaction (7-6) is not included, because compounds that sensitize singlet oxygen via their excited singlet state are uncommon.

Given in the following text are several selected conditions and the integrated rate equations which describe them.

Negligible Autooxidation of the Singlet Oxygen Acceptor. The rate equation for acceptor photooxidation under the preceding assumptions is

$$\frac{d[AO_2]}{dt} = -\frac{d[A]}{dt} = k_9[A]\left(\frac{k_{1,S}[S]\phi_{ISC,S}}{k_9[A] + 1/\tau_{^1O_2}}\right) \tag{7-11}$$

where $1/\tau = k_8[O_2]$ and ϕ_{ISC} is the quantum yield for intersystem crossing. Essentially, this describes the rate of photooxidation based on known and calculable parameters.

In the case where the decrease in [A] is small compared to the amount of A available (i.e., pseudo-first-order conditions) and $1/\tau \gg k_9[A]$, Eq. (7-11) can be integrated to obtain

$$\Delta[AO_2] = -\Delta[A] = k_9 k_{1,S}\phi_{ISC,S}\tau_{^1O_2}[A]_0[S]t \tag{7-12}$$

Under these conditions, the signal is expected to increase linearly with photon flux of the lamp, absorption coefficient of the analyte, quantum yield for intersystem crossing, rate constant for reaction with the acceptor, lifetime of singlet oxygen in the solvent, concentration of the acceptor, concentration of the analyte and reaction time.

In the case where $k_9[A] \gg 1/\tau$ the integration simplifies to

$$\Delta[AO_2] = -\Delta[A] = k_{1,S}\phi_{ISC,S}[S]t \tag{7-13}$$

which also predicts a signal linear in analyte concentration.

Autooxidation of the Singlet Oxygen Acceptor Occurs. In this case a second term must be added to Eq. (7-11). This term describes the autooxidation of A and is exactly analogous to the first term describing analyte-induced photooxidation. When the possibility of autooxidation is included, the result is

$$\frac{d[AO_2]}{dt} = -\frac{d[A]}{dt} = k_9[A]\left(\frac{k_{1,S}[S]\phi_{ISC,S}}{k_9[A] + 1/\tau_{^1O_2}} + \frac{k_{1,A}[A]\phi_{ISC,A}}{k_9[A] + 1/\tau_{^1O_2}}\right) \tag{7-14}$$

For the case when $1/\tau \gg k_9[A]$, Eq. (7-14) integrates to give

$$\Delta[AO_2] = -\Delta A = \frac{k_{1,S}\phi_{ISC,S}}{k_{1,A}\phi_{ISC,A}}\ln\left(1 + k_9 k_{1,A}\phi_{ISC,A}\tau_{^1O_2}[A]_0 t\right)[S] \tag{7-15}$$

For several of the systems discussed in this chapter where autooxidation of the acceptor is occurring, the preceding assumption will be approximately correct. For detection using 2,5-diphenylfuran, the acceptor discussed most thoroughly in this chapter, $1/\tau$ will range from 5 to 20 times greater than $k_9[A]$. The opposite case, $k_9[A] \gg 1/\tau$ is only encountered here with respect to systems having negligible autooxidation [Eq. (7-13)]. This is because much higher acceptor concentrations can generally be used when no autooxidation is occurring.

As noted at the beginning of this section, the photochemistry can be quite complex and may not fit well into the given categories. These are merely intended as rough guidelines for interpretation of the observed kinetic behavior.

Detection of Singlet Oxygen Sensitizers In HPLC

As discussed in the theory section, the basis of detection is the analyte-induced photooxidation of a singlet oxygen acceptor which has been spiked into the HPLC mobile phase. A change in concentration of the acceptor is detected by a suitable detector, the choice of which depends on the characteristics of the patent compound or those of the oxidized product. Because each analyte molecule can trigger the oxidation of many acceptor molecules (the degree varies widely from analyte to analyte), the integrated absorbance change of the acceptor may well be much greater than the absorbance of the analyte at its particular concentration. Data presented in the following sections will demonstrate this.

Many considerations are necessary in choosing a workable singlet oxygen acceptor. An ideal acceptor would not absorb light at the wavelength of analyte excitation, would have a fast rate constant for reaction with singlet oxygen and would undergo a dramatic spectral or other change upon oxidation, such that an appropriate detection wavelength or other principle could be chosen to monitor either the acceptor loss or product gain. Because all of these criteria are not easily met simultaneously, compromises must be made. The result is that each acceptor will be useful for a somewhat different excitation wavelength range. Filters are required in some cases, and optima exist in terms of acceptor concentrations, reaction time and lamp intensity. These factors are illustrated by the examples to follow.

Detection by use of 2,5-Diphenylfuran

Detection by Absorption. Shown in Fig. 7-1 are the oxidative changes in the absorption spectrum of a 3×10^{-5} M solution of 2,5-diphenylfuran (DPF) in acetonitrile upon irradiation in a cuvette by a low-pressure mercury lamp.[31] The oxidation reaction produces *cis*-dibenzoylethylene as shown in reaction (7-16)[32]:

DPF $\xrightarrow{^1O_2}$ *cis*-DBE (7-16)

Notice that the spectra in Fig. 7-1 show a decrease in absorbance at the λ_{max} of DPF at 320 nm and an increase in absorbance due to product appearance near 245 nm. There is also an isosbestic point at 270 nm, apparently because the reaction product has the same extinction coefficient as DPF. Theoretically, either the 245 or 320 nm region could be monitored for detection (absorbance gain at 245 nm or absorbance loss at 320 nm). The largest spectral change occurs at 320 nm, so this would seem to be the best choice, although many analytes absorb well at 245 nm, and thus their native absorbance would directly add to the oxidative signal at this wavelength. Detection of absorbance loss at 320 has, however, been determined to be more sensitive by a significant margin than absorbance gain at 245 nm.

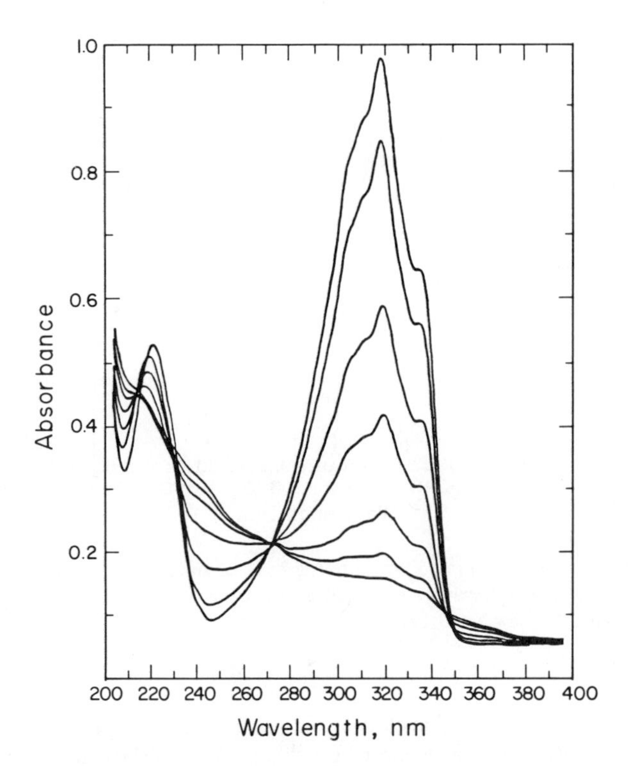

Figure 7-1 ■ Spectrum of a 3×10^{-5} M solution of 2,5-diphenylfuran (DPF) in acetonitrile after bulk irradiation in a cuvette with a low-pressure mercury lamp. Exposure times are 0, 30, 60, 90, 130, 170 and 210 s in order of decreasing absorbance at 320 nm.

Other considerations for detection must also be mentioned. One must choose a wavelength of excitation at which absorbance by the acceptor, and hence autooxidation, is minimized. With DPF this would be either in the 250 nm range or at wavelengths greater than 350 nm (see spectrum in Fig. 7-1). In the work described here, the 254 nm output from low-pressure mercury lamps was used in conjunction with a bandpass solution filter for the 254 nm region (see Fig. 7-2). Although 254 nm is at a minimum in the DPF spectrum, it still has an extinction coefficient of approximately 2000 $1 \text{ mol}^{-1} \text{ cm}^{-1}$. This absorption, although relatively low, results in significant autooxidation. For example, while working with DPF concentrations of 10^{-4} M and using two mercury pen lamps for excitation, the original DPF is approximately halved within 2 min of reaction time. Work with DPF without filtering the mercury lines lying in the 280 to 350 nm region is not feasible due to near complete DPF destruction.

The details of the photochemical reactor, instrumental configurations and solution filter preparation have been detailed elsewhere.[31] Figure 7-3 is a schematic diagram of the instrumental setup. Basically, the HPLC mobile phase is spiked with DPF at a concentration typically in the range of 0.5 to 1.0×10^{-4} M, the effluent passes through the photochemical reactor tubing with the associated mercury lamp and filter in place

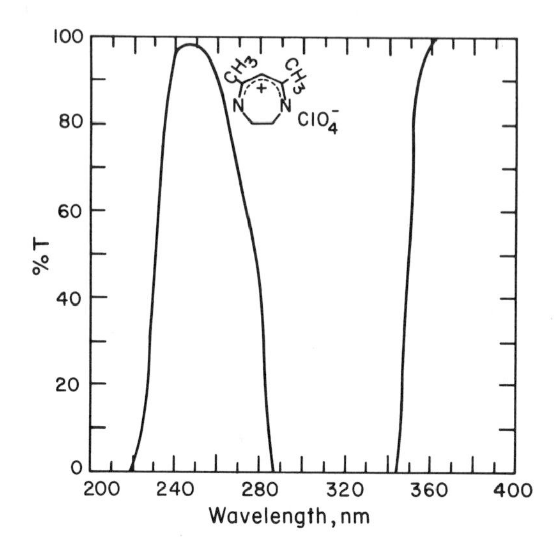

Figure 7-2 ■ Light transmission in a 1 cm path length cell of an aqueous solution of 2,7-dimethyl-3,6-diazacyclohepta-2,6-dieneperchlorate at 0.20 g/l. Adapted from S. L. Murov, *Handbook of Photochemistry*, Dekker, New York, 1973, p. 99.

and then passes into an absorbance detector set for 320 nm. The large background absorbance, typically in the range 0.5 to 1.0 a.u., is electronically subtracted by use of the zero offset feature of the UV detector, and the leads carrying the signal to the integrating recorder are reversed so that the chromatographic peaks are positive.

Many parameters have been studied for this system. Among them is the effect of reaction time on the signal-to-noise ratio. An example of the effect of reaction time is plotted in Fig. 7-4 for the three compounds anthraquinone, 1-nitronaphthalene and 4,4′-dichlorobiphenyl. The chromatography and detection were carried out in a solvent consisting of 95% acetonitrile spiked with 10^{-4} M DPF. Two mercury pen lamps

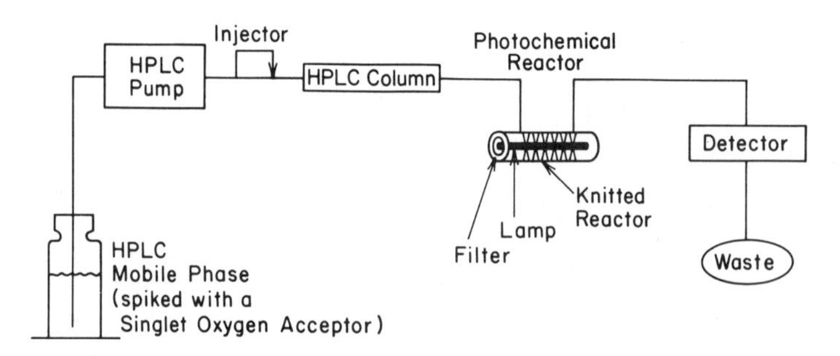

Figure 7-3 ■ Schematic of the general instrumental setup used in singlet oxygen sensitized detection.

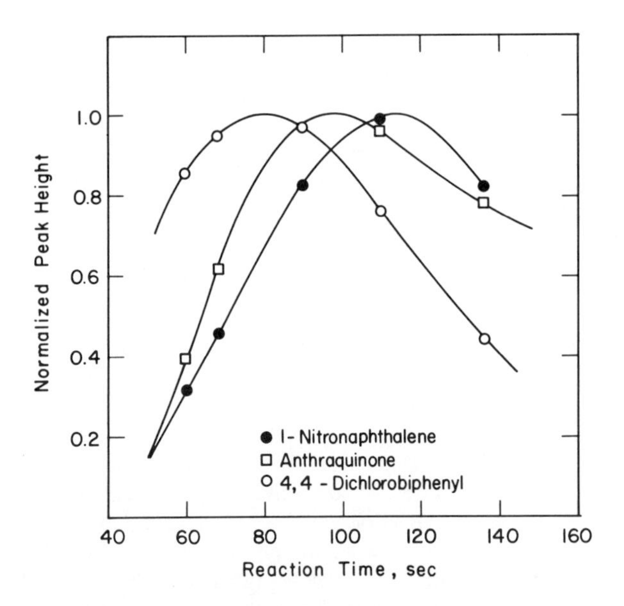

Figure 7-4 ■ Chromatographic peak heights as a function of reactor residence time for detection by DPF photobleaching at 320 nm. Mass injected: 0.25 ng anthraquinone, 1.3 ng 1-nitronaphthalene and 2.8 ng 4,4′-dichlorobiphenyl. Mobile phase is 10^{-4} M DPF in 95% acetonitrile.

served as the excitation source. The optimal residence time within the photochemical reactor was found to fall in the range 1 to 2 min, depending on the analyte and solvent conditions. Longer reaction times resulted in decreases in the signal and can be attributed to the continued depletion of DPF by autooxidation. In fact, as a result of autooxidation, the signal will decrease to zero at very long reaction times. This is an example of a tradeoff encountered in choosing a viable acceptor. A similar tradeoff exists for excitation lamp intensity. Signal increases with intensity (as is theoretically predicted) but eventually declines as an increasing percentage of DPF undergoes autooxidation. Other considerations, such as band broadening and back pressure of the lengthy crocheted reactor tubing, limit reaction time for poor sensitizers.

Detection limits for this DPF system were determined for six model compounds in HPLC with 95% acetronitrile as the solvent.[31] For these measurements, the HPLC flow rate was 0.7 ml/min, the photochemical reaction time was 75 s using two lamps and the DPF concentration was 10^{-4} M. The limits of detection are defined as the mass of analyte that results in a peak having a height equal to three times the peak-to-peak noise of the baseline. The results are summarized in Table 7-5. Detection limits range from 1.5 pg for anthracene to 40 pg for quinoline. Of course, detection limits are dependent on the quality of the UV absorption detector used. For this reason, detection limits were also determined for these compounds by direct UV absorption at their individual λ_{max} using the same detector. Table 7-5 compares the detection limits obtained by the two methods and gives the factors by which the detection limits are

Table 7-5 ▪ Detection Limit Comparisons, Diphenylfuran Absorbance Quenching

Compound	Optimized UV Absorbance	Photobleaching at 320 nm (pg)	Enhancement Factor
Anthracene	45 pg (253 nm)	1.5	30 ×
Anthraquinone	90 pg (250 nm)	8.0	11 ×
1-nitronaphthalene	160 pg (243 nm)	3.6	44 ×
Quinoline	900 pg (234 nm)	40	22 ×
Biphenyl	130 pg (260 nm)	20	6.5 ×
4,4′-dichlorobiphenyl	90 pg (270 nm)	10	9 ×

improved over direct UV absorption by use of the photochemical reaction. Sensitivity enhancement factors range from 6.5 for biphenyl to 44 for 1-nitronaphthalene.

Because this detection system is designed for chromatography where the solvent conditions must be varied to optimize separation, the effect of solvent composition on sensitivity was examined. It is difficult to predict in advance the overall effect of solvent composition, because altering the solvent can at the same time enhance and reduce the rates of some of the individual reactions contributing to DPF photooxidation. For example, increasing the water : acetonitrile ratio would decrease the lifetime of singlet oxygen, thereby slowing the reaction sequence, while increasing the rate of reaction of DPF with singlet oxygen, as it has been found that singlet oxygen reacts faster with substituted furans in more polar solvents.[33, 34] In fact, all of the relevant reactions are expected to be somewhat solvent-dependent.

Table 7-6 compares signals produced by analytes in acetonitrile containing 5, 25 and 50% water by volume using flow injection analysis (no column). The results are presented in Table 7-6. Because the flow dynamics in the knitted reactor were such that the flow injection peaks broadened when going to higher percentages of water, peak areas rather than heights were used for comparison. Although both increases and decreases in signal occurred, the signal did not change by more than a factor of 1.7 when the water content was varied from 5 to 50% by volume. These results suggest that

Table 7-6 ▪ Effect of Solvent Composition on DPF Photobleaching

	Relative Peak Area		
Compound	5% H_2O	25% H_2O	50% H_2O
Anthracene	1.0	0.98	1.6
Anthraquinone	1.0	0.68	0.71
1-nitronaphthalene	1.0	1.1	1.3
Quinoline	1.0	1.1	1.3
Biphenyl	1.0	0.80	0.68
4,4′-dichlorobiphenyl	1.0	0.70	1.0

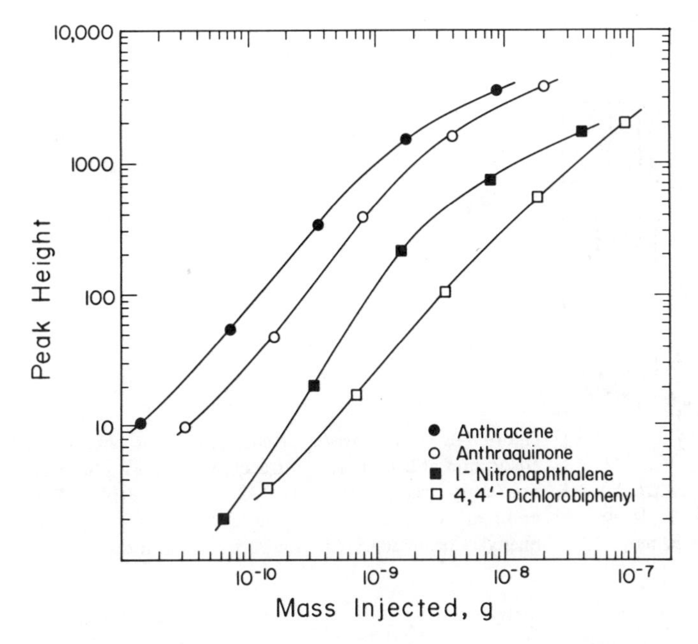

Figure 7-5 ■ Chromatographic peak heights as a function of quantity injected for anthracene, anthraquinone, 1-nitronaphthalene and 4,4'-dichlorobiphenyl. Mobile phase is 10^{-4} M DPF in 95% acetonitrile. Reactor residence time is 75 s.

this reaction system may be used in chromatography under a wide range of solvent conditions.

Working curves for detection of the model compounds in chromatography are shown in Fig. 7-5. Conditions are the same as for the detection limit determinations. On a log-log plot, the working curves are sigmoidal in shape. At lower masses injected, the sensitivity increases with increasing analyte mass. However, at sufficiently high concentrations the sensitivity begins to decrease with mass of analyte injected. The deviations from linearity at higher concentrations can be attributed to depletion of DPF; i.e., at high analyte concentrations 10^{-4} M DPF is not sufficient to maintain the kinetics as pseudo-first order in DPF. However, at analyte levels much above 1 ng there is no advantage of using a photochemical amplifier anyway. The explanation of the supralinearity observed at low analyte concentrations is not known with certainty; however, recall that Eq. (7-15) in the reaction kinetics section predicts a nonlinear system under these conditions.

To better illustrate the advantage of using this photochemical amplification technique in chromatography, chromatograms of a four component mixture were obtained using UV absorption at 254 nm and using DPF photobleaching at 320 nm. The two chromatograms are compared in Fig. 7-6. In both cases the mobile phase was 95% acetonitrile and the flow rate was 0.7 ml/min. The photochemical reaction conditions again were 75 s reaction time using two lamps and a DPF concentration of 10^{-4} M. The masses of the analytes injected were all in the subnanogram range and were chosen

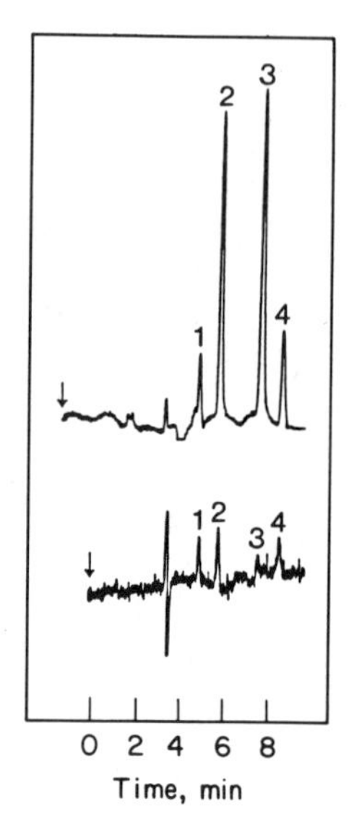

Figure 7-6 ■ Comparison of chromatograms obtained by direct UV absorption at 254 nm (lower trace) and by DPF photobleaching at 320 nm (upper trace). Peaks: (1) 0.32 ng 1-nitronaphthalene; (2) 0.16 ng anthraquinone; (3) 0.072 ng anthracene; (4) 0.70 ng 4,4′-dichlorobiphenyl. Chromatographic conditions given in text.

so that the peaks would be very near the detection limits for direct UV absorption. The advantage of the photochemical amplification, as shown in Fig. 7-6, is quite striking.

Detection Based on Fluorescence. DPF is a fluorescent compound with an emission maximum at 374 nm. The oxidized furan is not fluorescent and thus detection can also be accomplished by monitoring the quenching of a standing fluorescence signal at 374 nm, with excitation at 320 nm. In general, this has not been found to be as sensitive as detection by absorbance loss at 320 nm (discussed in the previous section). The analyte 1-nitronaphthalene, for example, is detected 10 times less sensitively by fluorescence loss than by absorbance loss. Anthracene is detected 13 times less sensitively. The fact that detection by fluorescence is not as sensitive correlates somewhat with data showing that DPF as an analyte is found to be detected a factor of 3 to 4 times more sensitively by absorbance than by fluorescence under optimized conditions. Also, there is more noise associated with fluorometric detection relative to absorbance detection when monitoring a standing signal. This probably accounts for the remainder of the difference between the two methods. Three fluorimeters were used and compared through the course of the work: the Kratos SF 950, Kratos FS 970 and the Shimadzu RF 530. The detection limits obtained using these instruments were all within a factor of 2.

Although the fluorescence quenching method did not result in detection limits that were substantially improved relative to conventional UV absorption, experiments using this system helped elucidate the general features of the photochemical system, and for this reason some results of these experiments will be described.

Figure 7-7 plots the logarithm of the DPF concentration, as measured by fluorescence, vs. residence time in the photochemical reactor for an initial DPF concentration of 10^{-4} M in methanol and in acetonitrile. The decay curve for DPF in methanol is

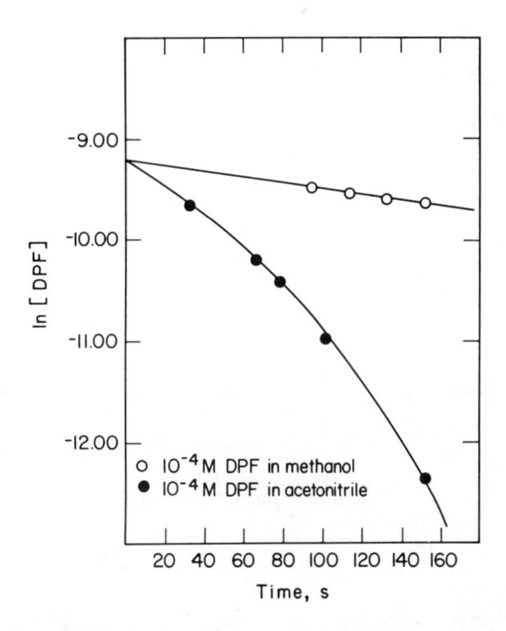

Figure 7-7 ▪ Concentration of DPF as a function of residence time in the photochemical reactor (due to autooxidation). Solvents are methanol and acetonitrile.

seen to be linear, consistent with a first-order photochemical reaction. In acetonitrile the reaction rate (negative slope of the line) increases with increasing time. This could possibly be explained by the appearance of a product that itself absorbs reactor light and adds to the rate of DPF destruction. When methanol is the solvent, however, the extent of reaction, and therefore product accumulation, is too small, even at 160 s reaction time, for an enhancement of the rate of photooxidation to become apparent. The initial rate of self-induced photooxidation in acetonitrile is 4.5 times greater than in methanol. This is consistent with the fact that the lifetime of singlet oxygen in acetonitrile is 4.3 times greater than in methanol (Table 7-3, 30 μs compared to 7 μs). Of course, other factors, such as the extinction coefficient of DPF, its quantum yield for intersystem crossing and its rate of reaction with singlet oxygen in the two solvents, also contribute to differences in the degree of self-photooxidation.

As shown in Fig. 7-8, the signal resulting from quenched DPF fluorescence increases linearly with reaction time up to about 150 s in acetonitrile when anthracene is the analyte and for an initial DPF concentration of 10^{-4} M. Similar behavior is found when using methanol as the solvent. Beyond ~150 s, peak heights level off and additional time spent in the photochemical reactor beyond ~3 min provides little enhancement of the signal, while band broadening continues to increase. Note that at infinite reaction time, all DPF would be destroyed as a result of self-photooxidation whether or not an analyte is present, and there would be no analyte peak. Peak heights will begin to decline after sufficient reaction time, as confirmed by results presented in the previous section where DPF is monitored by use of UV absorbance (Fig. 7-4).

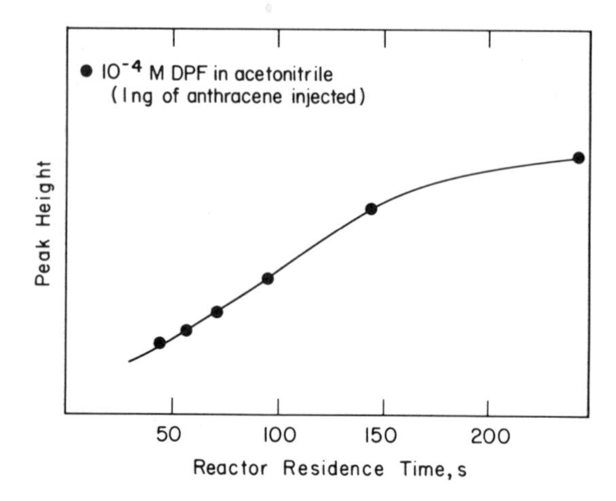

Figure 7-8 ■ Anthracene peak height as a function of residence time in the photochemical reactor. Initial [DPF] is 10^{-4} M in acetonitrile.

As a further test of the understanding of this reaction system, the relative responses were obtained for equal molar solutions of a group of substituted anthracenes. The mobile phase was 10^{-5} M DPF in acetonitrile and the photochemical reactor residence time was 60 s using one lamp. Their relative absorbances of 254 nm (the photoexcitation wavelength) were also measured in a Cary 219 spectrophotometer. Ratioing the peak heights for DPF fluorescence quenching to their relative absorbances allows one to factor out the effects of k_1, the absorption rate constant, on the signals obtained. These ratios were then normalized to the result for anthracene and are given in Table 7-7. Unless these compounds are undergoing a significant amount of photooxidation themselves, the ratios should differ only as a result of differences in the quantum yields for intersystem crossing (refer to discussion in the theory section). In Table 7-7, it is seen that the substituted anthracenes, in general, show the expected trend. Progression down the table is expected to be in the order of larger to smaller ϕ_{ISC}. There are only two flaws in this trend. The order of amino and methylanthracene are reversed from that expected, although misplacement is slight. Their fluorescence quantum yields are

Table 7-7 ■ **Effect of Substituents on Photochemical Amplification**

Compound	Peak Height/A_{254}
9-nitroanthracene	1.5
2-chloroanthracene	1.2
Anthracene	1.0
9-anthracenecarboxylic acid	0.66
2-aminoanthracene	0.59
2-methylanthracene	0.37

very similar. 9-anthracene-carboxylic acid is also out of place. This compound is expected to lie between 9-nitroanthracene and 2-chloroanthracene. The reason for this discrepancy is unknown, but it probably relates to the potential of that compound to undergo photooxidation itself. The altered analyte molecules will most likely have a different extinction coefficient at the excitation wavelength and may either slow down or speed up the rate of singlet oxygen production.

The DPF concentration was varied between 10^{-6} and 10^{-3} M for a reaction time of 80 s to determine the optimal concentration in fluorescence quenching. Anthracene (33 ng) served as the analyte for the optimization. A concentration of 10^{-4} M was found to give the best signal-to-noise ratio for both acetonitrile and methanol solvents. However, this optimum was very broad. For example in acetonitrile the signal-to-noise ratio did not vary by more than 20% in the range 10^{-3} to 10^{-5} M. Furthermore, the detection limit for anthracene (20 pg) was not significantly different in methanol and acetonitrile solvents. Although the lifetime of singlet oxygen is greater in acetonitrile, the noise in this system is greater because of a greater amount of background autooxidation.

For anthracene and several other polycyclic aromatic hydrocarbons, detection limits for DPF fluorescence quenching was a factor of 2 to 6 better than could be achieved by optimized UV absorption. However, PAH are more sensitively detected by fluroescence than by either UV absorption or the photochemical method of DPF fluorescence quenching. Also, fluorescent compounds can give signals from their own native fluorescence exactly opposite in sign to the DPF quenching signal. This would nullify much of the photochemical amplification. The detection limits for several anthraquinone derivatives were found to be a factor of 2 to 3 poorer using DPF fluroescence quenching as compared to UV absorption.

Detection using 2,5-Dimethylfuran

Figure 7-9 shows the photooxidative changes that occur when 2,5-dimethylfuran (DMF) is oxidized by singlet oxygen. The solvent in this experiment is acetonitrile. Note the appearance of a UV-absorbing product in the wavelength region 240 to 320 nm. A UV detector set to monitor a wavelength in the 250 to 280 nm region effectively detects the analyte-induced oxidation. The peaks are positive absorbance changes, as opposed to the negative peaks discussed with respect to DPF absorbance loss or fluorescence loss.

It is interesting to contrast the use of DPF and DMF. At first glance, DMF has several apparent advantages. Because it has little absorbance at the excitation wavelength of 254 nm, or at longer wavelengths for that matter, autooxidation of the acceptor is low. Its only source is the relatively weak mercury lines at wavelengths < 254 nm. Autooxidation from these lines can be inhibited by filtering.[31] These absorption characteristics allow the use of greater acceptor concentrations, more intense light sources and longer reaction times, all of which can improve sensitivity. The net result is a baseline which includes little background absorbance and more potential for improvement relative to DPF. Also, DMF has a rate constant for oxidation which is three to four times that of DPF (see Table 7-4 and references therein). Because higher concentrations can be used and because the singlet trapping rate is inherently higher, the DMF system can trap a significantly greater percentage of the singlet oxygen relative to DPF. Table 7-8 contains sample values for efficiencies of

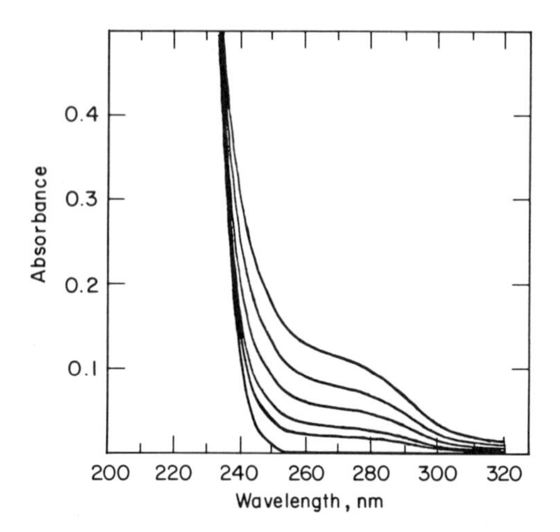

Figure 7-9 ▪ Absorbance of a 10^{-3} M solution of 2,5-dimethylfuran (DMF) in 95% acetonitrile after bulk irradiation in a cuvette with a low-pressure lamp. Exposure times are 0, 30, 60, 120, 180 and 240 s in order of increasing absorbance.

singlet oxygen trapping by DPF and DMF under normal operating conditions (recall the discussion on the index of reactivity in the theory section). Another potential advantage when using DMF is that one is monitoring a positive absorbance change. The analyte's native absorbance will thus have the potential to add directly to the photochemical signal. If monitoring at 254 nm, for example, detection sensitivity will be that available from standard UV detection at that wavelength plus the absorbance contribution from the catalytically oxidized DMF.

For all the seeming advantages of DMF, DPF has been found to allow more sensitive detection, at least for work to date. The reason lies in the much larger spectral change for DPF photooxidation relative to DMF photooxidation. At 320 nm, DPF has an extinction of 30,000 $l\ mol^{-1}\ cm^{-1}$ and the oxidized product has no absorbance. For DMF the extinction of the product is much smaller, estimated to be $< 500\ l\ mol^{-1}\ cm^{-1}$. Thus, the catalytic yield in the DPF system can be much smaller than that for DMF and still result in more sensitive detection.

Table 7-8 ▪ Fraction of Singlet Oxygen Trapped by 2,5-Dimethylfuran (DMF) and 2,5-Diphenylfuran (DPF)

Acceptor	Solvent[a]	Fraction of $^1O_2$ Trapped
DMF (10^{-3} M)	Acetonitrile	0.91
DMF (10^{-3} M)	Methanol	0.72
DPF (10^{-4} M)	Acetonitrile	0.33
DPF (10^{-4} M)	Methanol	0.022

[a]Singlet oxygen lifetimes in various solvents are listed in Table 7-3.

Table 7-9 ■ Detection Limit Comparisons, Dimethylfuran Product Absorbance

Compound	Optimized UV Absorbance	Acetonitrile: Absorbance at 280 nm (pg)	Acetonitrile: Enhancement Factor	Methanol: Absorbance at 280 nm (pg)	Methanol: Enhancement Factor
Anthracene	45 pg (253 nm)	10	4.5 ×	20	2.3 ×
Anthraquinone	90 pg (250 nm)	1.5	60 ×	5.0	18. ×
1-nitronaphthalene	160 pg (243 nm)	10	16 ×	50	3.2 ×
Quinoline	900 pg (234 nm)	45	20 ×	300	3.0 ×
Biphenyl	130 pg (260 nm)	40	3.3 ×	50	2.6 ×
4,4′-dichlorobiphenyl	90 pg (270 nm)	20	4.5 ×	60	1.5 ×

The detection limits for the same six model compounds examined by DPF were obtained by monitoring product absorbance at 280 nm. Table 7-9 summarizes the detection limits obtained in both 95% acetonitrile and 95% methanol using a 10^{-3} M concentration of DMF and a photochemical reaction time of 120 s with two mercury pen lamps. Again, the detection limits have been compared with those for optimized UV absorption and the sensitivity enhancement factors have been calculated. These enhancement factors are best in acetonitrile and range from 3.3 for biphenyl to 60 for anthraquinone. Except for anthraquinone, which is detected better in the DMF system, comparison of Tables 7-5 and 7-9 indicates better sensitivity when using DPF as the singlet oxygen acceptor.

The DMF system was characterized in a manner similar to that for DPF including the effects of reaction time, lamp intensity and solvent composition. Table 7-10 lists the relative peak areas for four model compounds detected in acetonitrile containing 5, 25 and 50% water by volume. Although the signal changes due to increasing water content are somewhat different from those found for DPF (Table 7-6), the variation is much less than that expected based only on consideration of the singlet oxygen lifetime. Again the data support previous work showing faster furan oxidation rates in solvents of greater polarity.[33,34] As in the DPF system, DMF can theoretically be used in chromatography over a wide range of solvent composition.

As noted in the theory section and elsewhere in this chapter, the signal is expected to increase with increasing reaction time, increasing lamp intensity and increasing

Table 7-10 ■ Effect of Solvent Composition on DMF Product Formation

Compound	Relative Peak Area: 5% H_2O	Relative Peak Area: 25% H_2O	Relative Peak Area: 50% H_2O
Anthraquinone	1.0	1.7	1.3
1-nitronaphthalene	1.0	0.83	0.59
Quinoline	1.0	0.82	1.1
Biphenyl	1.0	1.7	1.9

Table 7-11 ▪ Effect of Lamp Intensity on Signal for DMF Photobleaching

Compound	Mass Injected (ng)	Relative Signal[a]		
		1 Lamp	2 Lamps	3 Lamps
Anthraquinone	0.04	1.0	2.0	—
Anthraquinone	0.22	1.0	1.9	—
Anthraquinone	1.3	1.0	1.9	—
Chlorobenzene	50	1.0	1.7	3.7
Benzene	150	1.0	1.9	4.0

[a]Identical low-pressure mercury pencil lamps were used.

Table 7-12 ▪ Signal as a Function of DMF Concentration

DMF Concentration	Relative Signal	
	95% Methanol	95% Acetonitrile
10^{-5} M	1.0	1.0
10^{-4} M	1.9	3.3
10^{-3} M	6.8	15

acceptor concentration for acceptors that do not undergo appreciable autooxidation. All these were found to be true with DMF. Table 7-11 illustrates the effect of increasing excitation intensity for some example analytes. The data suggest the future use of more intense photochemical lamps for the DMF system. Work involving laser excitation is currently underway. As for increasing reaction time, after ~3 min both unacceptable band broadening and high back pressure on the PTFE reactor tubing prevent longer reaction times without slowing the flow rate to an impractical level. The signal as a function of DMF concentration is illustrated by data in Table 7-12 for the concentration range of 10^{-5} to 10^{-3} M in methanol and acetonitrile. The signal increases with increasing DMF concentration because a greater and greater percentage of the singlet oxygen can be trapped by the acceptor rather than undergo solvent-induced decay. The use of DMF concentrations higher than 10^{-3} M begins to become impractical because of the appearance of a significant autooxidative background absorbance.

Detection using 2-Methylfuran

Although 2-methylfuran (MF) has not been found to be a good acceptor for use in detection, it is discussed here for comparison with its relative, 2,5-dimethylfuran (DMF). The photooxidation of MF is shown in Fig. 7-10 for a concentration of 2×10^{-3} M in acetonitrile. It has an extinction coefficient at its λ_{max} of 211 nm of 740

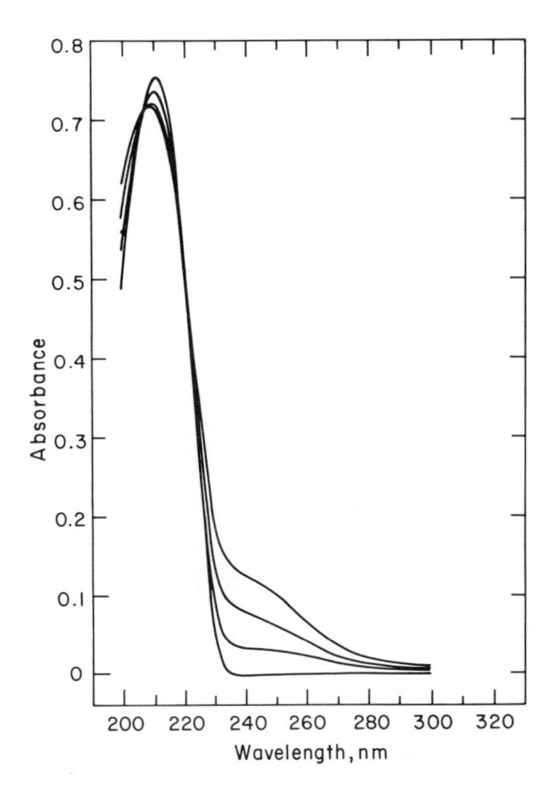

Figure 7-10 ■ Absorbance of a 2×10^{-3} M solution of 2-methylfuran (MF) in 100% acetonitrile after bulk irradiation in a cuvette with a low-pressure mercury lamp. Exposure times are 0, 90, 180 and 270 s in order of increasing absorbance.

$1\ mol^{-1}\ cm^{-1}$ compared to DMF with an extinction of ~ $7{,}000\ l\ M^{-1}\ cm^{-1}$ at its λ_{max} of 240 nm. A similar oxidation product appears to be produced, and indeed the products reported are analogous.[35,36] The rate constant for the reaction of MF with singlet oxygen is slower than that for DMF,[33] and the extinction coefficient of the oxidation product is likely to be smaller than the DMF oxidation product.

Detection using MF was investigated by monitoring absorbance at 245 nm and using a concentration of 10^{-2} M MF in 95% acetonitrile. The reaction time was 110 s using two mercury pen lamps. Sensitivity using MF was found to be slightly improved over that for normal UV detection (1.2 times better for anthraquinone and 3.4 times better for anthracene, for example), but significantly less sensitive than that with DMF.

Detection with 1,3-Diphenylisobenzofuran

Figure 7-11 illustrates the spectral change upon photooxidation of 1,3-diphenylisobenzofuran (DPBF) at a concentration of 5×10^{-5} in methanol. The reaction of DPBF

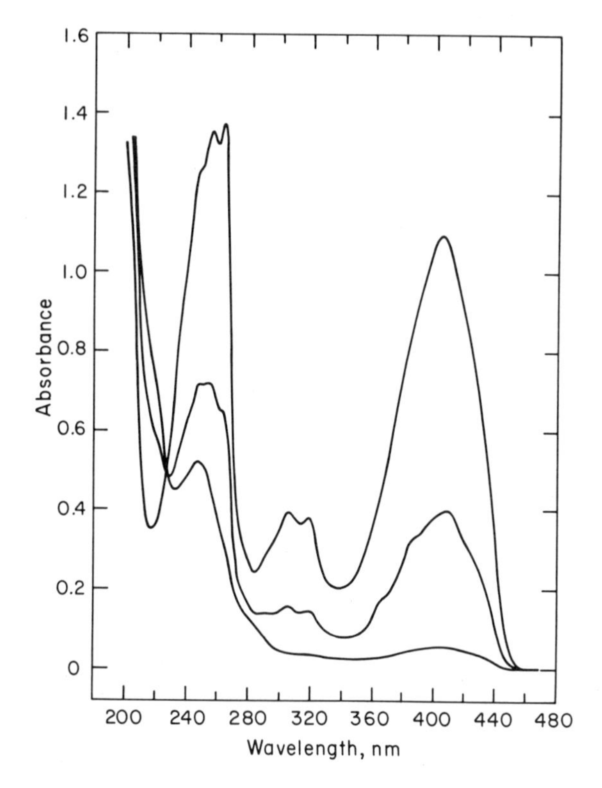

Figure 7-11 ■ Spectrum of a 5×10^{-5} M solution of DPBF in methanol after bulk irradiation in a cuvette with a low-pressure lamp. Exposure times are 0, 20 and 40 s in order of decreasing absorbance at 406 nm.

and singlet oxygen to form *o*-dibenzoylbenzene (DBB) is given by the reaction[36]

$$\text{DPBF} \xrightarrow{^1O_2} o\text{-DBB} \qquad (7.17)$$

DPBF is the most rapidly oxidized heterocycle reported in the literature. As can be seen in Table 7-4, it reacts with a rate constant of 9×10^8 mol l^{-1} s^{-1}, nearly an order of magnitude faster than DPF and several times faster than DMF. DPBF probably has been the most frequently employed acceptor in the multitude of singlet oxygen studies reported in the literature. Both fluorescence loss at 458 nm and absorbance loss at 406 nm have been used to monitor DPBF photooxidation. Because it has strong absorbance throughout the UV and into the visible it is very difficult to use in conjunction with photochemical lamps having outputs at wavelengths less than approximately 460 nm due to extremely fast autooxidation. At excitation wavelengths greater than this, DPBF would be the acceptor of choice. One would use a visible excitation source and filter out wavelengths shorter than 460 nm. This is discussed briefly in the following text.

Despite the problem of autooxidation, data has been obtained for UV absorbing compounds for comparison with the previously discussed systems. A mercury source was used and the detector was set to monitor absorbance loss at 406 nm. To prevent total autooxidation of the DPBF, very short reaction times were required (< 30 s). The result is sensitivity that is better than detection by UV absorption but not as good as that using DPF. Anthracene, for example, is detected with eightfold greater sensitivity than direct UV absorption, but DPF enables a 30-fold enhancement over direct UV detection (Table 7-5). Another difficulty with DPBF is that the baseline is very difficult to control. Monitoring a standing absorbance signal is not very feasible for a highly reactive acceptor such as DPBF under conditions where autooxidation is occurring. Any changes in lamp flux translate to enormous baseline drifts.

In the visible region, detection has been done using DPBF at concentrations ranging from 10^{-6} to 10^{-4} M. A white fluorescence lamp or a 100 W tungsten lamp was used for analyte excitation and a solution cutoff filter containing potassium chromate (Fig. 7-12) was used to protect the DPBF from autooxidation. Fluorescence loss at 458 nm was used for detection. Methylene blue was used to optimize and study the system. At a reaction time of 90 s in 10^{-5} M DPBF, it is detected at a level of 5×10^{-8} M, or 200 pg. All other visible absorbing compounds injected showed response in this system including a number of commercial dyes, food colorings and pigments extracted from spinach. In the latter experiment, spinach was extracted as described elsewhere[37] and the diluted extract was chromatographed on a C18 column in a mobile phase of 10^{-5}

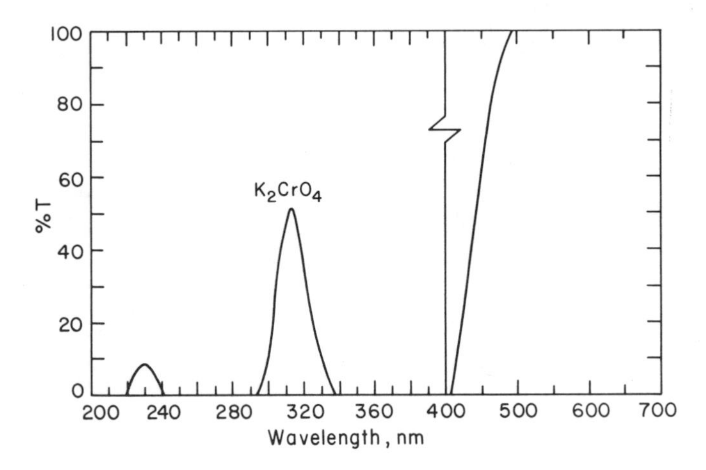

Figure 7-12 ■ Light transmission in a 1 cm path length cell of an aqueous solution of K_2CrO_4 (0.27 g/l) + Na_2CO_3 (1 g/l). Adapted from S. L. Murov, *Handbook of Photochemistry*, Dekker, New York, 1973, p. 99.

M DPBF in 100% methanol with a 40 s reactor residence time. Although the individual pigments were not identified, the 13 to 15 chromatographic peaks correlated with the number obtained by UV detection at 486 nm. The excitation source was modulated on and off to show that the components were due to the photochemical reaction rather than a mechanism of fluorescence quenching.

Detection using 2,4,5-triphenylimidazole

A final example of heterocyclic singlet oxygen acceptors is that of 2,4,5-triphenylimidazole (TPI). The bulk photooxidation of 1×10^{-5} M TPI in 95% methanol can be seen in Fig. 7-13. There are three potential modes of spectroscopic detection: absorbance gain in the 245 nm region, absorbance loss at 297 nm and fluorescence loss at 385 nm. Absorbance gain at 245 nm was found to provide the most sensitive detection.

Detection was generally much poorer than that with either DPF or DMF. For example, 1-nitronaphthalene has a detection limit of 100 pg, just slightly better than that by UV absorption. Anthraquinone has a detection limit of 60 pg, also only slightly better than UV absorption. Because the extinction coefficients for DPF and TPI are within a factor of 2, the probable reason for greatly differing sensitivities is simply a difference in reactivity in the acetonitrile/water mobile phase; i.e., DPF traps a greater percentage of singlet oxygen.

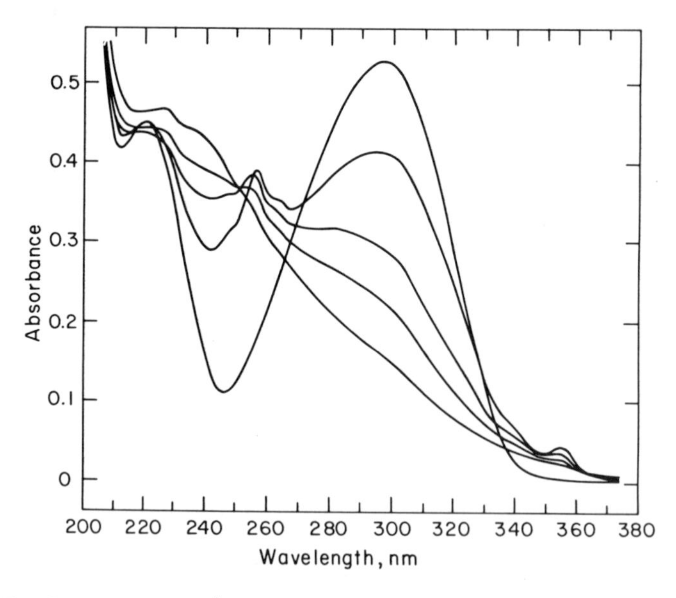

Figure 7-13 ■ Spectrum of a 10^{-5} M solution of 2,4,5-triphenylimidazole (TPI) in 95% methanol after bulk irradiation in a cuvette with a low-pressure lamp. Exposure times are 0, 30, 60, 90 and 150 s in order of decreasing absorbance at 297 nm.

Detection of Singlet Oxygen Quenchers in HPLC

There are several classes of compounds that quench singlet oxygen. One type is the carotenes and other compounds with extensive conjugation.[27,38-40] Other classes include amines,[41,42] phenols,[43,44] metal complexes,[45] sulfides,[46,47] nitroso compounds,[48] bilirubin and biliverdin[49] and some miscellaneous inorganic species including N_3^-, O_2^- and I^-.[50-52] Several reviews have dealt with singlet oxygen quenching.[28,53,54] Table 7-13 lists examples from these various classes and their rates of quenching and/or reaction with singlet oxygen.

Detection of singlet oxygen quenchers can be accomplished with little or no modification of the previously discussed methods. Some modification is necessary, however, in order to reduce or eliminate interference from singlet oxygen sensitizers.

Theory

The quenching of singlet oxygen by nonsolvent molecules is analogous to reaction (7-8) in the theory section and can be written

$$^1O_2 + Q \rightarrow {}^3O_2 + Q \qquad (7\text{-}18)$$

Because many quenchers also react with singlet oxygen (notably, amines with abstractable hydrogen atoms, alpha to the amine group, phenols, sulfides, and bilirubin),[28] a second reaction must be written

$$^1O_2 + Q \rightarrow QO_2 \qquad (7\text{-}19)$$

When quenchers undergo both reaction and quenching (pseudo quenching rate equal to $[Q](k_r + k_q)$, the relative contributions can usually be sorted out by careful kinetic

Table 7-13 ▪ Total of Rate Constants for Quenching ($k_Q + k_r$) of $O_2(^1\Delta_g)$

Compound	Solvent	$k_Q + k_r$ ($M^{-1}\ s^{-1}$)	k_Q/k_r	Ref.
β-carotene	C_6H_6	1.2×10^{10}	—	29
Diethylamine	MeOH	1.9×10^6	—	55
Triethylamine	MeOH	1.0×10^7	9.4	55, 56
Nicotine	AcN/D_2O (4:1)	5.9×10^7	5.4	57, 58
N,N,N′,N′-tetramethyl-p-phenylenediamine	MeOH	1.0×10^9	—	55
α-tocopherol	MeOH	5.3×10^8	14	43
2,4,6-triphenylphenol	MeOH	2.4×10^8	Very large	59
Ni(II) complexes	—	5×10^7–5×10^9	No reaction	45
2-nitroso-2-methylpropane	$CH_2Cl_2/MeOH$ (15:1)	9.3×10^9	—	48
NaN_3	MeOH	2.8×10^8	—	50
Bilirubin	CCl_4	2.3×10^9	10	49
Biliverdin	$CHCl_3$	3.3×10^9	Very large	49

analysis as discussed elsewhere.[28] As can be seen in Table 7-13, the rate constant for quenching is usually larger than that for reaction and many quenchers do not react at all.

The mechanisms of quenching fall into two major categories, depending on the identity of the quencher. These are charge transfer and energy transfer. The energy transfer mechanism is the most efficient and can have rate constants up to 2×10^{10} $M^{-1}\ s^{-1}$.[28] This is equal to or slightly less than that for diffusion control. Rate constants measured for charge transfer quenching are slower than those for energy transfer and generally fall into the wide range of approximately 10^{-4} to 10^{-9} $M^{-1}\ s^{-1}$.[24]

Energy Transfer Quenching. Energy transfer quenching was first linked to carotene[60] and has since become the accepted mechanism of quenching by many dyes with extensive conjugation. It is just the opposite of energy transfer from sensitizer to singlet oxygen. Just as sensitizers transfer energy to oxygen from excited triplet states of ≥ 22.5 kcal mol^{-1}, singlet oxygen transfers energy to compounds such as carotene with triplet energies very near or below 22.5 kcal mol^{-1}.

Another class of quencher thought to fall into this mechanistic class are nitroso compounds. Their rate of quenching is near diffusion controlled, and they have "effective" triplet energies of < 23 kcal mol^{-1}. This exceptionally low effective triplet energy is postulated to result from the coupling of the energy transfer to a relaxation of the normal excited state configuration of the nitroso group. The alternate configuration is of lower energy and is isoelectronic with ground-state oxygen.[48]

Quenching by solvent molecules, discussed previously, roughly falls into the category of energy transfer[23] but occurs by the relatively slow route of electronic to vibrational energy transfer, as opposed to electronic-to-electronic energy transfer.

Charge Transfer Quenching. This mechanism involves the formation of a charge transfer complex between the electron-poor (high electron affinity) singlet oxygen molecule and electron-rich (low ionization potential) donor molecule. Intersystem crossing restrictions are relaxed in the complex and it dissociates to ground-state oxygen and donor as[61-63]

$$D + {}^1O_2 \rightarrow [D^+ \cdots O_2^-]^1 \rightarrow [D^+ \cdots O_2^-]^3 \rightarrow D + {}^3O_2 \qquad (7\text{-}20)$$

The types of compounds postulated to quench by this mechanism include amines, phenols, some metal complexes, sulfides, iodide, azide, superoxide ion and various other electron-rich compounds. As already noted, some of these species can also react with singlet oxygen. The quenching : reaction ratios vary substantially and are somewhat solvent dependent.[22, 28]

Kinetics. One of the methods used to study singlet oxygen quenchers is to make use of acceptors such as 1,3-diphenylisobenzofuran (DPBF). The quencher competes with the acceptor for the same singlet oxygen molecules: Photooxidation conditions are set up and initial and final DPBF levels are quantified by absorbance or fluorescence. Discussion of the preceding method and various other methodologies can be found in numerous reports.[28, 38, 53]

Table 7-14 ▪ Rate Constants for the Quenching of Singlet Oxygen and the Triplet States of Rose Bengal and Methylene Blue by Anilines, X–ϕ–$N(CH_3)_2$[a]

		Triplet States	
X	Singlet Oxygen $k_Q \times 10^{-8}$ (s^{-1} M^{-1})	Rose Bengal $k_Q \times 10^{-8}$ (s^{-1} M^{-1})	Methylene Blue $k_Q \times 10^{-8}$ (s^{-1} M^{-1})
p-$N(CH_3)_2$	10 ± 2	26	37
p-OCH_3	1.8 ± 0.4	28	110
p-CH_3	1.2 ± 0.4	23	88
H	0.73 ± 0.08	7.5	86
m-OCH_3	0.48 ± 0.04	3.9	82
p-Br	0.17 ± 0.04	2.8	32
m-Cl	0.11 ± 0.01	1.2	4.2
p-CN	0.0057 ± 0.001	0.22	0.13
p-CHO	0.012 ± 0.006	—	—

[a] From R. H. Young, R. Kayser, R. Martin, D. Reriozi, and R. A. Keller, *Can. J. Chem.* **52**, 2892 (1974).

A fact that complicates the study of singlet oxygen quenchers is that many quenchers can also quench the triplet state of the sensitizer. Quenching the triplet prevents the formation of singlet oxygen, resulting in another mechanism of photooxidation inhibition. Essentially, the quencher and ground-state oxygen compete for the sensitizer triplet state. Because the rate constant for sensitizer-to-oxygen energy transfer is approximately 2×10^9 M^{-1} s^{-1} and because oxygen is at a concentration of 10^{-3} to 10^{-2} in solution, the quencher concentration must be very high to quench a significant portion of triplet sensitizers, even though the rate constant for triplet quenching is an order of magnitude faster. Table 7-14 contains quenching rate constants for both singlet oxygen and sensitizer triplet states. From values such as these and from the value of oxygen solubility and singlet oxygen lifetime in the particular solvent, one can calculate the concentration of quencher that will result in significant sensitizer quenching.

Many of the quenchers can also quench the excited singlet states of the sensitizer. Because of the very short singlet lifetime, however, the quencher concentration would have to be extremely high for this to be significant.

The kinetics of the photooxidation sequence in the presence of a quencher are thus quite complex at first glance. A quencher has the potential to quench the excited singlet or triplet states of the singlet oxygen sensitizer as well as the ability to quench singlet oxygen. Once singlet oxygen is formed, it may react with or be quenched by the quencher, or it may react with the singlet oxygen acceptor. It may also be quenched by the solvent, as discussed earlier. The degree of acceptor photooxidation thus depends on the relative concentrations and rate constants for all the species and reactions involved. Derivation and discussion of various scenarios and resulting rate equations have been reported elsewhere.[28]

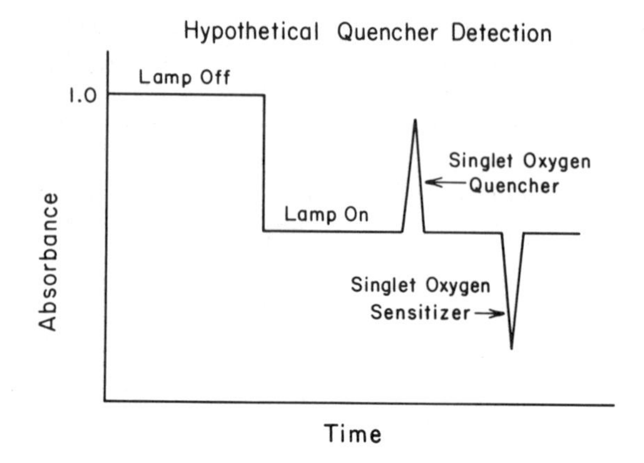

Figure 7-14 ■ Model for the detection of singlet oxygen quenchers using DPF or other suitable singlet oxygen acceptor. Discussion given in text.

Detection Methods for Quenchers

Detection of singlet oxygen quenchers can be accomplished with little or no modification of the previously discussed methods. The detection of quenchers in the simplest but most nonselective manner is modeled in Fig. 7-14. The detector is set to monitor the acceptor at its absorption maximum and conditions are chosen such that some autooxidation of the acceptor can occur in the photochemical reactor. This is shown in the diagram by the drop in absorbance when the excitation lamp is turned on. When a quencher flows through, quenching singlet oxygen as it goes, it protects the acceptor and a positive chromatographic peak results. In other words, the standing level of autooxidation is decreased. As indicated in the model, if absorbing singlet oxygen sensitizers are present in the sample, they will give signals as well as quenchers, and these peaks will be opposite in direction.

Some modification is thus necessary in order to make the detection more selective for quenchers, and there are several possible means presently under study. One possibility is to adjust the acceptor concentration to one which discriminates against sensitizers. This has not been very successful. Examination of Fig. 7-15 shows optimal DPF concentrations to be between 10^{-5} and 3×10^{-5} M for the quenchers nicotine and *N*-nitrosodimethylamine. Although the optimal DPF concentration for the detection of sensitizers is higher, the preceding concentration will still result in efficient sensitizer response. Working at a DPF concentration that is a factor of 10 lower than the optimum and sacrificing some quencher sensitivity will, however, result in some discrimination against sensitizers.

Another method of achieving a degree of selectivity over singlet oxygen sensitizers is to shift to longer excitation wavelengths. As the excitation wavelength moves out of the UV toward the visible wavelengths, the number of sensitizers that can interfere will drop dramatically. Sensitizers must absorb light in order to interfere, while quenchers need not absorb light in order to give a signal. They need only quench singlet oxygen. In fact, this illustrates the true potential of this mode of detection; no chromophore in

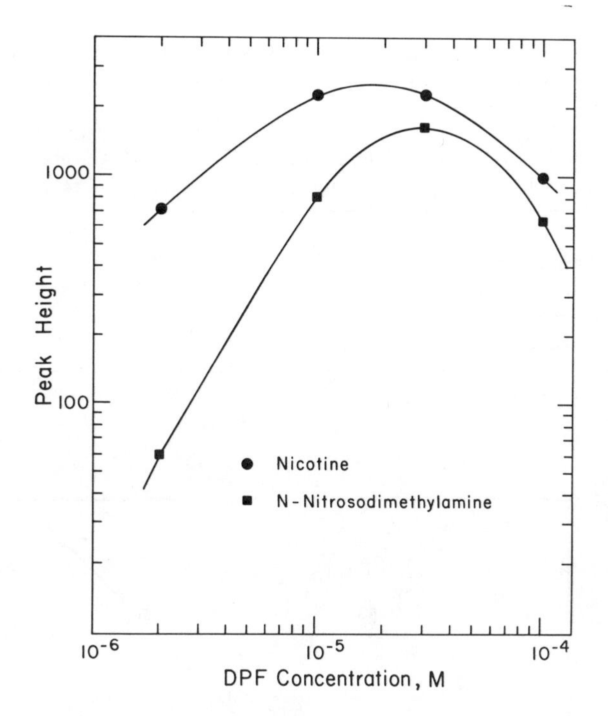

Figure 7-15 ■ Signal as a function of DPF concentration for the singlet oxygen quenchers nicotine and *N*-nitrosodimethylamine. Conditions given in text.

the UV or visible spectral region is necessary for the detection of quenchers. One can shift to the visible wavelength region by using an acceptor that absorbs in the visible (and thus undergoes autooxidation under visible light) or by spiking the mobile phase with a visible absorbing singlet oxygen sensitizer. Either of these methods can cause a standing level of acceptor oxidation necessary for detection while not exciting UV absorbing sensitizers. The farther one shifts into the visible, the more selective the detection system will be for quenchers.

To date, the acceptors utilized have been 2,5-diphenylfuran (DPF) and 1,3-diphenylisobenzofuran (DPBF). As can be theoretically predicted, the use of DPBF results in poorer detection limits. The reason is DPBF's faster rate constant for reaction with singlet oxygen (Table 7-4). DPBF more effectively competes with the quencher for singlet oxygen, thus reducing the sensitivity. For the test analyte nicotine, detection was an order of magnitude less sensitive than that for DPF under identical conditions. Acceptors with very slow rate constants are not expected to work well either. More research is necessary in order to define the ideal choice of acceptor for detection of singlet oxygen quenchers.

Table 7-15 contains detection limits for four quenchers detected by both photochemistry and UV absorption. The chromatography made use of an amino column with DPF as the singlet oxygen acceptor at a concentration of 2.5×10^{-5} M in a

Table 7-15 ■ Detection Limits for Singlet Oxygen Quenchers[a]

Compound	Detection Limit (ng)	Photochemical Sensitivity Relative to UV Absorbance ($\lambda = 254$ nm)
Nicotine	1.6	1.3 ×
N-nitrosodimethylamine	0.30	3.7 ×
N,*N*,*N*′,*N*′-tetramethyl-*p*-phenylenediamine	0.33	2.1 ×
1-nitroso-2-naphthol	0.46	1.3 ×

[a] Conditions given in text.

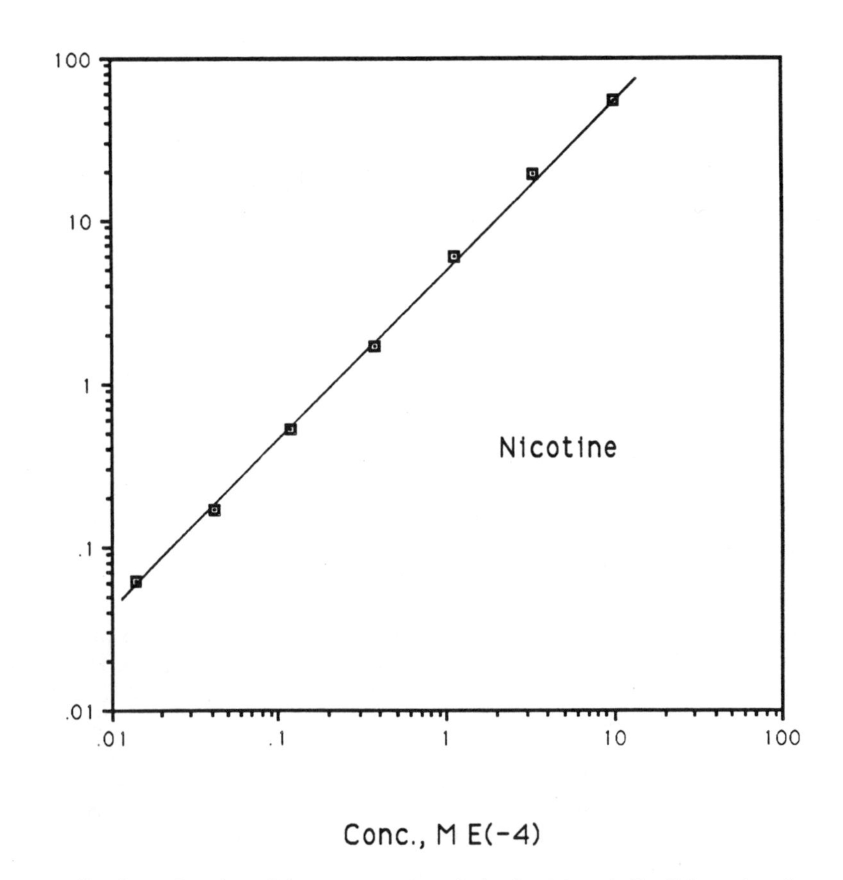

Figure 7-16 ■ Signal as a function of the concentration of nicotine injected. Conditions given in text.

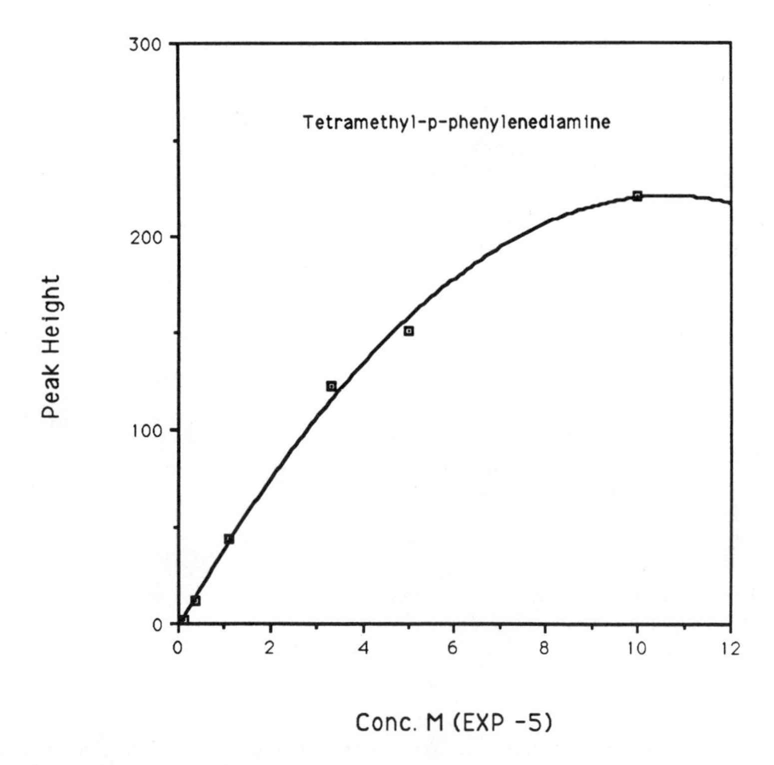

Figure 7-17 ▪ Signal as a function of the concentration of tetramethyl-*p*-phenylenediamine injected. Conditions given in text.

mobile phase of 85 : 15 acetonitrile : water buffered at a pH of 6. A mercury excitation source was used with a reactor residence time of 40 s and the absorbance of DPF was monitored at 320 nm. These data show that sensitivity by photochemical detection is somewhat better in each case than detection by absorbance at 254 nm, although many more example compounds need to be studied in the future.

Figures 7-16 and 7-17 illustrate the signal as a function of quencher concentration for the quenchers nicotine and tetramethyl-*p*-phenylenediamine, respectively. A mobile-phase DPF concentration of 10^{-5} M was used in both cases and the reaction times for the two studies were 50 and 80 s, respectively. Other factors were the same as those used for the detection limit experiment in Table 7-15. For nicotine, the response was linear over the entire concentration range tested, 1.4×10^{-6} to 10^{-3} M. For tetramethyl-*p*-phenylenediamine, the response is linear for the lower four concentrations, 1.2×10^{-6} to 3.3×10^{-5} M, and begins to level off by 5.0×10^{-5} M.

The nonlinearity at higher concentrations of tetramethyl-*p*-phenylenediamine may well have to do with significant absorption of the excitation radiation by the analyte. It may also be due to quenching of the sensitizer triplet state. As the quencher concentration is increased, it will eventually reach a level where it can effectively compete with molecular oxygen for the excited triplet states.

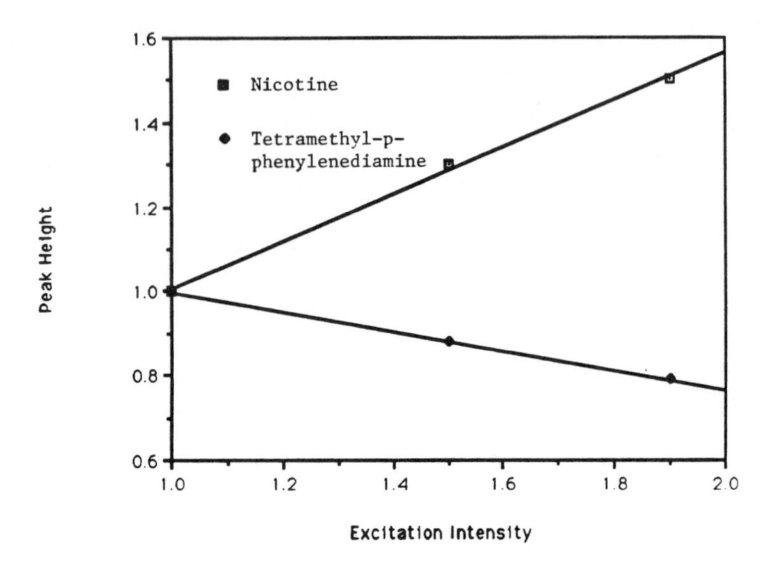

Figure 7-18 ■ Signal as a function of excitation intensity for nicotine and tetramethyl-*p*-phenylenediamine. Conditions given in text.

Figure 7-18 is a plot of signal as a function of light intensity for nicotine and tetramethyl-*p*-phenylenediamine. The conditions are the same as those used in Figs. 7-16 and 7-17. Note that the signal increases with intensity for nicotine and decreases for tetramethyl-*p*-phenylenediamine. This could be due to a greater tendency toward photooxidation for tetramethyl-*p*-phenylenediamine than nicotine. Reported reaction rates are available for nicotine but not tetramethyl-*p*-phenylenediamine.

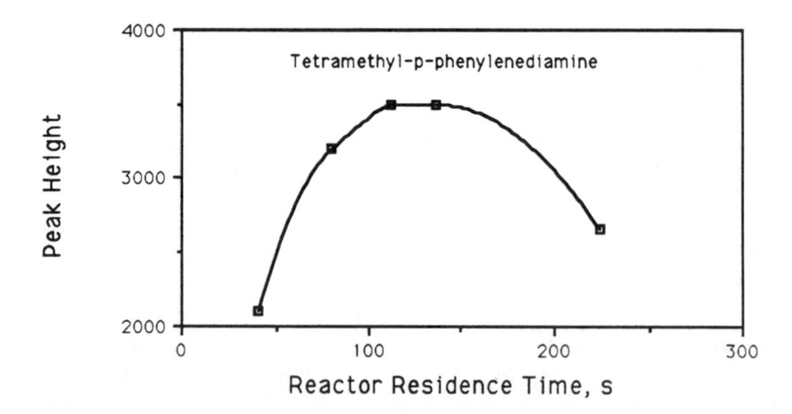

Figure 7-19 ■ Signal as a function of reactor residence time for tetramethyl-*p*-phenylenediamine. Conditions given in text.

Figure 7-19 shows the effect of reactor residence time (reaction time) on the signal from tetramethyl-*p*-phenylenediamine. The conditions are the same as those used before. Other quenchers were also studied in the same manner. As with the singlet oxygen sensitizers, there is always an optimum reaction time, and this optimum is very analyte- and condition-dependent. The optimal reaction time can be varied by changing excitation intensity, acceptor concentration, solvent and other parameters.

Summary. Although quenchers can be detected with sensitivity slightly better than absorbance detection, more work needs to be done to ascertain whether this is a viable technique. To make the technique useful would require either significantly greater sensitivity or good selectivity while maintaining adequate sensitivity. The detection process needs to be applied to real samples where the question of selectivity can be directly evaluated. Quenchers without strong UV chromophores also need to be evaluated.

Future Trends

The success already achieved using singlet oxygen sensitization in liquid chromatography (sensitivity enhancements of 1 to 2 orders of magnitude) suggests that further exploration of the post-column chemistry of singlet oxygen could be fruitful. Future studies should include investigations coupling singlet oxygen sensitization with chemiluminescence and electrochemical detection in HPLC, as well as the development of GC detectors where mass spectrometric and photoionization techniques are applicable. Here we review some techniques that have not yet been evaluated or have been only partially evaluated, and show how these detection principles could possibly be applied to HPLC and GC.

Detection Based on Chemiluminescence

Singlet oxygen is involved in a variety of chemiluminescent reactions. The mechanisms fall roughly into three categories, which are outlined in reactions (7-21) through (7-23), where S is the designation for a singlet oxygen sensitizer, A is a singlet oxygen acceptor, and Flr is an aromatic fluorescer:

$$S + h\nu \rightarrow {}^1S^* \rightarrow {}^3S^* \tag{7-21}$$

$${}^3S^* + {}^3O_2 \rightarrow S + {}^1O_2({}^1\Delta_g) \tag{7-22}$$

$$2{}^1O_2({}^1\Delta_g) + A \rightarrow 2{}^3O_2 + {}^1A* \rightarrow A + h\nu \tag{7-23a}$$

$${}^1O_2({}^1\Delta_g) + A \rightarrow [AO_2] \rightarrow \text{product}^* \rightarrow \text{product} + h\nu \tag{7-23b}$$

$${}^1O_2({}^1\Delta_g) + A \rightarrow AO_2 \xrightarrow{\text{Flr}} [\text{complex}] \rightarrow {}^1\text{Flr}^* + \text{products}$$

$${}^1\text{Flr}^* \rightarrow \text{Flr} + h\nu \tag{7-23c}$$

Reaction (7-23a) is the direct energy transfer from singlet oxygen dimers (${}^1\Delta_g, {}^1\Delta_g$) to a dye capable of fluorescence. The formation of the dimers is well studied, and numerous investigators have reported chemiluminescence by this mechanism. Acceptors have included violanthrene,[64] methylene blue[65] and chlorophyll.[66, 67] The issue has been clouded by the fact that many of the dyes also undergo reaction with singlet oxygen to chemiluminesce as shown in (7-23b). Long-lived luminescent intermediates are formed.[66] The use of time and wavelength resolution is often needed to distinguish between these emission mechanisms. Direct energy transfer luminescence can only occur when the energy of the dye's singlet state is below that of the singlet oxygen dimer, 45 kcal mol^{-1}.

A detection scheme has already been developed for HPLC to detect singlet oxygen acceptor dyes.[68] The system (Fig. 7-20) involves the construction of a chemilumines-

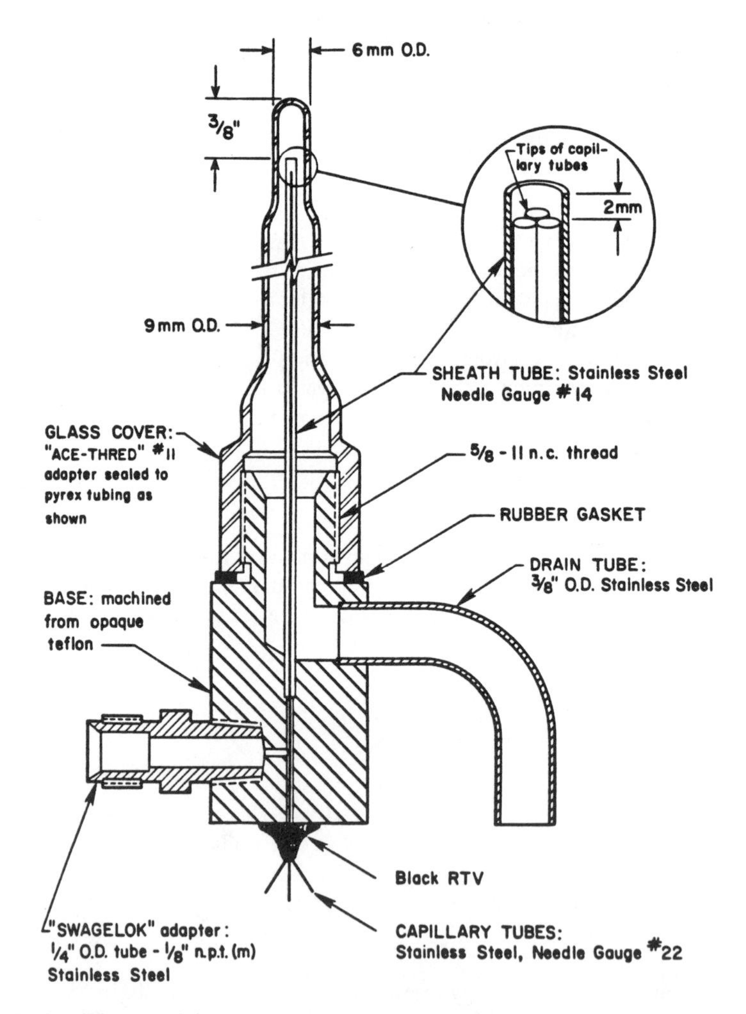

Figure 7-20 ■ Diagram of the chemiluminescence spray cell. From B. Shoemaker and J. W. Birks, *J. Chromatogr.* **209**, 253 (1981).

cence spray cell, where the column effluent and chemiluminescence reagents are mixed at the exit of a pneumatic nebulizer and the emission is monitored by a PMT. Filters are used to discriminate against oxygen dimer emissions at 634 and 706 nm and also to distinguish energy transfer emission [reaction (7-23a)] from the emission by oxidized intermediates [reaction (7-23b)]. In this system singlet oxygen is produced by the reaction between sodium hypochlorite and hydrogen peroxide[69] rather than by sensitizers:

$$OCl^- + HOOH \rightarrow \rightarrow H_2O + ClOO^- \quad (7\text{-}24)$$

$$ClOO^- \rightarrow Cl^- + {}^1O_2({}^1\Delta_g) \quad (7\text{-}25)$$

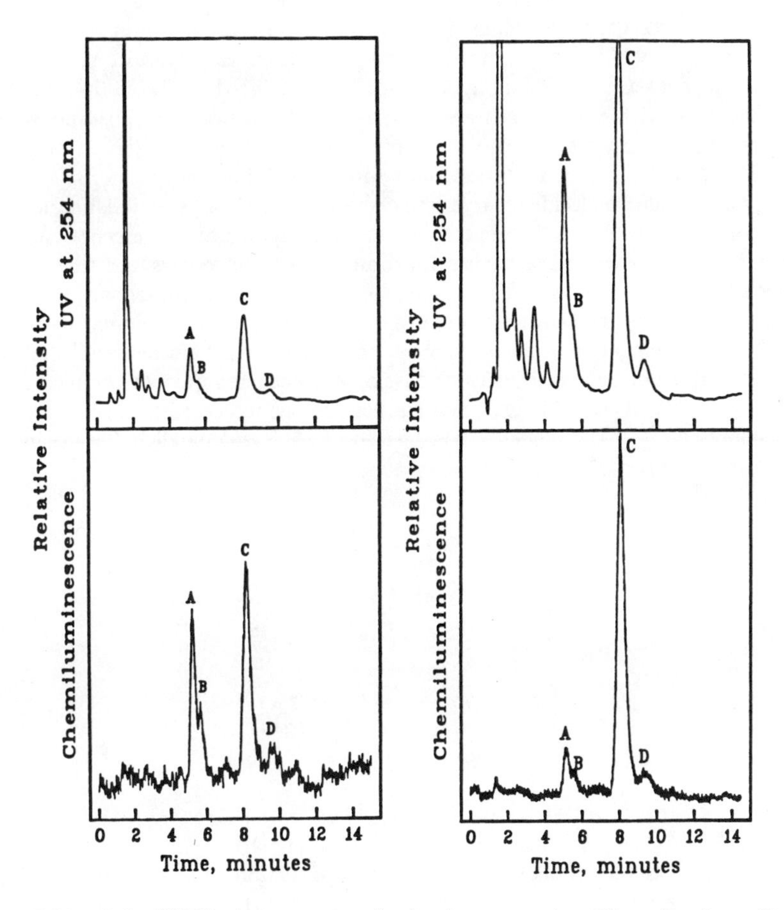

Figure 7-21 ▪ Left: UV-CL chromatogram of spinach extract using 671 nm bandpass filter for wavelength selection. Peaks: A = chlorophyll b; B = chlorophyll a′; C = chlorophyll a; D = pheophytin a. Right: Spinach extract chromatogram using 600 nm long wavelength cutoff filter. Peak identification as noted above. From B. Shoemaker and J. W. Birks, *J. Chromatogr.* **209**, 260 (1981).

The singlet oxygen acceptor is the analyte. The detector was applied to selectively detect chlorophyll in an extract of spinach and also to distinguish chlorophyll a from b. The resulting chromatograms can be seen in Fig. 7-21. The UV traces at 254 nm are included for comparison.

Although this system is set up to detect chemiluminescent acceptors, it may well be possible to detect sensitizers as well. In this case, the acceptor would be a chemiluminescent reagent and the sensitizer would be the source of singlet oxygen rather than hypochlorite and hydrogen peroxide.

In reaction (7-23b) the acceptors form dioxetane intermediates that can decompose to yield chemiluminescence. Although most dioxetanes decompose in solution to yield predominantly triplet excited states affording little chemiluminescence, there are acceptors that form dioxetanes having good excited singlet state yields. These include lophine (2,4,5,-triphenylimidazole) and some of its derivates,[70, 71] various 1,4-dioxenes,[72, 73] various vinyl pyrenes[74] and many dyes, as previously noted.[66] The reaction rate constant for 2,3-diphenyl-1,4-dioxene is listed in Table 7-4. In these classes the most viable candidates are those with easily oxidizable substituents. The preceding cited references indicate a chemiluminescent decomposition involving an intramolecular redox reaction. A transient is formed with positive and negative charge residing on the same molecule (charge transfer complex). Charge annihilation leaves a fraction of the molecules in the excited singlet state, capable of fluorescence emission.

The chemiluminescence mechanism illustrated in (7-23c) occurs with the greatest efficiency for species that form dioxetanones upon oxidation. Dioxetanones have been identified as the key intermediate in many biogenic emissions, including that of the firefly.[75] Their chemiluminescent reactions are among the most efficient ever studied.

The oxidation of ketenes by singlet oxygen has been reported to produce dioxetanones, and efficient chemiluminescence has been shown to occur in the presence of an added fluorescer.[76, 77] The sequence is shown in (7-26)[76] and parallels the mechanism outlined in (7-23c):

$$\begin{aligned}
&R_2C{=}C{=}O + {}^1O_2 \longrightarrow \begin{matrix} O-O \\ | \quad\; | \\ R_2C-C{=}O \end{matrix} \\
&\begin{matrix} O-O \\ | \quad\; | \\ R_2C-C{=}O \end{matrix} + Flr \longrightarrow [Complex] \\
&[Complex] \longrightarrow {}^1Flr^* + R_2C{=}O + CO_2 \\
&{}^1Flr^* \longrightarrow h\nu + Flr
\end{aligned} \tag{7-26}$$

This mechanism is very similar to that for reaction (7-23b), but it is intermolecular rather than intramolecular. The mechanism has been termed *chemically initiated electron-exchange luminescence* (CIEEL).[78] It involves a transient electron transfer from the fluorescer to the high-energy dioxetanone (for R = H the estimated decomposition energy of the dioxetanone is -197 kcal mol^{-1}, for example)[76] and subsequent charge annihilation, which leaves the fluorescer in the excited singlet state. The aromatic fluorescer actually acts as a catalyst for the decomposition reaction which can proceed orders of magnitude faster than the unimolecular uncatalyzed decomposition.[79] As with the mechanism in (7-23b), the efficiency of this process is directly correlated with the oxidation potential of the aromatic fluorescer.[79]

Detection utilizing singlet oxygen acceptors that produce dioxetanes and dioxetanones [(7-23b) and (7-23c)] has not been experimentally demonstrated as yet. It may be possible, but there are many uncertainties. Besides the requirement for efficient chemiluminescence, there are several requirements that must be met in order to use these compounds in the post-column photochemical system. One of the most critical factors is the stability of the dioxetane or dioxetanone intermediate. In detection, after the sensitizer induces singlet oxygen formation and singlet oxygen reacts to form the high-energy compound, it must be stable enough to make it to the entrance of the light detection cell where thermal-, solvent- or fluorescer-catalyzed decomposition will be triggered. Although there are dioxetanes and dioxetanones that can be formed and maintained at ambient temperature and in polar solvents, many of the reactions cited in the literature involve synthesis at low temperatures and in nonhydroxylic solvents. Ideally, the oxidized compounds would have stability under the conditions of formation, and decomposition could be induced at the correct point in time upstream from the detector.

Another consideration in choosing a viable acceptor is the absorption characteristics of the compound. Compounds that undergo autooxidation by acting as sensitizers will cause a background of standing chemiluminescence upon which the analyte signal would be superimposed. The background noise would definitely reduce sensitivity. One of the big advantages of chemiluminescence detection lies in the fact that there is no light source in the detector and ideally no light arriving at the PMT in the absence of an analyte. Were this advantage lost, it is questionable whether chemiluminescence would be any more sensitive than some of the detection methods already discussed.

Detection Based on Electrochemistry

The two major requirements for detection in the photooxidation system are an efficient singlet oxygen acceptor and a means to measure the change in the acceptor concentration. Only spectroscopic changes have been discussed to this point, but because electrochemical detectors possess equal or greater sensitivities, they should be considered as well.

The major question is whether there exist good acceptors that are redox active or have products that are redox active. An added bonus would be realized if the acceptor did not absorb at the wavelength of excitation and undergo autooxidation. There are some potential candidates, and work is underway to screen them for their electrochemical characteristics.

As with the spectroscopic detection systems, one can either monitor the decrease in the concentration of the acceptor or the appearance of the photooxidized product. Detecting product appearance would at first seem to be the most straightforward. Certainly, many of the oxidized products (organic peroxides, carbonyls, dioxetanes, etc.) would be reducible at an electrode surface. Unfortunately, the reduction of oxygen would often be an interference. Oxygen begins to reduce at -0.1 V vs. SCE. Hydroperoxides (a common oxidation product), for example, would be impossible to detect, as

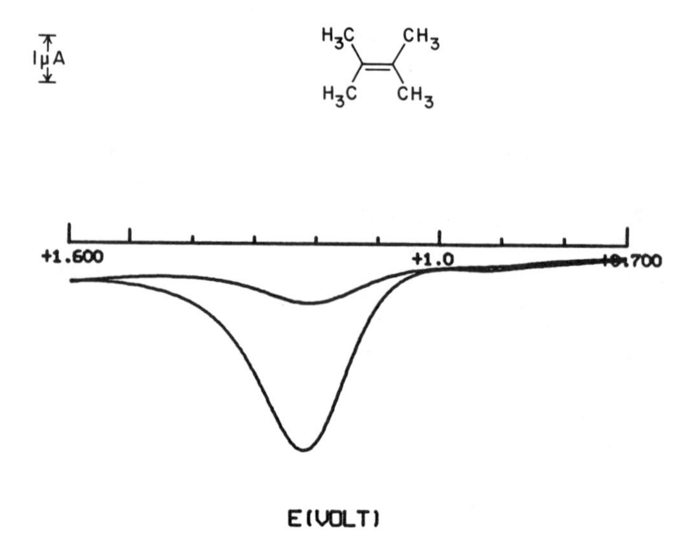

Figure 7-22 ■ Square wave voltammetry of a 10^{-3} M solution of tetramethylethylene before (lower trace) and after (upper trace) 60 s irradiation in a cuvette with a low-pressure mercury lamp. The solvent is acetonitrile containing 0.1 M tetrabutylammonium tetrafluoroborate as an electrolyte and 10^{-5} M anthracene as a singlet oxygen sensitizer.

they have reduction potentials in the range of -0.6 to -0.9 V vs. SCE. Some dioxetanes, on the other hand, may reduce at potentials more positive than oxygen, and if the dioxetane is stable enough to reach the electrochemical detector, acceptors of this nature would have potential.

If the singlet oxygen acceptor is readily oxidizable, its disappearance can be monitored. To date, experiments have been done with two different acceptors fitting into this category. The first of these is tetramethylethylene (TME). The singlet oxygen mediated reaction for substituted ethylene is

$$R_1R_2C{=}CR_3R_4 \xrightarrow[O_2]{\text{Sens},\ h\nu} R_1C(=O)R_2 + R_3C(=O)R_4 \qquad (7\text{-}27)$$

where Sens is the organic singlet oxygen sensitizer.[80] For TME, R_1–R_4 are methyl groups and the photooxidation product is acetone.

Figure 7-22 illustrates this photooxidation. The lower trace is a square wave voltametric sweep of 10^{-3} M TME in acetonitrile/0.1 M tetrabutylammonium tetrafluoroborate spiked with 10^{-5} M anthracene serving as a singlet oxygen sensitizer. The upper trace is the same solution after 60 s irradiation by a low-pressure mercury lamp. One can see that 90% of the TME is photooxidized in this time frame. The oxidation potential is 1.22 V vs. Ag/Ag^{+}. For detection using TME in the mobile phase, current would be continuously sampled at this potential. An analyte would produce a negative chromatographic peak corresponding to oxidation of the acceptor and consequent drop in current.

A second singlet oxygen acceptor which has been tested is *N*-allyldimethylamine. This compound is an example of an enamine (an alkene with an amine group), and the general structure and photooxidation reaction is

$$R_1R_2C{=}CH{-}NR_2 \xrightarrow[O_2]{\text{Sens}, h\nu} R_1R_2C{=}O + O{=}CH{-}NR_2 \qquad (7\text{-}28)$$

For *N*-allyldimethylamine, R_1 and R_2 are hydrogens and the R groups on the nitrogen are methyls. The reaction with singlet oxygen has no solvent dependence.[81] The enamine double bond is electron rich and highly reactive, and these compounds possess some of the most rapid rates of oxidation by singlet oxygen. From Table 7-4, for example, it is seen that *N,N*-dimethylisobutyleneamine (R, R_1 and R_2, are methyl groups) has a rate constant of 5×10^9 M^{-1} s^{-1}, only slightly slower than that of 2,3-diphenylisobenzofuran.

Figure 7-23 illustrates the photooxidation of the *N*-allyldimethylamine under solvent conditions identical to those for the TME experiment (Fig. 7-22). The bottom

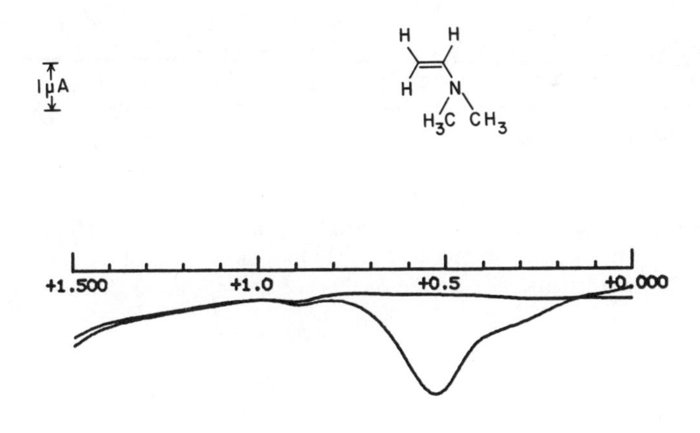

Figure 7-23 ■ Square wave voltametry of a 10^{-3} M solution of *N*-allyldimethylamine before (lower trace) and after (upper trace) 20 s irradiation in a cuvette with a low-pressure mercury lamp. The solvent is acetonitrile containing 0.1 M tetrabutylammonium tetrafluoroborate as an electrolyte and 10^{-5} M anthracene as a singlet oxygen sensitizer.

trace is before irradiation and the top trace is after 20 s irradiation. Under these conditions the acceptor is completely oxidized and comparison to TME shows a significantly faster reaction. The oxidation potential is 0.540 V vs. Ag/Ag^+. Detection using this acceptor would be exactly analogous to detection using TME, only current would be sampled at a lower potential.

This mode of detection (monitoring acceptor loss) would be quite free from oxygen interference, but one would have to contend with the strong possibility of electrode poisoning. The electrode would be set at a potential to monitor a constant standing oxidation of the organic acceptor, potentially leading to adsorption and accumulation on the electrode surface. The use of pulsed amperometric detection,[82] however, may very well prevent this problem. This involves a triple step potential waveform. First there is a positive pulse to the potential appropriate to sample oxidative current from the species of interest. This is followed by a more positive pulse which oxidizes the electrode surface and effectively desorbs adsorbed organics and inorganics. The final pulse is large and negative and serves to reduce the oxidized surface and clean the metal in preparation for the next pulse train.

If the potential problem of electrode surface contamination can be avoided, anodic electrochemical detection using acceptors of this nature has definite potential. Besides favorable electrochemical characteristics, TME and *N*-allyldimethylamine possess low UV cutoffs (< 250 nm). This would minimize or eliminate the possibility of autooxidation. The question of sensitivity and overall practicality cannot be answered, however, until an actual chromatographic system is utilized.

Detection in Gas Chromatography

Although work on detection in gas chromatography is only beginning, it is worthy of discussion here. The potential for ultrasensitive detection is at least equal to and probably greater than that in solution. Type II photooxidation is equally applicable to the gas phase. The overall quantum yield for production of singlet oxygen from an absorption event is very favorable and the efficiency of singlet oxygen trapping can theoretically reach 100%.

The considerations for a detection system will be somewhat different in the gas phase. Because the rate of diffusion is so much faster and the carriers (He, H_2, N_2, etc.) are much different than the solvents used in solution, the kinetics of the reaction will be quite different. Particulars, such as the method of adding the singlet acceptor to the

Table 7-16 ▪ Singlet Oxygen Lifetimes in Quartz Capillaries[a]

Capillary i.d. (mm)	τ (s)
0.1	0.04
0.25	0.1
0.50	0.2

[a] Based on $k_{wall} = \gamma\bar{u}/2r$ where k_{wall} is the first-order rate constant for wall loss, $\bar{u}$ is the mean molecular velocity ($\simeq 4.3 \times 10^4$ cm s^{-1}) and γ is the deactivation coefficient ($\simeq 1.3 \times 10^{-5}$).

mobile phase, the method of adding oxygen, the choice of a detector and so on, will also be different.

The deactivation of singlet oxygen is quite slow in the gas phase leading to an inherently long singlet oxygen lifetime. By utilizing quenching rate constants for various GC carrier gases and comparing their quenching rates to the wall collision rates in the photochemical reactor tubing, we expect that wall loss would be the dominant deactivating factor. Table 7-16 is a compilation of calculated results for different sizes of capillary reactor tubing. Note that the singlet oxygen lifetime is up to 10^5 times longer than that encountered in solution (see Table 7-3). This will dramatically increase the fraction of singlet oxygen that can be trapped. As will be shown, it is theoretically possible to trap all of the singlet oxygen produced.

Once the singlet oxygen is produced by analyte sensitization, it must be detected. Two methods of detection will be discussed. The first is the trapping of singlet oxygen with an acceptor, exactly analogous to that discussed in LC detection. The second is the direct detection of singlet oxygen by a photoionization detector. The merits and potential drawbacks of each will be discussed.

Detection By Use of a Singlet Oxygen Acceptor. Many investigators have demonstrated the feasibility of trapping singlet oxygen in the gas phase. Some of the same acceptors that are used in solution can be used in the gas phase, although the relative reaction rates may be different. Two acceptors with rapid rate constants in both the liquid and gas phase are 2,5-dimethylfuran (DMF) and 2,3-dimethyl-2-butene (tetramethylethylene, TME). Both have been used for gas-phase singlet oxygen studies[83] and would appear to be ideal for detection of singlet oxygen sensitizers. They do not absorb beyond 250 nm, and thus, any excitation source with output greater than this can be used without the occurrence of autooxidation. Both DMF and TME have been shown to react in a simple pressure-independent bimolecular reaction and each yields one major product.[83] The second-order rate constant for the reaction with DMF is 2.4×10^{-14} cm^3 mol^{-1} s^{-1} and that for TME is 1.4×10^{-15} cm^3 mol^{-1} s^{-1}. The rate for DMF is 7 times faster than its reaction rate in solution and that for TME is equal to its solution phase value.[83] Given the lifetime of approximately 0.1 s for singlet oxygen and the rate constant previously listed, a DMF concentration could easily be met such that essentially all singlet oxygen produced could be trapped. This would be true for TME as well. For DMF, the trapping of greater than 98% of the singlet oxygen would require doping the carrier gas with about 1 torr of DMF. For TME, approximately 20 torr would be required. These are both easily obtainable concentrations for work in GC, given boiling points of 93 and 73°C for DMF and TME, respectively.

To utilize these acceptors for detection would be quite simple. The appropriate amount of O_2 and DMF would be added to the effluent of a GC (perhaps from a permeation tube) and the spiked effluent mixture would then pass through a section of transparent tubing irradiated by an intense source of UV radiation. The reaction tubing would either be a short section of fused silica, having no internal or external coatings, or a coil of polytetrafluoroethylene. The latter is permeable to oxygen, which could have the advantage that it may not be necessary to add oxygen from a tank; however, its temperature range is limited and would not be applicable to all analyses. The oxidized DMF or TME resulting from reaction with singlet oxygen could be detected by a bench-top mass spectrometer. The masses of both DMF + O_2 $(m/e = 128)$ and

TME + O_2 ($m/e = 116$) have been shown to correlate in a 1:1 manner with the amount of acceptor consumed and the amount of oxygen consumed.[83]

It is expected that this method could be a very sensitive means of detecting UV absorbing organics in gas chromatography. When using DMF in HPLC, amplifications over absorption detection have been obtained which range from 3.3 × to 60 × (see Table 7-9). This is despite the fact that the DMF oxidation product, as detected by UV absorption, has a very small extinction coefficient (< 500 l mol^{-1} cm^{-1}). The relative sensitivity per molecule of oxidized DMF would be much greater by mass spectrometric detection in the gas phase than UV absorption in solution. Also, as previously noted, DMF in the gas phase would potentially trap every singlet oxygen produced. At 10^{-3} M in methanol, DMF traps approximately 70% (Table 7-8). Because of these factors, there is every reason to believe that excellent sensitivity could result.

Because one would have a choice of photoexcitation sources, another possible strength of this system would be a certain degree of selectivity. Especially desirable for many analyses would be the ability to discriminate against hydrocarbons and many other common matrices. Essentially, this detection method would provide a selectivity similar to that of UV absorption.

Detection by the Use of Photoionization. The use of a photoionization detector to detect the singlet oxygen is also under examination in our laboratory. A photoionization detector (PID) has been developed and marketed by HNU Systems, Inc. for gas chromatography. Applications of this detector have been made to aromatic hydrocarbons in water (EPA methods 503, 602, 8020), chlorobenzenes and other halogenated hydrocarbons, pharmaceuticals, fatty acid methyl esters, flavor and odor analysis, inorganics such as H_2S and NH_3, nitrosamines, pesticides, polyaromatic hydrocarbons, priority pollutants and various sulfur compounds. The detector has a lower limit of detection for benzene of 2 pg and a dynamic range of 10^7.

The principle of detection is that of photoionization. Absorption of a vacuum UV photon by the analyte leads to ion formation as

$$A + h\nu \rightarrow A^+ + e^- \tag{7-29}$$

where A is the analyte and $h\nu$ is a photon with an energy greater than or equal to the ionization potential of the molecule. The ions are collected at electrodes, as in other ionization detectors, and the current is measured as the signal. Sealed vacuum UV lamps are available with photon energies of 11.7, 10.9, 10.2, 9.5 and 8.3 eV. The hydrogen Lyman α (10.2 eV) lamp is standard. Lamps emitting lower energy lines are more selective in that fewer compounds can absorb light to be ionized.

For detection of singlet oxygen, the PID would be placed immediately downstream from the photochemical reactor. The length of reactor tubing would be chosen so as to deliver the singlet oxygen to the detector before it decayed to ground state. As already noted, the lifetime of singlet oxygen in the post-column reactor is expected to fall within the limits of 0.04 and 0.2 s depending on the diameter of the tubing employed in the post-column reactor. This calculation is based on quartz but could increase or decrease for different materials or for different coatings applied to the reaction tubing.

The singlet oxygen exiting the photochemical reactor would enter the PID and be detected using an argon lamp. The argon emission lines lie at 104.8 nm (11.82 eV) and 106.7 nm (11.62 eV), i.e., the 11.7 eV lamp of HNU. This energy is high enough to photoionize $O_2(^1\Delta_g)$, which has an ionization potential of 11.08 eV, but would not photoionize ground-state O_2, whose ionization potential is 12.06 eV. If organic analytes are ionized at this energy, their signal would simply add to the singlet oxygen signal. Photoionization has often been used as a sensitive means of direct detection of singlet oxygen in gas-phase studies of singlet oxygen reactions.[84, 85]

Preliminary work has confirmed the potential use of the PID for singlet oxygen detection.[86] Singlet oxygen produced from solid-phase rose bengal excitation was selectively detected with the argon lamp over a large excess of oxygen. The use of rose bengal-based singlet oxygen generators has been previously described.[87,88] The sensitivity of singlet detection has not yet been evaluated.

Of the two modes of gas-phase singlet oxygen detection described here, photoionization would probably be the simplest to implement. A PID detector would be simpler than a mass spectrometer and a singlet oxygen acceptor, such as DMF, would not have to be doped into the effluent. However, the trapping of singlet oxygen is potentially more sensitive. The residence time in the photochemical reactor would be limited to less than 0.2 s when using photoionization (the singlet oxygen lifetime in the reaction tubing), whereas longer reaction times could be used if singlet oxygen were being trapped as it was produced. Additional research will be necessary to judge the efficacy of both of these systems.

References

1. Kautsky, H., and deBruijn, H. *Naturwiss.* **19**, 1043 (1973).
2. Kasha, M., and Khan, A. U. *Ann. N.Y. Acad. Sci.* **171**, 5 (1970).
3. Khan, A., and Kasha, M. *J. Am. Chem. Soc.* **92**, 3293 (1970).
4. Kearns, D. R. *Chem. Rev.* **71**, 395 (1971).
5. Schaap, A. P. *Singlet Molecular Oxygen*, Dowden, Hutchinson, and Ross: Stroudsberg, PA, 1976.
6. Wasserman, H. H., and Murray, R. W., eds., *Singlet Oxygen*, Academic, New York, 1979.
7. Frimer, A. A., ed., *Singlet O_2*, CRC Press, Boca Raton, FL, 1985.
8. Turro, N. J. *Modern Molecular Photochemistry*, Benjamin/Cummings, Menlo Park, CA, 1978, p. 183.
9. Turro, N. J., *Modern Molecular Photochemistry*, Benjamin/Cummings, Menlo Park, CA, 1978, p. 186.
10. Rosenthal, I. In A. A. Frimer, ed., *Singlet O_2*, CRC Press, Boca Raton, FL, 1985, p. 15.
11. Wu, K. U., and Trozzolo, A. M. *J. Phys. Chem.* **83**, 3180 (1979).
12. Shold, D. M. *J. Photochem.* **8**, 39 (1978).
13. Kawaoka, K., Khan, A. U., and Kearns, D. R. *J. Chem. Phys.* **46**, 1842 (1967).
14. Gijzeman, O. L. J., Kaufman, F., and Porter, G. *J. Chem. Soc. Faraday Trans. II* **19**, 708 (1973).
15. Gollnick, K., Franken, T., Schade, G., and Dörhöher, G. *Ann. N.Y. Acad. Sci.* **171**, 89 (1970).
16. Kawaoka, K., Khan, A. U., and Kearns D. R. *J. Chem. Phys.* **47**, 1883 (1967).
17. Duncan, C. K., and Kearns, D. R. *J. Chem. Phys.* **55**, 5822 (1971).
18. Kearns, D. R., Khan, A. U., Duncan, C. K., and Maki, A. H. *J. Amer. Chem. Soc.* **91**, 1039 (1969).
19. Arnold, S. J., Kubo, M., and Ogryzlo, E. A. *Adv. Chem. Ser.* **77**, 133 (1968).
20. Rosenthal, I. In A. A. Frimer, ed., *Singlet O_2*, CRC Press, Boca Raton, FL, 1985, p. 17.

21. Kearns, D. R., and Kahn, A. U. *Photochem. Photobiol.* **10**, 193 (1969).
22. Monroe, B. M. In A. A. Frimer, ed., *Singlet O_2*, CRC Press, Boca Raton, FL, 1985, p. 177.
23. Kearns, D. R. In H. H. Wasserman and R. W. Murray, eds., *Singlet Oxygen*, Academic, New York, 1979, p. 115.
24. Wilkinson, F., and Brummer, J. G. *J. Phys. Chem. Ref. Data* **10**, 809 (1981).
25. Gollnick, K., and Kuhn, H. J. In H. H. Wasserman and R. W. Murray, eds., *Singlet Oxygen*, Academic, New York, 1979, p. 287.
26. Wasserman, H. H., and Lipshutz, B. H. In H. H. Wasserman and R. W. Murray, eds., *Singlet Oxygen*, Academic, New York, 1979, p. 430.
27. Poulsen, J. R., Birks, J. W., Gübitz, G., van Zoonen, P., Gooijer, C., Velthorst, N. H., and Frei, R. W. *J. Chromatogr.* **366**, 371 (1986).
28. Foote, C. S. In H. H. Wasserman and R. W. Murrary, eds., *Singlet Oxygen*, Academic, New York, 1979, p. 139.
29. Wilkinson, F. In B. Ranby and J. F. Rabek, eds., *Singlet Oxygen-Reactions with Organic Compounds and Polymers*, Wiley, New York, 1978, p. 27.
30. Kearns, D. R. In H. H. Wasserman and R. W. Murray, eds., *Singlet Oxygen*, Academic, New York, 1979, p. 115.
31. Shellum, C. S., and Birks, J. W. *Anal. Chem.* **59**, 1834 (1987).
32. Krinski, N. I. In H. H. Wasserman and R. W. Murray, eds., *Singlet Oxygen*, Academic, New York, 1979, p. 602.
33. Young, R. H., Wehrly, K., and Martin, R. L. *J. Am. Chem. Soc.* **93**, 5774 (1971).
34. Young, R. H., Chinh, N., and Mallon, C. *Ann. N.Y. Acad. Sci.* **171**, 130 (1970).
35. Wasserman, H. W., and Lipshutz, B. H. In H. H. Wasserman and R. W. Murray, eds. *Singlet Oxygen*, Academic, New York, 1979, p. 432.
36. Foote, C. S., Wuesthoff, M. T., Wexler, S., Burstain, I. G., Denny, R., Schenck, G. O., and Schulte-Elte, K.-H. *Tetrahedron* **23**, 2583 (1967).
37. Eskins, K., Scholfield, C. R., and Dutton, H. J. *J. Chromatogr.* **135**, 217 (1977).
38. Foote, C. S., Chang, Y. C., and Denny, R. W. *J. Am. Chem. Soc.* **92**, 5216 (1970).
39. Merkel, P. B., and Kearns, D. R. *J. Am. Chem. Soc.* **94**, 1029 (1972).
40. Smith, W. F., Jr., Herkstroeter, W. G., and Eddy, K. L. *J. Am. Chem. Soc.* **97**, 2764 (1975).
41. Young, R. H., Martin, R. L., Feriozi, D., Brewer, D., and Kayser, R. *Photochem. Photobiol.* **17**, 233 (1973).
42. Monroe, B. M. *J. Phys. Chem.* **81**, 1861 (1977).
43. Foote, C. S., Ching, T.-Y., and Geller, G. G. *Photochem. Photobiol.* **20**, 511 (1974).
44. Stevens, B., Small, R. D., and Perez, S. R. *Photochem. Photobiol.* **20**, 515 (1974).
45. Monroe, B. M., and Mrowca, J. J. *J. Phys. Chem.* **83**, 591 (1979).
46. Foote, C. S., and Peters, J. W. 23rd IUPAC Congr., *Special Lect.* **4**, 129 (1971).
47. Kacher, M. L. Ph.D. Dissertation, University of California, Los Angeles, 1977.
48. Singh, P., and Ullman, E. F. *J. Am. Chem. Soc.* **98**, 3018 (1976).
49. Stevens, B., and Small, R. D., Jr. *Photochem. Photobiol.* **23**, 33 (1976).
50. Hasty, N., Merkel, P. B., Radlick, P., and Kearns, D. R. *Tetrahedron Lett.* **1**, 49 (1972).
51. Rosenthal, I. *Isr. J. Chem.* **13**, 86 (1975).
52. Rosenthal, I., and Frimer, A. *Photochem. Photobiol.* **23**, 209 (1976).
53. Foote, C. S., Denny, R. W., Weaver, L., Chang, Y., and Peters, J. *Ann. N.Y. Acad. Sci.* **171**, 139 (1970).
54. Bellus, D. In B. Ranby and J. F. Rabek, eds., *Singlet Oxygen-Reactions with Organic Compounds and Polymers*, Wiley, New York, 1978, p. 61.
55. Young, R. H., Brewer, D., Kayser, R., Martin, R., Feriozi, D., and Keller, R. A. *Can. J. Chem.* **52**, 2889 (1974).
56. Smith, W. F., Jr. *J. Am. Chem. Soc.* **94**, 186 (1972).
57. Peters, G., and Rogers, M. A. J. *Biochem. Biophys. Acta* **52**, 2889 (1974).

58. Schenck, G. O., and Gollnick, K. *Z. Naturforschung.* **28c**, 302 (1973).
59. Thomas, M. J., and Foote, C. S. *Photochem. Photobiol.* **27**, 683 (1978).
60. Foote, C. S., and Denny, R. W. *J. Am. Chem. Soc.* **90**, 6233 (1968).
61. Toung, R. H., and Martin, R. L. *J. Am. Chem. Soc.* **94**, 5138 (1972).
62. Young, R. H., and Brewer, D. R. In B. Ranby and J. F. Rabek, eds., *Singlet Oxygen-Reactions with Organic Compounds and Polymers*, Wiley, New York, 1978, p. 36.
63. Ogryzlo, E. A. In B. Ranby and J. F. Rabek, eds., *Singlet Oxygen-Reactions with Organic Compounds and Polymers*, Wiley, New York, 1978, p. 17.
64. Ogryzlo, E. A., and Pearlon, A. E. *J. Phys. Chem.* **72**, 2915 (1968).
65. Stauff, J., and Fuhr, H. *Ber. Bunsenges. Physik. Chem.* **73**, 245 (1969).
66. Stauff, J., and Fuhr, H. *Z. Naturforschung.* **26b**, 260 (1971).
67. Fuhr, H., and Stauff, J. *Z. Naturforschung.* **28c**, 302 (1973).
68. Shoemaker, B., and Birks, J. W. *J. Chromatogr.* **208**, 251 (1981).
69. Kahmn, A. U., and Kasha, M. *J. Chem. Phys.* **39**, 2105 (1963).
70. Sonnenberg, J., and White, D. M. *J. Am. Chem. Soc.* **86**, 5685 (1964).
71. White, E. H., and Harding, J. J. C. *J. Am. Che. Soc.* **86**, 5686 (1964).
72. Zaklika, K. A., Kissel, T., Thayer, A. L., Burns, P. A., and Schaap, A. P. *Photochem. Photobiol.* **30**, 35 (1979).
73. Zaklika, K. A., Burns, P. A., and Schaap, A. P. *J. Am. Chem. Soc.* **100**, 318 (1978).
74. Thompson, A., Lever, J. R., Canella, K. A., Miura, K., Posner, G. H., and Seliger, H. H. *J. Am. Chem. Soc.* **108**, 4498 (1986).
75. Koo, J.-Y., Schmidt, S. P., and Schuster, G. B. *Proc. Natl. Acad. Sci.* **75**, 30 (1978).
76. Bollyky, L. J. *J. Am. Chem. Soc.* **92**, 3230 (1970).
77. Turro, N. J., Ito, Y., and Chow, M. *J. Am. Chem. Soc.* **99**, 5836 (1977).
78. Koo, J.-Y., and Schuster, G. B. *J. Am. Chem. Soc.* **100**, 4496 (1978).
79. Schmidt, S. P., and Schuster, G. B. *J. Am. Chem. Soc.* **102**, 306 (1980).
80. Bartlett, P. D., and Landis, M. E. In H. H. Wasserman and R. W. Murray, eds., *Singlet Oxygen*, Academic, New York, 1979, p. 245.
81. Foote, C. S., Dzakpasu, A. A., and Liu, J. W.-P. *Tetrahedron Lett.* **14**, 1247 (1975).
82. Johnson, D. C., Polta, J. A., Polta, T. Z., Neuburger, G. G., Johnson, J., Tang, A. P.-C., Yeo, I.-H., and Baur, J. *J. Chem. Soc., Faraday Trans. 1* **82**, 1081 (1986).
83. Herron, J. T., and Huie, R. E. *Ann. N.Y. Acad. Sci.* **171**, 89 (1970).
84. Wayne, R. P. *Ad. Photochem.* **7**, 400 (1969).
85. Clark, I. D., and Wayne, R. P. *Chem. Phys. Lett.* **3**, 93 (1969).
86. Wells, K. M. Personal communication.
87. Patterson, R. C., Kalbag, S. M., and Irving, C. S. *Ann. N.Y. Acad. Sci.* **171**, 133 (1970).
88. Schaap, A. P., Thayer, A. L., Blossey, E. C., and Neckers, D. C. *J. Am. Chem. Soc.* **97**, 3741 (1975).

Compound Index

Subject Index